Controlled Release Pesticides Formulations

Author

Nate Cardarelli

Professor, General Technology
Director, Environmental Management Laboratory
Community and Technical College
Division of Engineering and Science
University of Akron
Akron, Ohio

Editor-in-Chief
CRC Pesticide Series

Gunter Zweig

U.S. Environmental Protection Agency
Washington, D.C.

CRC Press
Taylor & Francis Group
Boca Raton London New York

CRC Press is an imprint of the
Taylor & Francis Group, an **informa** business

First published 1976 by CRC Press
Taylor & Francis Group
6000 Broken Sound Parkway NW, Suite 300
Boca Raton, FL 33487-2742

Reissued 2018 by CRC Press

Library of Congress Cataloging in Publication Data

Cardarelli, Nate F.
Controlled release pesticide formulations.

Bibliography: p.
Includes index.
1. Pesticides, Controlled release. I. Title.
SB951.C3 628.9 75-46632
ISBN 0-8493-5114-6
Former ISBN 0-87819-080-5

Publisher's Note
The publisher has gone to great lengths to ensure the quality of this reprint but points out that some imperfections in the original copies may be apparent.

Disclaimer
The publisher has made every effort to trace copyright holders and welcomes correspondence from those they have been unable to contact.

ISBN 13: 978-1-315-89178-1 (hbk)
ISBN 13: 978-1-351-07088-1 (ebk)

Visit the Taylor & Francis Web site at http://www.taylorandfrancis.com and the
CRC Press Web site at http://www.crcpress.com

PREFACE

In recent years there has been conflict between the absolute need for pesticidal chemical usage in agriculture and public health domains and humanity's desire for a quality environment free of toxic materials. Whereas there probably can never be complete reconciliation, the recent advances in controlled release pesticide technology can be ameliorative. This new technology provides a method of substantially decreasing the amount of pesticide added to the environment, vastly reduces the health hazard to those handling the toxicants involved, and, most importantly, allows the substitution of nonpersistent materials for those having an undesirably long environmental life. Furthermore, controlled release methods appear essential for the practical utilization of insect growth regulators and pheromones.

Relatively few controlled release products have reached the market place. A few years ago there was only an antifouling rubber material; now we have six products generally available and nine others ready for marketing. Additionally, several hundred governmental groups and industrial concerns are presently investigating this area. The obvious environmental and economic advantages inherent in the controlled release methodology have provided impetus to this rapidly growing field of technical endeavor.

The main goal of this book is to present a summary of the state of the art with historical background. Conventional chemical pest control methods are mentioned only as a means of comparison to controlled release. Research endeavor with biological weapons, potential usage of such controls, and the few instances of success are likewise brought into focus with the same motive. Formulations and methods of preparing controlled release pesticides are discussed in some detail as concerns the antifouling and molluscicide areas, where the compounding methodology has been well developed. The mathematical basis of controlled release has been developed to an extent and is presented in an abbreviated form.

This text covers most of the subject area work published in the United States up to March 1975, plus unpublished data from the author's laboratory. A great deal of the information presented is derived from the Proceedings of the 1974 Controlled Release Pesticides Symposium held at the University of Akron.

Although there are now many investigators examining the controlled release concept, I wish to acknowledge those who pioneered the area and whose scientific works have been of great value in the composition of this book: Dr. William Barnes, now at TVA, for his basic work with insecticides incorporated in plastics; Capt. James Nelson and his colleagues at the United States Army Environmental Hygiene Agency for bringing that work to fruition; Dr. Frank Harris (Wright State University) for his seminal effort with appendent herbicides, and also for reviewing this book; Dr. Charles Stockman of the B. F. Goodrich Company who encouraged the development of controlled release molluscicides; and Drs. John Engelhart and Al Ross at the M & T Chemical Company who have demonstrated both foresight and humanity in making such materials commercially available. Likewise, I am grateful that the International Copper Research Organization and the United States Army Corps of Engineers have sponsored projects leading to the development of controlled release aquatic herbicides as well as the discovery of the chronicity phenomenon. The assistance of Miss Katherine Walker of my staff and Miss Joyce Greene who patiently typed this manuscript are gratefully acknowledged.

Lastly, I wish to thank Dr. Gunter Zweig of the Environmental Protection Agency for suggesting this project and editing the manuscript, and the efficient and courteous staff of CRC Press.

Nate Cardarelli
Akron, Ohio
December 1975

FOREWORD

This book on controlled release pesticides by Professor Nate Cardarelli represents the third work in the CRC Press series on pesticides. The other two books already published are G. T. Brooks' two-volume *Chlorinated Insecticides* and M. Eto's *Organophosphorous Pesticides: Organic and Biological Chemistry.* Controlled release pesticide formulations, relatively new in pesticide application technology, have already found wide use in the chemical control of snails causing the serious, debilitating disease of schistosomiasis in many parts of the world. Another important application of the controlled release technique has been the development of antifouling paints and formulations for the protection of ship hulls and other marine structures. Professor Cardarelli has been a pioneer in these important developments. Recently, controlled release formulations of parathion (an insecticide) and a mosquito larvicide (methoprene or Altosid®) have been registered by the U.S. Environmental Protection Agency. The usefulness of these products lies on greater safety to applicators (in the case of parathion) and the extended life of environmentally labile compounds (methoprene). It is hoped that Professor Cardarelli's provocative theories will encourage other researchers to develop controlled release pesticide formulations, especially in the area of aquatic weed control.

Gunter Zweig
Editor
CRC Pesticide Series

THE AUTHOR

Nathan F. Cardarelli is Professor, General Technology, Community and Technical College, and Director of the Environmental Management Laboratory, Division of Engineering and Science at the University of Akron, Ohio.

Professor Cardarelli's degrees, all from the University of Akron, include a B.S. in natural science, M.A. in philosophy, M.A. in history, and M.S. in physics.

After service in the U.S. Air Force as a meteorologist, Professor Cardarelli spent 15 years in the rubber industry, holding various positions in research and development. As the inventor of controlled release antifouling rubber, molluscicides, and herbicides in the 1964–66 period, he is regarded as one of the pioneers of this rapidly expanding field. In 1968 he became an associate professor in general technology at the University of Akron, and in 1970 was one of the founders of the Creative Biology Laboratory.

Professor Cardarelli holds over 50 patents, mostly in the controlled release area, and has written a sizable number of publications. He chaired the first Controlled Release Pesticide Symposium (University of Akron, September 1974) and has served as an expert consultant to the World Health Organization, various public health agencies, and private industry.

TABLE OF CONTENTS

Chapter 1

INTRODUCTION

Since his advent on earth man has engaged in a continuous struggle for survival with many thousands of predatory or competing species of plant and animal life. Malnutrition, starvation, disease, suffering, and untimely death were and remain the lot of the majority of mankind.

Various species of insects serve as vectors of pathogenic microbes. Others, along with snails, act as vectors or hosts of parasites that infect millions of humans. Weeds compete with cultivated crops for soil nutrients, sun, and water. Aquatic weeds block irrigation channels, retard fish production, harbor undesirable insects, and hamper navigation. Rats and other rodents destroy crops in the field and in storage as well as serve as disease vectors. Birds and insects often reap what man has sown. Vampire bats destroy millions of head of cattle annually. The upsurge in coyote population has increased the economic risks of sheepherding in the western United States. Malaria and yellow fever have returned to previously "clean" areas. Bubonic plague cases are rising in the United States. The list of public health and agricultural problems arising from pest organisms is long indeed and the outlook pessimistic.

Pest control through the use of chemical agents is tenuous at best. Rapidly rising world population creates a global imperative to increase food production. Agricultural surplus in the United States constitutes about 6%. Any significant decrease of available foodstuffs in this nation would lead to an unprecedented crisis.

Approximately 100,000 different species of weeds, insects, trematodes, and fungi must be contained in the United States. The control costs amount to billions of dollars per annum, and the actual economic losses sustained because of the inefficiency of control techniques are somewhere between 13 and 30 billion dollars per year. In order to meet the worldwide hunger challenge it cannot be doubted that U.S. food production must be increased. If the challenge is not met by the food exporting nations the resulting social upheavals will lead to economic and political instability, revolutions, counterrevolutions, and wars of possibly unprecedented magnitude.

Pest control costs are tremendous and bearable by only a few prosperous nations. Even so, the increasing tolerance of many pest species to chemical agents results not only in the increased substitution of more toxic materials for less toxic substances, but also in the use of greater amounts of the selected agent for a specific control application. Concommitant widespread soil, water, and air contamination result with increasingly visible and unfavorable effects on nontarget biota, including man.

The public health and agricultural problems involving pest control are intertwined in most of the developing nations. Low agricultural productivity in many areas of Asia, Africa, and South America is grounded not only in the general lack of materials and know-how to combat pest life, but also in the general debility of the agricultural worker, who is physically unable to labor at performance levels common to those in good health.

In general, human resources in the developing lands are not effectively utilized. A tragic cycle is apparent. Riddance of disease requires medical assistance, sanitation, and pest control measures. However, the nations involved, due mainly to low agricultural productivity, lack the economic base necessary to tackle the public health problem.

A critical need exists for improved pest control methodologies. Chemical agents have been the focal point of pest management in the past along with environmental restructuring through engineering, improved tillage practices, and so on. Present techniques have led, in some instances, to serious ecological damage, and as ever-increasing amounts of toxicants are added continuously to the environment a critical point may be reached where human survival itself is threatened.

Three new control methods are undergoing rapid development, spurred on by the threat of an impending ecological crisis and a critical need to expand agricultural productivity. The much-discussed panacea supposedly lies in "biological control" wherein a species harmless to man, his crops, domestic animals, and artifacts is pathogenic to, or a predator of, a pest species, and, is reared and released in the infested area.

Biological control, however, must be approached with extreme caution. Pragmatically there have been but a few relatively unimportant successes and a large number of failures. A great deal of time and money concerning research is necessary. A biological control must be found, evaluated, and proven harmless to man's health or economic well being. Ecological impact must be ascertained, and methods of rearing and releasing developed. This effort must, of course, be repeated for each biocontrol species, and predictably will become increasingly difficult as the interactions between control species must also be assessed in all their ecological implications.

Effective biological control of major pest species, where the amount of chemical agents used can be appreciably decreased, is probably decades away.

The concept of target biocides, or "baits," promises a much faster method of achieving useful control of at least many insect pests. Specific attractants are used that entice the pest target to discover and contact a poisonous bait. Nontarget biota are protected to a much higher degree than that observed with the use of conventional control techniques. However, target biocides are in their infancy and a great deal of research remains in most areas before useful control can commence.

The third concept, and the main subject of this text, is "controlled release." The methodology involved requires the long-term continuous loss of a chemical agent at ultralow environmental concentrations. It is likely that target biocides, to be cost effective, must be of a long-term control nature, and thus they constitute a proper subject for this document.

Controlled release methods are postulated as a means of increasing pest control efficacy while vastly decreasing the quantity of toxic materials added to the general ecology, thus reconciling chemical control with the need for environmental quality and serving as a stopgap measure until biological control techniques can be developed and exploited.

DEFINITIONS

Prior to a detailed account of controlled release and all its ramifications it is appropriate to offer a brief description of terms.

"Conventional control methodology" as used herein refers to the normal methods of applying pest control agents. Such chemicals are toxic to the target organisms and almost always toxic to various members of the nontarget biota. They are prepared as emulsions in oil or water, wettable powders, granules, dusts, and so on through well-known formulation techniques. Application is made to the pest-infested area by spraying, dusting, or other methods at dosages necessary to destory the pest or retard its growth.

Two major disadvantages arise from the use of conventional methods. Dosages must be very large, in that natural detoxification processes limit target exposure to but a few hours or at most a few days for adequate mortality. The use of nonpersistent agents means continuous retreatment with a relatively short interval between each treatment, whereas the use of persistent agents, for example DDT, may lead to grave long-term ecological damage. Secondly, the total environment in a given area is exposed to the toxicant even though the pest organism inhabits only a minute fraction of that environment. This factor coupled with the need for massive dosages often leads to destruction of nontarget life and undesirable effects on the food chain. Conventional pest control by its nature relies on acute intoxication of the target.

A "target biocide" in most cases is synonymous with a bait. An attractive element such as a pheromone or a desirable foodstuff is released at a slow, continuous rate into the environment. The target, be it a fruit fly, boll weevil, rat, etc., is attracted to the bait. Kill occurs through ingestion of the bait or by contact and absorption of an incorporated toxic element. Conceptually, the perfect bait releases only a nontoxic, generally innocuous, chemical into the environment wherein only the specific target is affected. The toxic element in this optimal situation is not lost into the external environment but remains ecologically unavailable until the target pest reaches the bait.

In more recent developments it appears likely that what is essentially a bait can be made containing no toxic ingredient whatsoever. For instance, a pheromone can be released over a wide area as a confusant, thus vastly decreasing the ability of male and female to locate each other. Or a juvenile hormone mimic can be released that does not affect the environment but prevents the maturation of a specific target species.

The terms "controlled release" or "slow release" are often used interchangeably although their meanings are not precisely the same. Also, both terms have been used in the past by pesticide manufacturers to promote formulations of a few days' release life. For instance, pest control agents are adsorbed or absorbed on various clays, silicas, or organically derived materials and then desorbed slowly upon contact with soil or water. The toxicant involved provides superior control by increasing target exposure time. Various polymeric additives to emulsifiable concentrates can likewise prolong toxicant release.

Controlled release as used in this book will apply to certain formulations that slowly and continuously lose the toxic element into the environmental interface over a period of time measured in months or years. Biological efficacy arises through a chronic intoxication mechanism or in several special cases through repellency. Controlled release implies that the formulations involved are so created as to deliver the proper dosage for effective long- or short-term pest control. In contrast, slow release suggests continuous toxicant loss but at variable rates over a long period of time.

CONTROLLED RELEASE CATEGORIES

Controlled release materials can be conveniently classified in accordance with the mechanism of release.

Diffusion-dissolution

This type of controlled release is based upon the solution properties of various elastomers. Many pesticidal materials are soluble in natural rubber and other polymers possessing the thermodynamic characteristics of elastomers. Upon incorporation in the elastomeric matrix at levels below the solution threshold, a condition of solution equilibrium is established. If the toxic element is volatile, molecules on or near the surface will desorb into the exterior air environment. Likewise, if the toxicant is water soluble, molecules on or near the elastomer surface will, through dissolution, pass into the water interface. In both cases internal solution pressure will lead to the migration of interior solute molecules towards the depleting surface, and a continual toxicant loss is established at the interface.

Three-phase Carrier Systems

In instances where the toxicant of choice is not soluble in an elastomeric or more commonly, a plastic matrix, a third necessary element is added. This "carrier material" is usually soluble in the matrix, whereas the toxicant is matrix insoluble but soluble or dispersable in the carrier. As the carrier moves via solution pressure or thermodynamic instability towards the surface it carries with it the pesticide. Surface loss through volatilization or other means occurs.

Leaching

Nonmatrix soluble materials can be incorporated in elastomers or plastics through mechanical bonding. Upon exposure to water toxicant molecules near the polymer surface slowly pass into solution. As the surface layers are depleted a pore structure is formed, provided toxicant loading is sufficiently high. Water thus penetrates the pore structure and washes out freshly exposed toxicant molecules. The process is continuous although rate of loss usually varies with tortuosity of the growing pore structure and other factors.

Exfoliation

Various elastomers and plastic substances will degrade upon environmental exposure. Such degradation may arise from physical, chemical, or biological attack. If the matrix incorporates a nonsoluble or soluble toxicant, its destruction leads to the loss of the pesticide. Rate of release depends upon a number of environmental factors, and as a rule, such materials do not provide a continuous, *constant*, toxicant loss.

Microencapsulation

In this instance the pest control agent is enclosed within a permeable or semipermeable polymeric envelope. The toxicant moves through the polymer wall by diffusion processes and is desorbed from the external surface, passing into the exterior environment.

Volatility Retardation

Toxic agents or repellants can be incorporated in various film-forming polymeric solutions. Upon evaporation of the solvents an adherent film is laid upon a substrate, this film containing through solution or mechanical bonding the active agent. The agent then passes from the film surface into the surrounding medium, usually air, through diffusion and volatility processes. The film acts to retard volatility, thus increasing lifetime.

Ingestible Baits

An attractant and a toxicant are bound together in an ingestable matrix, which may be an elastomer or other substance. For proper function the attractive element slowly escapes, affecting various chemoreceptors of the target. The pest organism locates the bait, ingests a lethal dose, and succumbs.

Contact Baits

The matrix element contains an attractant that is released slowly and a contact toxicant which, optimally, is not released at all.

Nontoxic Baits

The matrix element contains a biologically active material which by nature is nontoxic to the target or other species, but alters target behavior in such a manner as to reduce pest populations.

Controlled Release Predacides

A toxic element is held within a polymeric enclosure releasing only at the instant that the predator target attacks the protected nontarget organism.

HISTORICAL SURVEY

Pest Control

Pest control is undoubtedly an art as ancient as civilized man. The earliest available reports describe the Egyptians in ca. 2000 B.C. using the cat for biological control of rodents in granaries. A little later we find the Egyptians and Persians using torsion-style rodent traps. Probably the earliest forms of chemical pest control lie in the use of arsenic compounds in rodent poisons (described by Aristotle in 350 B.C.), and in the development of ship-bottom coatings to prevent worm damage and retard fouling attachment.

Ship hulls were treated as early as the 8th Century B.C. Aristotle (384–321 B.C.) described the barnacle problem in the 4th Century B.C. and Publius Ovidius (43 B.C.–17 A.D.) discussed marine boring organisms. Early antifouling devices consisted of copper sheathing, an effective but expensive method, and crude formulations of toxic coatings. Such forerunners of antifouling paints consisted of pitch-, wax-, and oil-based preparations usually containing sulfur and one or more salts of lead, tin, mercury, and arsenic. Christopher Columbus used a toxic pitch compound on his three famous vessels. As paint technology improved over the centuries cuprous oxide in a shellac or resin base, first used in the 1860s, gradually became the standard antifouling preparation. However, paints containing other copper compounds, oxides and salts of mercury, and various arsenicals were still in widespread use until the 1940s.

In the insect control area the first known bait composition consisted of arsenic/honey combinations used in ant control in 1681. Nicotine was in use as an agricultural insecticide 9 years later. Carbon disulfide was probably the first soil sterilant, reported in use in France in 1854. Paris green was used in 1865 for the control of the potato beetle, Bordeaux mixture for mildew protection in 1882, and the first synthetic organic pesticide, potassium dinitro-*o*-cresylate, in 1892. DDT was tested as a mothproofing material in 1939 and soon became the universal insecticide.

At this time there are approximately 700 pesticidal materials in commercial use in the United States. Over the last 30 years agricultural yield has more than doubled on a world-wide basis, due in no small part to the use of pest control agents as well as fertilizers. Malaria, typhus, plague, encephalitis, and other insect-borne diseases have been nearly eradicated in large areas of the world. Both factors have led in turn to dramatic population growth in Africa, Asia, and South America, and to the beginnings of industrialization. However, the widespread use of pesticides has led to a number of problems.

In accordance with natural selection, pests have developed increasing resistance. Codling moth resistance to arsenic was reported in 1928, housefly resistance to DDT in 1946, and mosquito tolerance towards DDT in 1947. As acquired resistance increases in a given species effective control can only be achieved by the use of larger amounts of toxic materials or increasing the treatment dosage. Sometimes a decrease in the between-treatment interval is also necessary.

The concept of biological control became a popular topic after 1955 when it was successfully demonstrated that field populations of the screw worm could be destroyed by sterilizing and dispersing males of the species. This was an outstanding and much-publicized success, but unfortunately has proven to be a very difficult act to follow.

In 1962 Rachel Carson's *Silent Spring* alerted scientists and laymen alike to the dangers of indiscriminate use of chemical pest control agents.

Controlled Release Formulations

In the 1930s Bungenburg de Jong published his studies on coacervation. Barrett Green used this work as a foundation for the preparation of the first microcapsules. In the early 1960s polar materials were encapsulated inside polymeric substances through phase separation techniques. So-called "time release" pharmaceuticals were one result of these developments. Many commercial applications of microencapsulation were brought about in the 1950s. In all cases an agent is contained within a polymeric membrane and released through any of a number of processes: application of heat, pressure, dissolving of the retaining capsular wall, and so. However, the contained liquid might also be released at a slow and perhaps uniform rate if it could diffuse through the polymeric membrane. In recent years pest control agents have been microencapsulated and a degree of controlled release experienced, as will be discussed in detail in later chapters.

An improved method of slowly releasing a biologically active agent arose from the discovery in 1964 that a very long-term release of antifouling agents can be realized by their incorporation in an elastomeric matrix. Longevity is measured in years. This in turn led to the formulation of controlled release larvicides wherein halogenated hydrocarbons, carbamates, and various organophosphorus agents are added to an elastomeric or plastic matrix, and the release rate controlled by proper compounding. In 1965 the first controlled release molluscicides based on organotins were developed, followed in 1967 by the incorporation of niclosamide, a "conventional" molluscicide, in rubber. By 1973 inorganic copper salts had been successfully formulated in slow release matrices and evaluated as molluscicides.

In 1969 an ester of 2,4-D was formulated in natural rubber and the first long-term controlled release herbicide developed.

The history of controlled release pesticides is a strange one indeed. The initial flurry of activity with rodent baits and antifouling compositions occurred in prehistoric times with relatively little technological advance until the early 1960s. The following chapters follow the same basic chronology, beginning with a consideration of antifouling compositions, followed by molluscicide, cercariacide, herbicide, and insecticide developments that were generated by the discovery of antifouling elastomers. Microencapsulation is examined as a parallel control method of considerable promise.

Baits are separately classified in that such materials release an attracting agent rather than a toxicant, and consequently several sections are devoted to snail and slug baits, rodent baits, insect baits, and so on. The release of pest repellents is examined, and the recent advances in triggered release predator control agents are discussed in the last chapters.

CONVENTIONAL PESTICIDE FORMULATION

Historically the first pesticide formulations were simple baits made by mixing toxicant and attractant. Once it was discovered that chemical control of insects and other pests was possible, man sought effective means of delivering the agent to the target. This was relatively easy with water-soluble materials, but much more difficult with insoluble ones. Later, as scientists investigated the mechanisms of toxicity in plants and animals, it became obvious that the properties of the formulation played a vital role. To be biologically effective the control agent must contact the pest and overcome various physiological barriers. From both economic and environmental standpoints proper formulation leads to easier application and greater efficacy, thus reducing the total amount of toxicant needed.

By the late 19th Century the pesticide industry discovered that prolonging the target-population agent contact time was very important to efficient control. Thus pyrethrum, nicotine, and other toxicants were mixed with various clays, talc, waxes, silicates, etc. as binding elements. The agent was adsorbed or absorbed by the binder. Once applied the toxicant molecule was gradually released through dissolution or mechanical destruction of the binder. Biological efficacy improved as persistence increased from a few hours to perhaps a few days, thus greatly increasing target contact probability.

Most modern pesticides are not water soluble, so application from a water system requires the use of an emulsifier. Such materials are characterized by a chemical structure in which the molecule is divided into two moieties. One part of the structure is hydrophilic (water soluble) and the other part is hydrophobic (water insoluble). Usually the hydrophobic moiety is soluble in organic liquids. Essentially, the emulsifier allows the homogeneous dispersion of nonsoluble control agents in water.

Surface-active agents usually show adjuvant effects when used with herbicides. For effective control of weeds the herbicide must be in contact with the entry site for a period of time. Such factors as spreading of the emulsion on the leaf surface, run-off due to gravity, and the like are vital considerations. By improving the wetting property of the emulsion the contact time can be extended. Surface-active agents affect wettability by influencing both the surface tension and the contact angle.

Emulsion stability is important if the material is to be stored for any length of time prior to use. Various additives, usually blends of anionic and nonionic surface-active agents, are used. Salts of alkyl aryl sulfonates are popularly used as anionic agents.

In the early 20th Century water-insoluble organic pesticides were often solubilized in petroleum distillates and used as sprays. However, this practice has lost popularity in favor of emulsified water systems, where damage to the protected organism is much less. Kerosene, for instance, causes leaf burn in many agricultural crops, and had to be used with care to adequately destroy the target without unacceptable damage to the desired plant.

Whereas emulsified systems are popular because of being more biologically effective and economical in application compared to other methods, other dispersal systems are in widespread use. These include dusts, fogs, wettable powders, and granules. Nonemulsified systems almost always depend upon some carrier material to allow proper dilution and dispersal of the active agent. Carriers are by definition inert materials, but in reality play an important active role in pesticide-pest interaction.

Carriers can be classified as inorganic minerals and botanicals. Minerals consist of such materials as various clays, talcs, carbonates, sulfates, and phosphates. Ground corn cobs, nut shells, wood dust, etc. fall under the botanical category.

The most important characteristic of any carrier is its sorptive capacity (adsorption and/or absorption). The carrier must "hold" a given amount of the toxicant, usually a liquid, while retaining physical properties such as flowability.

A granular carrier is defined as one of particle size range from 4 to 80 mesh (U. S. Standard Sieve Series). Dusts are by definition materials having particle sizes smaller than 80 mesh.

Granules release the sorbed toxicant by mechanical disintegration, dissolving of toxicant at a water interface, displacement of the toxicant by water, and volatilization. Each process takes a certain finite time so that release is slow, often taking several weeks for the granule to completely release all the retained agent.

Carrier properties are often vital to the proper activity of the toxicant. For instance, alkaline carriers tend to decompose many carbamate pesticides and thus must not be used.

Dusts are finely powdered carriers which contain one or more toxicants at concentrations up to 50% or more. Usually the toxicant can only be added to the dust by the use of an impregnating agent, aliphatic and aromatic petroleum solvents, alcohols, ethers, ketones, esters, and vegetable oils, for instance. Important dust properties are bulk density, flowability, and particle size range.

Wettable powders are dust concentrates that contain surface-active agents. When diluted with water such materials become stable, usually sprayable, emulsions.

Chapter 2

CONTROLLED RELEASE ANTIFOULING FORMULATIONS

Marine fouling is defined as the attachment and growth of various living organisms to objects immersed in a marine (salt-water) environment. Fresh-water fouling also exists but does not show the severe undesirable effects seen in the oceans and seas. The fouling community varies with geographical location, water temperature, water depth, salinity, and other factors. All natural waters harbor fouling species; however, abundance is very small in arctic waters and great in tropical waters.

Fouling is characterized by a biotic progression. Although the species differ and the rate of attachment and growth varies with location, the general scheme is this: Primary fouling, which begins at the time of exposure, consists of unicellular slimes. Diatoms, bacteria, algae, and protozoa compose the major portion of slime films. Next, various sessile organisms attach as larval forms and metamorphose into adults. This secondary community is composed mainly of barnacles, hyozoans, mussels, and various calcareous helminths. Organisms causing secondary fouling are probably, in some form, attracted by the food value of the primary foulants. The secondary forms continue to attach, often completely overrunning the first community. Barnacle attached to barnacle four or five layers thick are commonly observed. Each wave of encrustation tends to deprive prior attached organisms of contact with foodstuffs, thus destroying them. Multicellular algaes now appear along with copepods, amphipods, and other free-living forms that find the fouling community a source of nutrition and protection.

Although not strictly speaking members of the fouling community, certain other organisms that destroy exposed structures by boring or consuming are generally thought of as part of the fouling problem. Since protection against fouling generally leads to protection against the "ship worm" and other detrimental organisms, they will be considered in this chapter.

FOULING PROBLEM

General Considerations

The attachment and growth of fouling organisms leads to a number of economically serious problems. Fouling accumulation adds to the weight of the exposed object, which can be considerable; buoys, for instance, will gradually sink as fouling builds up. On ship hulls even small amounts of fouling destroy streamlines and leads to increasing drag effects. This along with increased weight means that the vessel must burn more fuel in order to maintain speed. For instance, if friction at the hull/water interface increases at a rate of 0.25% per day, at the end of 6 months the vessel must consume 35 to 50% more fuel to maintain speed, and, furthermore, the maximum speed is reduced by 1 to 2 knots.[1] Ships have been known to run out of fuel prior to reaching port due to fouling.

Fouling attachment is minimal when the ship is moving on the high seas. Barnacle attachment does not apparently occur at speeds greater than 2 knots, and the limiting velocity for borers lies somewhere between 1.4 and 1.9 knots.[2] The barnacle and other organisms that attach by secreting an adhesive material need relatively calm water so that the larva are not mechanically washed away before adhesion is complete. The problem is particularly serious for naval vessels that spend a significant proportion of time stationary in harbors, or in the case of nuclear submarines laying motionless at an underwater duty station for several months.

Commercial freighters, tankers, and liners have a rapid turn-around time so that the stationary period in harborage is relatively small. However, all vessels foul to some extent and the economic loss due to greater fuel consumption and slower speed is enormous. A reduction of only 0.15 knots in cruising speed due to fouling can lead to a direct cost of $12,000 per annum.[3]

Fouling leads to severe mechanical damage to marine instruments and equipment, rendering movable parts inoperative and causing other dysfunctions. For instance, fouling of the vanes of current measuring equipment has led to considerable loss of scientific data and replacement expense.

Some types of fouling greatly accelerate underwater corrosion. The growing barnacle shell knives under and lifts protective coatings from materials, exposing the metal substrate to salt water. Furthermore, the barnacle shell edge actually penetrates steel, as shown in Figure 2–1. Usually the greatest deterioration of the substrate is localized with extreme pitting observed below the shell base of the barnacle.[4] An oxygen-rich galvanic cell is probably created within the pit.

Calcareous Fouling

Calcareous fouling organisms are those that attach to the immersed surface and then develop a shell. Barnacles, tube worms, and coral are of major importance. However, encrusting bryozoans and mussels are properly included, even though the mussel is not sessile but can change location.

It is likely that the calcareous acorn barnacle is the major fouling pest. This is a general statement in that other foulants may play a leading role in specific locations. For instance, tube worm fouling is more severe around the Hawaiian Islands. However, vast areas of the Atlantic, Mediterranean, Indian, and Pacific Oceans find the acorn barnacle in great abundance. Detachment of the acorn barnacle is generally more difficult than removal of other fouling organisms. With a density somewhat near that of calcium carbonate (2.7 g/cm^3), the barnacle and other calcareous organisms add true weight to the object whereas the specific gravity of nonshelled foulants varies little from that of sea water.

Barnacles

Barnacles are crustaceans of the phylum *Arthropoda.* There are four orders of barnacles; however, *Thoracica* is the only group of importance to fouling. The genera *Balanus* (acorn barnacles) and *Mitella* (goose barnacles) are germane to this discussion.

Adult barnacles are sexually hermaphroditic although cross fertilization is the rule. Eggs develop in the parents' mantle cavity, hatching into the microscopic larval form termed "nauplius." The nauplii swim free of the parent, undergo five to seven molts depending on the species, and finally become the morphologically distinct cyprid form. The cypris seeks an attachment site. Various surfaces are explored by a creeping motion, making use mainly of antennae for locomotion. When a suitable site is found attachment is made by secretion of an adhesive from the antennae. Within a few minutes the adhesive bond is firm. The body reorients, resulting in the ventral surface contacting the substrate. Decortication occurs as the cyprid carapace is shed. At about the 12th to 14th hr after attachment shell formation appears.[5]

Barnacles (Figure 2–2) are hardy creatures able to undergo long periods of starvation,[6] survive over a wide temperature range,[7] and tolerate large salinity changes.[8] Fast-growing species mature and reproduce within 7 weeks.[9]

The cementum secreted by the barnacle is of great interest for several reasons. It adheres to all tested substrates including polished glass, Teflon,® and other objects where man-made adhesives fail; cross-linking occurs rapidly underwater; it is nonsoluble in concentrated acids and bases at any temperature save for hot glacial acetic acid; the tensile strength is much greater than that found in any commercial adhesive.[10]

Unlike acorn barnacles, the goose barnacle adheres to firm substrates by means of an organic stalk in which are located the cement glands. This pest is not as universal in habitat as the acorn barnacle and thus, save in rather small geographical areas, does not pose the serious problem noted with the genus *Balanus.*

Tube Worms (Hydroides)

Polychaetes of the Serpulidae family include a number of marine worms that live in calcareous tubes. The larval form attaches to a surface by means of a so-far unknown cementation mechanism and builds a calcareous tube. Such tubes may be 7 or more cm in length and 2 cm or more in diameter at the distal end. Like barnacles they are firmly attached, and death of the worm does not alter the unwanted effects of the tube. Figure 2–3 shows both barnacles and tube worms on a fouled panel.

Coral

Coral build-up on immersed objects constitutes a serious fouling problem in various areas of the tropical Pacific Ocean. Coral encrustations several inches thick have been noted on oil drilling vessels stationary for several years off New Guinea.

Oysters

Many bivalve molluscs are fouling organisms. Adults are sessile and once attached grow rapidly, creating problems of weight and drag on submersed objects. This type of fouling is present along both coasts of the United States. However, compared to the tube worm and barnacle, oysters and scallops are of minor importance as fouling organisms.

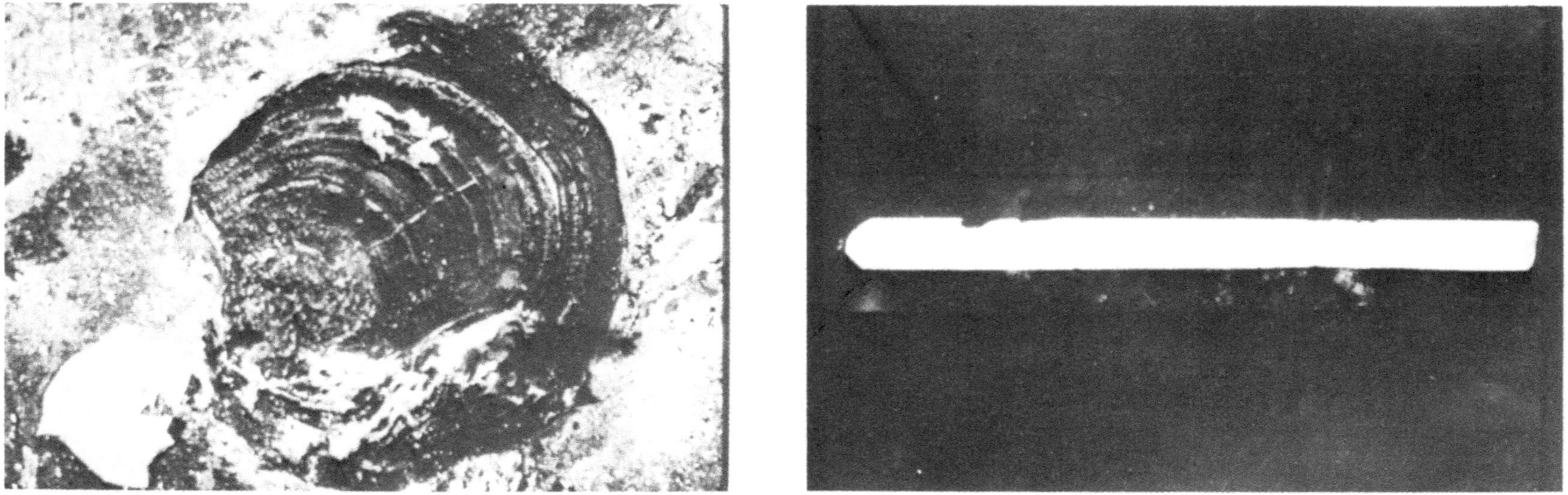

FIGURE 2–1. Barnacle shell peripheral penetration pits; top view (left) and lateral view (right). With each growth period the leading shell edge penetrates deeper into the metallic substrate. Two growth periods are observed. The deeper penetration is in the rostral direction. Top view shows pitting of steel substrate. (Photographs by the author.)

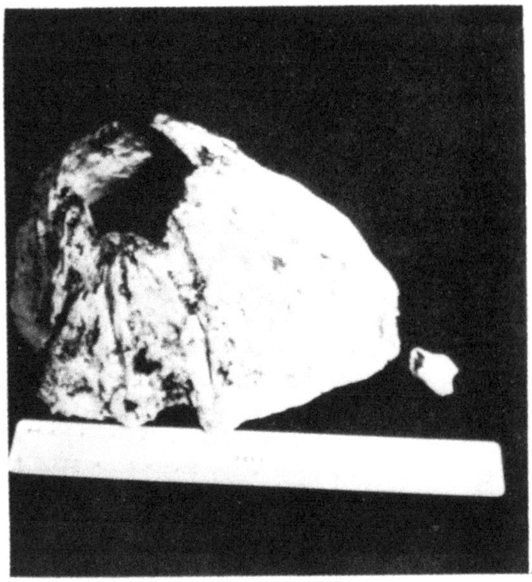

FIGURE 2–2. Growth extremes between *Balanus nubilus*, a U.S. Pacific Coast giant, and one of the common Floridian species. (Photograph by the author.)

Mussels

Mussel fouling is usually not a serious problem save in water intakes such as those used in electrical generation stations. Continuous chlorination of intake waters alleviates the problem.[11] Mussels (*Mytilus edulis*) are found from North Carolina and San Francisco northward; they attach by means of a byssal thread which is organic in nature. Slow movement from site to site is possible and once the living organism dies the attachment decays and the shell will eventually drop.

Encrusting Bryozoa

The Bryozoa (phylum Ectoprocta) are sessile colonial animals that live in secreted cases. Most are salt-water dwellers, and marine species are numerous. Members of the order Ctenosomata form brown and red colonies, usually encrusting. Such growth will overlay and destroy barnacles, oysters, and other calcareous forms as the fouling community ages. The genus *Bugula* (order Cheilostomata) is a sessile, common form resembling moss.

Noncalcareous Fouling

Nonshelled fouling organisms, although they add little weight to the substrate, can vastly increase drag on hulls and block ports, valves, tubing, pipes, and various marine equipment.

Tunicates

Tunicates are invertebrate members of the Phylum Chordata. Sea squirts are the most common fouling form. Although members of the fouling community, they are seldom numerous and are of minor importance. They are usually not found as primary or secondary foulants, but typically arrive after barnacles and other organisms have completely covered the immersed object.

Algae

Algae constitute the major nonshelled fouling class. Single-cell algae such as diatoms, which are primary foulants, are not in themselves a problem. However, it is likely that they attract sessile forms that feed on them.

1. Blue-green algae (Cyanophyta) – Blue-green algae are usually small, seldom growing to more than a millimeter in length, and slimy. They are primary members of the fouling community.
2. Green algae (Chlorophyta) – Marine green algae vary from flagellated planktonic, colonial forms of microscopic size to filamentous forms of various size. *Cladophora* grows to but a few millimeters in length, sea moss (*Bryopsis*) to 3 in., some species of *Enteromorpha* (sea "grass") to several feet, and *Ulva* (sea lettuce) to 4 or more feet.
3. Brown algae (Phaeophyta) – These are the largest of all algae, some such as *Pelagophycus* growing to lengths of 100 or more feet. Only the smaller kelps are very often found as members of the temperate water fouling community. *Sargassum* and *Alaria* of several feet in length are sometimes seen on fouled objects.
4. Red algae – Red algae, usually small and filamentous forms, are often seen as secondary or tertiary foulants. The larger plants such as *Porphyra* are seldom found in fouling communities.

Marine Borers

In a strict sense the boring worms and clams are not foulants in that they are not sessile. However, the protective measures used against fouling organisms also serve to prevent boring. In general, borers attack immersed wooden objects, vessel hulls, pilings, etc.; however, the pholad (Pholadidae) clam bores into concrete and often causes severe damage to sea walls and pilings.

The teredo (Teredinidae), popularly called a "ship worm" due to its appearance, is actually a

FIGURE 2–3. Fouled panel after 2-month immersion at Miami Beach. Note tubeworm in foreground. (Photograph by the author.)

FIGURE 2–4. *Limnoria* ingesting a test block. (Photograph by the author.)

mollusc related to the clam. The shells do not protect the organism; rather, the bored dwelling in wood protects the long wormlike body. Shells are found at the anterior end of the burrow and are used as boring instruments. At the posterior end are found siphons protected against predation by pallets. In case of attack the pallets draw close to seal off the burrow entrance. Teredoes are free swimming only in the larval form, and once penetration of wood has been initiated the animal cannot leave its location. The 45 or more species of economic importance are able to withstand considerable variation in temperature and salinity. Widespread dispersal of many of the species may have occurred due to their releasing of larvae at various ports of call. The teredo can grow to 3 or more feet in length, and some of the larger forms are eaten as a sea food in Thailand and the Fiji Islands.[11]

Unlike most fouling forms, save serpulid worms, boring organisms are found at depths as great as 6000 ft.[12] Materials at risk include wood, plastics, rope, and fibers, and even lead has been penetrated.[13]

Another wood-boring organism is the isopod *Limnoria* ("gribble"). Unlike the teredo, *Limnoria* ingest wood and are picturesquely called "marine termites." Being air breathers they cannot work at any great depth, but rather cause extensive damage to wood structures in the intertidal zone. Pilings are especially prone to attack and unless treated will not last a year in heavily infested areas. Even creosoted pilings are cut through within 5 or 6 years in such areas. Figure 2–4 illustrates gribbles eating within a wood block and Figure 2–5 shows piling damage.

Often gribble and teredo attack occur simultaneously, leaving a spongelike wood residue with little if any strength (Figure 2–6). The teredo will not or cannot bore when current velocity is over 1.4 knots, and the gribble is limited to velocities less than 1.8 knots.[14]

FIGURE 2–5. Piling damage due to gribbles. (Photograph by the author.)

FIGURE 2–6. Untreated wood damaged by teredo and gribble attack; 3-month exposure in subtropical Atlantic Ocean. (Photograph by the author.)

Other Organisms

The oceans teem with bacteria, although relatively little is known concerning their biology or ecology. Muraoka reports that water samples taken at 5300-ft depths contain 6000 bacteria per milliliter, and sediments at the same depth 25,000 organisms per gram (wet).[15] Bacteria are foulants, although in themselves probably do little harm, save possibly to increase drag; however, it can be shown under laboratory conditions that marine bacteria will ingest plastic coatings, and they almost certainly attract other much more detrimental fouling organisms.

Sponges will settle and grow on immersed objects, but so far as is known constitute a very minor foulant.

A number of nonsessile life forms are attracted to the fouling community, as it offers both foodstuffs and some degree of protection. Various isopods, decapods, lobsters, octopi, and squids, for instance, are found on and around heavily fouled stationary objects.

CONVENTIONAL TECHNOLOGY

Survey

Over the centuries man has devised various methods of combating fouling and boring organisms. Copper ion is repellent and/or toxic to all foulants and borers except possibly bacteria, and consequently ancient man covered ship hulls with copper plating.[16]

Copper sheathing is probably the best conventional method of preventing fouling and boring attack. However, it has many disadvantages besides that of cost. It cannot be used on metal hulls or other objects due to galvanic activity leading to rapid corrosion. Lead sheathing was also used in ancient times though probably not to as great an extent as copper. Lead does not prevent most fouling attachment though it was probably useful against worm penetration.

Antifouling paints became generally available during the 19th Century.[17,18] Many were based on copper since it was well understood that the gradual release of copper ion from a copper surface prevented fouling. Arsenic and mercury paint formulations were also available. McIness[17] based his formulation on copper sulfate, while Tarr and Wonson[18] based theirs on copper oxide. Some others were rather complicated: the Rahtjen plastic paint of 1863, for instance, contained copper carbonate, copper acetoarsenic, copper arsenite, arsenious oxide, and mercuric oxide in a resin base.[19] By 1943 the U.S. Navy standardized a formulation with cuprous oxide as the antifoulant. Reference 19 provides various U.S. Navy paint formulations and a rather optimistic view of their longevity.

Application of antifouling paints prevents fouling for a short period of time. In spite of extravagant claims, a 6 month foul-free life in tropical waters and 18 months in temperate waters is about all that can be expected. A few paints will exceed 8 months tropical.[20] Unfortunately, one can be very elaborate in applying paints to test panels, thus achieving relatively good longevity, while conventional shipyard practices do not permit the same degree of precision. Dry-docking costs for vessels of any size range from $5000 per day upwards. Paints are applied by large spray equipment as rapidly as possible. Twenty-one-coat hot plastic paints can be applied to test panels over a period of a week or so and perhaps give a year or more tropical fouling protection. Such systems have yet to be demonstrated as being equally effective on commercial or navy vessels.

The procedure is costly. A fouled vessel must be dry docked and defouled. To clean up a hull, high-pressure steam can be used to wash off noncalcareous growth. Barnacles, tube worms, coral, and so on require sand blasting for removal. Once fouling is cleared a primer coat is applied, allowed to dry, then followed by one or two anticorrosive coatings, and then one or more antifouling paint layers. Inadequate application at any point can lead to rapid spot corrosion and/or rapid fouling. Good antifouling paints often cost $70 or more per gallon and none are available at less than $18 per gallon as of this writing.

Antifouling paints function through several mechanisms. The most common type employed is leaching, wherein the toxicant molecule upon contact with sea water releases copper ions. If the molecule is very water soluble leaching proceeds rapidly (as with copper sulfate) and paint life is short. Water-insoluble copper compounds such as Cu(II) carbonate or copper (II) oxide release too slowly. Consequently, copper (I) oxide, which is slightly soluble, provides the longer lasting activity. A leaching mechanism, by its nature, requires the continual growth of a pore structure so that water can penetrate the paint film and contact fresh copper. System dynamics therefore require very high copper oxide loadings. Below about 70%, cuprous oxide concentration (dry weight) leaching is inadequate, and paint formulations foul immediately. Most copper paints contain 85 to 92% cuprous oxide (dry weight). The necessary high loadings affect detrimentally such physical properties as tensile strength, peel resistance, and abrasion resistance. Poor antifouling paint performance is as likely to arise from paint loss through mechanical action as from depletion of the toxicant.

A second type of paint functions through an exfoliating mechanism. In this case the paint film dissolves slowly in sea water or is mechanically lost through erosion, in both cases releasing the toxicant into the water-paint-film interface.

Copper (I) oxide paints suffer a number of deficiencies. Temperature change greatly affects performance: a 1°F increase may lead to a 3% increase in copper loss per day.[21] Leach rate increases with increasing salinity, pH, and speed.[22] The effect of large copper ion loss on nontarget biota in harbors is unknown. Also, copper ion readily forms a galvanic couple with iron, so a copper-impermeable paint must separate the antifouling coating from metallic ship hulls, etc. The reader is referred to standard texts which describe the corrosion problem, methods of prevention, anticorrosive undercoats, and so on.[23]

Effective antifouling paints must protect against all fouling organisms as far as use on vessels is concerned. Perhaps of primary concern is the prevention of microfouling which is the initial step in attracting calcareous organisms.[24] The slimes of diatoms, bacteria, algae, etc. possess the highest tolerance to toxic systems.[25] The toxicant chosen must be active against slimes as well as the secondary foulers. Turner has maintained, "The only satisfactory system for the prevention of fouling is one which works for all organisms anywhere in the world at all times of the year."[26]

Other systems for prevention of fouling exist besides the use of toxic coatings. Electrical systems have been explored.[25] The toxion method, where an antifoulant is released from a tube network around a ship's hull, has been an ecological disaster.[26] Also, poisonous gases such as chlorine can be continuously added to conduits such as condenser cooling water lines and cooling water intake tunnels for hydroelectric facilities.[26,27] Application must be continuous for effective control. Chlorination systems are more expensive than antifouling paints, and present a far greater potential hazard to users. Other methods under development are keyed to modifying the surface tension of the slime films so that foulants will find adherence difficult.[28] Also, new copper alloys are intrinsically antifouling and may be useful in hull design, conduits, and other areas.[29]

Preservation of wood structures such as pilings from borer attack is usually affected by impregnation with toxic materials. A common way of preserving wood is through pressure impregnation using creosote, coal tars, and water-borne preservatives. Methods are well documented in various publications.[30] Other methods of preserving wood exposed to the marine environment are much less effective in terms of longevity.[31] Many water-borne preservatives contain copper salts, arsenite compounds, or both.[32] Such materials are sooner or later leached out of the wood by the sea water. Other preservative methods have been suggested, as, for instance, impregnation with silver nitrate after first removing all natural chlorides (silver chloride, a toxicant, being formed upon contact with ions in the ocean),[33] or impregnation with copper formate, followed by autoclaving to form an insoluble copper-cellulose bond.[34]

Borer activity can also be stopped by application of plastic coatings after penetration, which deprives the organism of contact with the sea, its source of food and oxygen.[35]

Modern Antifouling Paints

It is not the purpose of this text to discuss the art and science of paint formulation. Many adequate volumes exist on this subject, both practical[36] and theoretical.[37] A paint has three essential elements: a binding polymer, a solvent or vehicle, and various pigments that provide color, improved endurance, etc.; antifouling pigments have another property, biological activity. Various additives such as plasticizers, drying oils, and antioxidants are used to enhance desirable physical or chemical properties or eliminate undesirable ones.

An effective antifouling coating contains a toxic agent, active against all fouling forms, bound in a dried polymeric film, and designed to provide a steady state release from that film at a rate above the attachment threshold.

The toxicant of choice is now cuprous oxide. Mercurials have been found to be somewhat less toxic to fouling, more costly, and perhaps more hazardous in application. Cyanides, arsenic salts, and oxides were too dangerous. Most insecticides such as DDT lacked universality.

The mechanism of toxic action involved is not well understood. The presence of copper does not deter the barnacle cyprid from attempting to settle, but apparently poisons the larva before it can develop a shell.[38]

Early U.S. Navy antifouling paints contained several toxicants. Formula 72 is shown in Table 2–1 as an example, and may be compared with Formulations 15RC (1935), 15HP hot plastic (1942), and 121/63 (1963).

Relatively few improvements in cuprous oxide paints have been made. The general trend seems to be (1) to increase the toxicant level by using polymers such as chlorinated rubber that will accept the higher loadings,[39] or (2) addition of a synergistic toxicant such as an arsenic compound or an insecticide.

New Toxicants

One major development in fouling prevention has been the introduction of new chemical agents. The organotins, comprising one such group, first appeared in the patent literature in 1960. Marine fouling was claimed to be reduced by the use of 0.01 to 0.2% of various triphenyl and tributyl tins.[40] S. M. Miller, in examining several hundred potential antifouling agents, demonstrated that bis(tri-*n*-butyltin) oxide, TBTO®, inhibited all forms of marine fouling attachment indigenous to the Miami Beach, Florida area.[41]

Antifouling paints containing TBTO first appeared on the commercial market in 1962. Generally, a vinyl or epoxy polymer was used with a 40 to 55% TBTO concentration in the dried film. These early paints were clearly inferior to cuprous

TABLE 2–1

U.S. Navy Paint Formulations

Formula 72[19]

Ingredient	Parts by weight
Rosin	120
Copper carbonate	15
Water	50
Linseed oil	15
Lead acetate	15
Petrolatum	15
Arsenic trioxide	30
Mercurous chloride	30
Pine tar	42
Japan wax	15
Carnauba wax	15
Glycerine	15

Formula 15RC, coal tar rosin[19]

Ingredient	Weight (pounds)
Rosin WW	248
Zinc oxide	238
Silica	78
Cuprous oxide	145
Mercuric oxide	45
Coal tar naphtha	47
Coal tar	6.4
Pine oil	5.2

Formula 15HP, hot plastic[19]

Ingredient	Weight (pounds)
WW gum rosin	525
Phenolic condensate[a]	157
Paraffin wax	374
Cuprous oxide	437
Magnesium silicate	113

Formula 121/63[19]

Ingredient	Weight (pounds)
Cuprous oxide	1440
Vinyl resin VYHH	55
Rosin	215
Tricresylphosphate	50
Methyl isobutyl ketone	165
Xylene	115
Antisettling agent	5–9

[a]Phenolic condensate:

Ingredient	Weight (pounds)
Phenol	600
Lead acetate	67.5
Formaldehyde (40%)	414

oxide paints based upon tropical lifetime, all those tested by the author providing only 3 to 4 foul-free months at Miami Beach. However, they did possess several advantages. Being nonionic in character, anticorrosive undercoatings were not necessary, and, unlike the red-brown to brown colors of the cuprous oxide paints, TBTO paints could take any color – brilliant blues, whites, yellows, greens, and reds.

Other organotin materials were also examined. bis(Tributyltin) adipate was found useful in antifouling formulations, 18 months' immersion lifetime being claimed.[42] Fouling protection of 2 years has been reported for a coating containing tributyltin pentachlorophenate.[43] Other organotin materials reported as being antifouling are tributyltin carboxyethylmercaptals,[44] trialkyltin oxymercuric alkyls,[45] bis(tributyltin) tetrachlorophthalate,[46] tributyltin sulfide,[47] and tributyltin fluoride.[48] Paints containing triphenyl or trialkyl tins in combination with other toxicants have been described; for instance, Beers combines tributyltin fluoride with 10,10-oxybisphenoarsine.[49]

The use of organotins in preserving immersed wood structures shows much promise. As previously considered creosote and coal tar have long been used. Creosote loadings as high as 20 to 30 lb/ft^3 of wood are often required, and, although this gives adequate control of the teredo, it does not prevent *Limnoria* attack. Vind and Hochman showed that select antifouling organotins were quite toxic to *Limnoria* and teredo larvae.[50] Also, tests carried out at the U.S. Naval Civil

TABLE 2–2

Vinyl-rosin Paint Formulations

Ingredient	Navy paint 121	ILZRO formula[a]
Cuprous oxide	800 lb	400 lb
TPLA	–	200 lb
PVC/AC copolymer[a]	80 lb	80 lb
WW gum rosin[b]	80 lb	80 lb
Tricresyl phosphate	30 lb	30 lb
Methyl isobutyl ketone	275 lb	275 lb
Xylene	185 lb	185 lb
Barytes	–	400 lb
Red iron oxide	–	200 lb
Total weight	1450 lb	1850 lb
Total yield	102 gal	125 gal

[a]Vinylite VYHH, Union Carbide Corp.
[b]Hercules, Inc.

From Carr, D. S., *Paint Varn. Prod.*, 64, 19, 1974. With permission.

Engineering Laboratory showed that 0.25 lb of TBTO impregnant per cubic foot of wood provided 25 to 30 months wood life against teredo and *Limnoria* attack, as compared to 15 months for 34 lb/ft^3 of creosote.[51]

Oldfield and Sansom report that organotin paints applied to wood panels showed poor antifouling properties, with tributyltin fluoride showing superiority over TBTO and TBTAc (tributyltin acetate).[52]

One new discovery allows the elimination of solvents in the use of antifouling and antiboring coatings. Epichlorohydrin bisphenol epoxy resins are used with extending pigments and TBTO. The reported combinations provide a sprayable or brushable paint having 100% total solids, thus eliminating a number of human health hazards associated with conventional paints.[53]

A number of organolead materials have shown promising performance in antifouling paint. Triphenyllead sulfide, triphenyllead chloride, triphenyllead 2-ethyl hexanoate, triphenyllead laurate, triphenyllead imidazole, and triphenyllead benzyl sulfide hexaphenyldilead have reportedly demonstrated 18 months' antifouling life in coatings.[19] Extensive antifouling evaluation of various paint bases containing TPLA (triphenyllead acetate), TBLA (tributyllead acetate), TPLF (triphenyllead fluoride), and TBLF (tributyllead fluoride) have been performed.[54,55]

Up to 5 years' antifouling protection in tropical waters has been reported where both cuprous oxide and TPLA were used in combination (ILZRO Formula 1),[56] a modification of U.S. Navy-type 121 vinyl-rosin paint, as shown in Table 2–2.

The long-term antifouling property observed in immersion tests is undoubtedly due to a synergistic mechanism since cuprous oxide alone or TPLA alone in the same paint base does not provide anywhere near the noted 5-year longevity.

Organolead paints based upon various elastomers have also been prepared and evaluated.[56,57] Elastomers used are Natsyn® 400 (Goodyear Co.), a styrene-butadiene block polymer; Kraton® 1101 (Shell Chemical Co.), a chlorosulfonated polyethylene; Hypalon® 30 (DuPont) also a chlorosulfonated polyethylene; Hydrin® 100 (B. F. Goodrich Chemical Co.), a poly-epichlorohydrin; and various combinations such as a butadiene-acrylonitrile polymer with a polyurethane. Figures 2–7 and 2–8 illustrate exposure studies.

A number of nonmetallic organic toxicants have been evaluated. Various carbonates, polychlorinated hydrocarbons having the methanobenzene structure, and others have been studied,[41,58] although such substances are almost always effective against a few categories of fouling. Probably the superior nonmetallic organic antifoulant discovered and in use today is the Diamond Shamrock Corp. DS2787, 2,4,5,6-tetrachloroisophthalonitrile.[59] This very stable, noncorrosive material has provided excellent antifouling properties to various paint bases.[60]

CONTROLLED RELEASE ANTIFOULING SYSTEMS

Historical

The first controlled release antifouling coverings were created in 1964 in response to a rubber sonar dome problem. Sonar systems are a vital part of a modern navy's ability to detect and respond to submarine attack. In order to protect sonar transmitters from the open sea they are encased in a rubber or metallic dome. Like any object immersed in the ocean such domes are prone to fouling attachment when the vessel is moving very slowly or is still. Antifouling paints applied to either type of sonar dome have an extremely short life due to surface cavitation

FIGURE 2–7. Test panel at left coated with U.S. Navy Type 121/63 vinyl high-rosin paint. Photo at right shows International Lead Zinc Research Organization Formula 1 organolead antifouling paint. Both paints exposed 31 month at Hawaii. (Photo courtesy of International Lead Zinc Research Organization.)

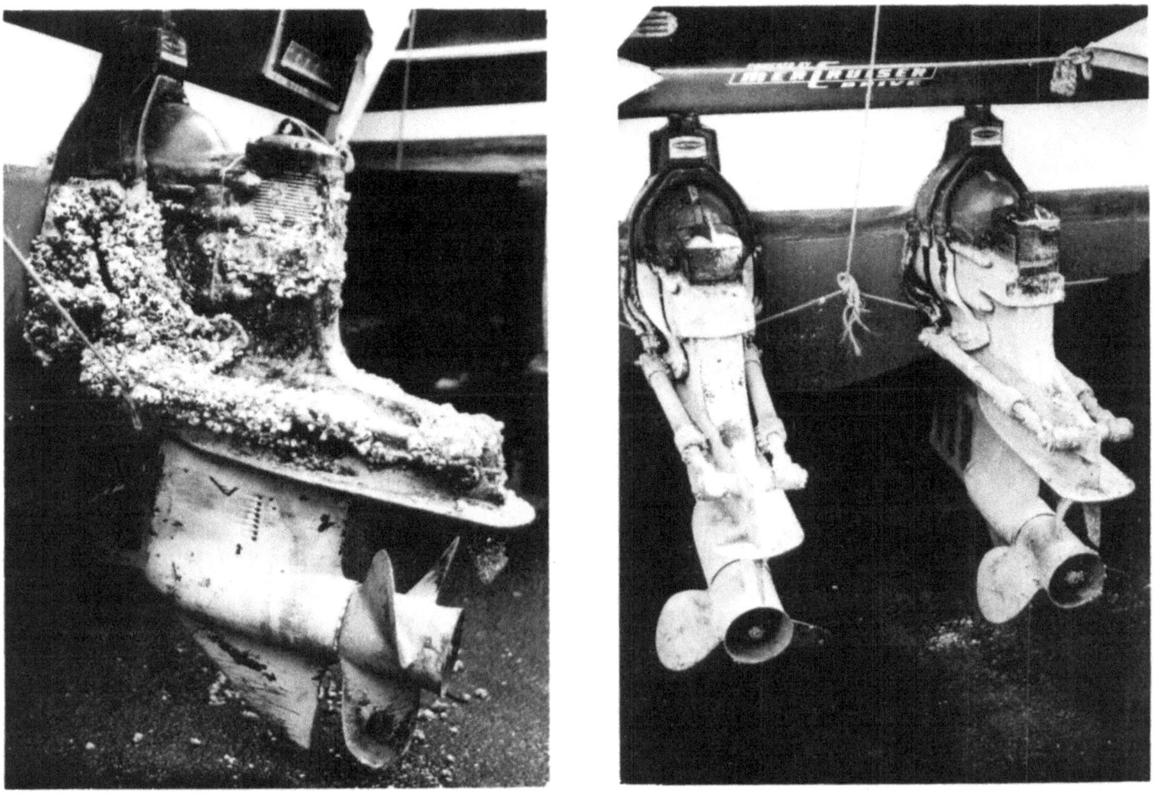

FIGURE 2–8. Outboard engines after exposure in Connecticut coastal waters from May to late fall 1972. Lower portion of engine on left coated with elastomeric antifouling paint containing TPLA. Entire immersed portions of engines on right were coated with organolead antifouling paint. (Photo courtesy of International Lead Zinc Research Organization.)

arising from transmission of the acoustical signals. In fact, domes may be denuded in spots after just a few hours' transmission. Calcareous fouling organisms will rapidly locate and attach to such unprotected spots. Development of shelled foulants on the dome surface leads to very serious interference and distortion of the sonar signal. Sonar becomes inoperable with a relatively small amount of calcareous fouling.

Fouled domes may be defouled by divers. However, this is tedious work that must be frequently repeated, and during this operation the sonar detection unit is not operable, so that the military function is impaired.

In order to meet this problem scientists at the B. F. Goodrich Company took several simultaneous approaches. One consisted of determining how the barnacle attached and then developing a surface whose chemical and/or physical nature prevented attachment.[4,10,61] This approach was abandoned in late 1964 (although several groups have recently revived it) when another concept, that of adding a toxicant directly to the molded rubber dome, was found feasible.[62] Once success was certain in the dome area, it was hypothesized that a sheet rubber product could be developed that would contain the toxic element and release it at an efficacious rate so as to provide long-term antifouling. This conceptual sheet product could then be adhesively attached to ship hulls, buoys, pilings, instruments, and various underwater structures to provide fouling protection.

A number of materials were selected from Miller's compendium[41] and evaluated in various rubber matrices. TBTO was found to satisfy the following criteria:

Universal antifoulant
Solubility to some degree in the elastomer
Adequate diffusion rate in the matrix
Relative insolubility in water
No interference with crucial physical properties of the elastomeric base.

TABLE 2–3

Elastomeric Types Suitable for Antifouling Rubber

Chloroprene (neoprene)
Styrene-butadiene polymers
Nitrile rubber
Acrylonitrile rubber
Polyisobutylene
Polyurethane
Ethylene-propylene polymers
cis-Polybutadiene
Natural rubber
Synthetic natural rubber

Very early in the development program it was realized that only materials soluble in the elastomer base would provide a sustained release. Cuprous oxide, for instance, can be incorporated in rubber, but will not release unless the binding matrix contains about 80% Cu_2O or better and a leach-type mechanism can be established. A few materials have slight solubility, in chloroprenes for example, and thus give some degree of protection. Copper pentachlorophenate and phenyl mercury oleate are examples.

Table 2–3 illustrates the elastomer categories found suitable to hold and release the toxicants listed in Table 2–4. The agents and generic categories given in these two tables cannot be used in random combination. For instance, TPTL works well in several chloroprenes, but not in styrene-butadiene polymers. TBTO is excellent in most chloroprenes but is very poor in the polyurethanes. The key is to properly match the matrix geometry with the toxicant. Knowledge of solubility of the agent in the elastomer (preferably high) and solubility in water (preferably low) and diffusion rate of the toxicant in rubber is a prerequisite. Since the physical characteristics of the elastomer are controllable, it is possible to optimize the antifouling character of a given system by:

1. Selective polymerization to create the proper matrix geometry
2. Vulcanization conditions which control cross-linking and thus the geometry of the interstices
3. Compounding techniques wherein additives are used to increase or decrease the diffusion rate (within limits) by extending or decreasing the effective diffusion path length, or to alter free volume.

TABLE 2–4

Antifouling Agents

TBTO®	bis(Tri-*n*-butyltin) oxide
TBTS	bis(Tri-*n*-butyltin) sulfide
TBTAc	Tributyltin acetate
TBTA	Tributyltin adipate
TBTF	Tributyltin fluoride
TPTC	Triphenyltin chloride
TBTR	Tributyltin resinate

The initial static test program was performed by the placement of candidate rubber specimens on various wood, metal, or fiberglass substrates and immersion in ocean waters (Figures 2–9 through 2–12). Miami Beach, Duxbury, Mass., Los Angeles, and Honolulu were the initial test sites. Nearly 3,000 test panels were immersed in this study as well as a number of small buoys, pilings, cables, chains, cement blocks, and other objects.

These shoreline immersions were at 6- to 10-ft depths. However, in a series off Bermuda, panels were exposed to 150-ft intervals to 6,6000-ft depths. At this time antifouling rubber has been successfully used in covering buoys,[63] sonar domes,[64] ship hulls,[65] and other objects.

Formulations and Evaluation

Elastomer Compounding

It is not within the scope of this text to detail the methodology and technique of elastomer compounding. A brief survey of this very extensive field is included here; the reader is referred to the many excellent publications in this area.[66,67]

Elastomers

The starting point in the creation of a rubber formulation is the selection of an appropriate elastomer. An elastomer is basically a polymeric material that will stretch under tension, and, when the force is released, will return to its original shape and size. This "elastic" condition arises from weak interchain forces and crosslinking characteristics. Thus, when the polymeric backbone is "straightened out" by application of a tensile force, the macromolecules involved lack sufficient attraction for each other to maintain the stretched state once tension is released. In contrast, plastic materials have interchain forces of moderate strength. If the polymeric chains are stretched the molecules retain the new configuration even after the tension is released. Of greater importance to controlled release, however, is the fact that most plastic materials are solids and all elastomeric materials are liquids at room temperature from the standpoint of thermodynamic and kinetic behavior. Thus, when considering a toxicant dissolved or dispersed in an elastomer we have solid-in-liquid or liquid-in-liquid solution or suspension.

FIGURE 2–9. Immersion test rack with panels. (Photograph by the author.)

Elastomers in their original state are macromolecules of essentially no strength or useful physical properties. They are spoken of as "gums." In order to change the gum form into a usable material (for almost all applications) it is necessary to cross-link the smaller macromolecules, creating extremely large macromolecules. This is accomplished by use of a cross-linking agent ("curative") and vulcanization ("curing"), i.e., heating the elastomer and the curative to some temperature for a given period of time in order to induce chemical reaction. Table 2–5 provides the common and chemical names of major elastomers used in controlled release work, along with a common curative.

Other than natural rubber, there are many forms of the categories shown in Table 2–5. For instance, there are more than 40 SBR materials in large-scale usage, and around 20 different neoprenes. Variation is usually with molecular weight of the macromolecule prior to vulcanization, but can be with the degree of unsaturation in the backbone, side branching, etc.

FIGURE 2–10. Test station view. (Photograph courtesy of Miami Marine Test Station.)

A.

FIGURE 2–11. Covered sonar dome. A: Front view. B: Side view. (Official Navy photographs.)

Curing Agents

Sulfur remains the principal curing agent, but other sulfur-bearing chemicals may also be used. Red lead oxide can be used to cross-link butyl rubber. Various peroxides serve as vulcanizing chemicals for SBR, and certain dioximes can be used in some instances.

Accelerators

In order to improve the rate of cure and often to enhance physical characteristics of the cured material certain active chemicals are used in conjunction with the curative. Accelerators can be classified in accordance with their rate of reaction. Slow (inorganic) accelerators such as lime, litharge, and magnesia are not much used. Medium accelerators such as diphenylguanidine (DPG) are usually used in conjunction with faster acting accelerators to retard cure rate and prevent "scorch" (premature vulcanization). Of much greater importance are fast accelerators such as mercaptobenzothiazole (MBT) and ultra-fast materials such as tetramethylthiuram disulfide (TMTDS).

Activators

An accelerator may provide too slow a cure rate when used alone. The reaction can be speeded up by adding a second (usually fast or ultra-fast) accelerator or some other materials that are poor accelerators by themselves. Accelerators such as various dithiocarbamates or thiuram sulfides may be employed; when used as such they are often termed "secondary accelerators." Usually the primary-to-secondary accelerator quantitative ratio

Figure 2–11 B.

FIGURE 2–12. Nofoul® covered vessel hull. (Photograph courtesy of Miami Marine Test Station.)

TABLE 2–5

Elastomers and Vulcanizing Agents

Name	Chemical name	Curative
Natural rubber	cis-1,4 Polyisoprene	Sulfur
SBR	Poly(styrene-butadiene)	Sulfur
Butyl	Poly(isobutylene-isoprene)	Sulfur
Neoprene	Polychloroprene	Magnesium oxide, zinc oxide, or both
Nitrile	Polyacrylonitrile	Sulfur
Urethane	Polyester urethane	Isocyanates
Urethane	Polyether urethane	Isocyanates
CB	cis-Polybutadiene	Sulfur

lies between 4 to 1 and 10 to 1. Zinc oxide and stearic, palmitic, or oleic acids are also used as activators.

Peptizing Agents

In order to mix rubber compounds it is necessary to "break down," i.e., lower the molecular weight of, the gum elastomer. Such action is performed on the rubber mill or in external mixers. Breakdown is accelerated by the use of a peptizing agent, often termed a lubricant, which acts by coating the broken molecular chains and preventing the reestablishment of bonds between molecules. Lubricants are complex organic sulfur compounds or various fatty acids.

Fillers

Fillers may be reinforcing, i.e., increasing tensile strength, decreasing elongation, increasing tear resistance, and in general enhancing favorable physical properties, or nonreinforcing materials or extenders (used to increase bulk, or to economically cheapen the material).

Carbon black is the reinforcing filler in general use, the particle size being vital to the amount of reinforcement imparted. These carbon blacks of large particle size such as FT (fine thermal – 200 nm) and MT (medium thermal – 400 nm) provide very little strength to the rubber and can be considered as nonreinforcing materials.

There are nonblack reinforcing fillers, a necessity if the product is to have any color but black. Various silicas of fine particle size are used, as well as zinc oxide, magnesium carbonate, and certain clays.

Extending fillers are numerous. Large-particle-size clays, talc, limestone, chalk, and barytes are used, along with many other materials.

Softeners

Softeners act through the physical lubrication of the elastomer molecules (although the effect may be similar, peptizing agents act chemically). Softeners increase the "tack" (stickiness) of the elastomer mix, facilitating the incorporation of various additives; also, by rendering the mixed stock stickier, they allow easier attachment to other surfaces prior to vulcanization. Pine tar, wood rosin, various oils, waxes, and fatty acids are used as softeners.

Protective Agents

Elastomers deteriorate upon exposure to various environments. Chemical degrading occurs mainly due to oxidation or contact with ozone, although some materials may slowly hydrolyze in the presence of water. Most solvents also affect rubber products. Bacteria and fungi are the major culprits in biological deterioration. Unprotected SBR sheet materials exposed to the combined action of sea water and tropical marine bacteria degrade to a strengthless lacelike substance in a few months. Proper choice of elastomer will overcome a part of the chemical attack problem. Usually, however, it is necessary to add an antioxidant such as phenyl-β-naphthylamine (PBNA) and/or an antiozonant. Various waxes, especially microcrystalline ones, can offer substantial protection against oxidation. Protection against microbe attack can be gained by adding a toxic agent.

Special Additives

Chemical additives are used in many ways: pigments to impart color, pumice to give an abrasive surface, low density materials to decrease the product specific gravity, blowing agents to

TABLE 2–6

Antifouling Rubber Compositions (SBR)[68]

Ingredient	Composition (by parts)				
	1	2	3	4	5[a]
Styrene-butadiene	100[b]	100[c]	100[d]	100[d]	100[d]
Zinc oxide	5	5	–	–	1
Sulfur	2	2	–	–	1.75
BTDS[e]	1.75	3	–	–	–
Stearic acid	–	1.5	–	–	1
HAF black	40	40	40	40	–
TBTO®	0.02–10	0.02–10	0.02–10	0.02–10	0.02–8
TMTDS[f]	–	–	3	–	–
Dicumerol peroxide	–	–	–	5	–
Silica	–	–	–	–	39
MBT[g]	–	–	–	–	0.75
DPG[h]	–	–	–	–	1.5
Hexamethylene – tetramine	–	–	–	–	1
Diethylene glycol	–	–	–	–	3

[a]Transparent antifouling SBR.
[b]NBS 387 styrene-butadiene (National Bureau of Standards).
[c]NBS 386 styrene-butadiene (National Bureau of Standards).
[d]Ameripol® 1007.
[e]Benzothiazyl disulfide.
[f]Tetramethylthiuram disulfide.
[g]Mercaptobenzothiazyl.
[h]Diphenyl guanidine.

decrease specific gravity, retarders to slow down vulcanization, antitack materials to remove stickiness, and substances that impart special physical qualities such as low temperature resistance, abrasion resistance, puncture resistance, and so on.

Nofoul® Rubber

Antifouling elastomers containing organotin toxicants were invented by B. F. Goodrich Company scientists in 1964.[26,58,62,68] The earliest (French) patent[68] provides the elastomeric recipes shown in Tables 2–6, 2–7, and 2–8. Acrylonitrile and polyurethane compositions are also included in the patent, though not depicted here. These recipes illustrate the range of workable materials. Both neoprene and natural rubber compounds were used in sonar dome coverings; however, the all-purpose antifouling sheet material for buoys, ships, pilings, etc. has been a neoprene composition.

In order to properly review the recent (and superior) recipes, the information in Tables 2–9 and 2–10 is derived from several later patents,[69,70] available B. F. Goodrich literature,[71,72] and publications.[61,62,73] Table 2–9 shows the standard 351 neoprene formula and Table 2–10 illustrates fouling results on 0.062-, 0.032-, and 0.125- in.-thick sheet panels immersed in subtropical waters. Rubber panels are secured to metal backing. Actual sizes are either 8 × 10 or 10 × 12 in. 500+ barnacle coverage is essentially complete fouling coverage. Both the amount of toxicant present and the panel thickness determine lifetime. Table 2–11 presents further evaluation of the variation in thickness.

It appears that long-term antifouling activity from neoprene-based materials requires at least a thickness of 0.05 in. and a TBTO content of 2 parts (based on 100 parts of elastomer) or better. The buoys of the first commercial application[63] remain essentially foul free after 92 months' immersion in the heavily fouled water off Key West, Fla.

TBTS at 7 parts/100 parts neoprene on test panels has also shown over 40 months' complete antifouling activity in subtropical immersion.

TABLE 2–7

Antifouling Rubber Recipes (Natural Rubber, Butyl Rubber)[68]

Ingredient	Composition (by parts)			
	1	2	3[a]	4
Natural rubber	100	100	100[b]	–
Zinc oxide	5	5	1	5
Sulfur	2.5	3	1	2
Stearic acid	2	3	1	3
HAF black	–	50	–	50
BTDS[c]	1	1	–	0.5
PBNA[d]	1	1	2	–
TBTO®	0.02–10	0.02–10	0.02–8	0.12–14
Silica	–	–	25–39	–
MBT[e]	–	–	1	–
Methyl zimate	–	–	0.25	–
Hexamethylene – tetramine	–	–	1	–
Sulfasan® R	–	–	1	–
Ammonium acetate	–	–	1	–
Butyl rubber[f]	–	–	–	100
TMTDS[g]	–	–	–	1

[a]Transparent recipe.
[b]Pale crepe.
[c]Benzothiazyl disulfide.
[d]Phenyl-β-naphthylamine.
[e]Mercaptobenzothiazyl.
[f]NBS No. 388 (National Bureau of Standards.)
[g]Tetramethylthiuram disulfide.

TABLE 2–8

Antifouling Neoprene (Chloroprene) Recipes[68]

Ingredient	Composition (by parts)				
	1	2	3	4[a]	5[b]
Neoprene GN	100	100	–	–	–
Neoprene WRT	–	–	100	100	100
Zinc oxide	5	5	5	5	5
Magnesium oxide	4	4	4	4	4
Stearic acid	0.5	1	–	–	–
HAF black	–	20–40	–	–	–
Petroleum wax	–	–	0.5	0–2	0–2
FEF black	–	–	12–20	–	–
PBNA[c]	–	–	2	2	2
MBT[d]	–	–	1	1	1
Lauric acid	–	–	0–4	0–4	0–4
TBTO®	0.02–12	0.02–12	0.02–20	0.02–8	0.02–8
Titanium dioxide	–	–	–	12–30	12–30
Pyrazoline red	–	–	–	–	2–5

[a]White composition.
[b]Red composition.
[c]Phenyl-β-naphthylamine.
[d]Mercaptobenzothiazyl.

TBTS is more expensive and somewhat more difficult to handle in processing, so it is unlikely that this substance will find use in antifouling materials, especially since its activity appears to be no greater than that of TBTO.

Antifouling elastomers must release TBTO at a certain rate to prevent fouling. By immersing a great number of test panels, taking periodic cores from such panels, and analyzing for residual tin it was possible to determine the fouling threshold for various fouling forms; Table 2–12 illustrates the results.

There are slight variations among the species that fall in each category, and the values in Table 2–12 hold only for Floridian barnacles and East Coast (U.S.) algae.

Accumulation of fouling organisms is gradual even after the dissolution rate falls below the threshold value. Figure 2–13 illustrates algae and barnacle attachment in terms of residual tin concentration and percent of the test panel covered. Curves are not as exact as shown due to specie variation, but are averages. Note the far greater resistance to TBTO by algae.

Algae seems to attach just above threshold, accumulate organotin for several days or weeks, and eventually die.

Carbon black content is crucial to the control of release rate. For instance, the following base recipe, B151, was loaded with TBTO or TBTS at varying levels and immersed, with results shown in Table 2–13.

TABLE 2–9[70]

Neoprene 351 Formulation

Ingredient	Formulation (parts)
Neoprene WRT	100
FEF carbon black	14.5
PBNA[a]	2.0
Zinc oxide	5.0
Lauric acid	3.0
Magnesium oxide	4.0
Ethylene thiourea	0.75
MBTS[b]	1.0

[a]Phenyl-β-naphthylamine.
[b]Mercaptobenzothiazyl disulfide.

TABLE 2–10[70]

351 Neoprene Series Stationary Fouling Evaluation

	TBTO® concentration			Fouling status after 40 months' immersion	
	Parts	Percent by weight	Thickness (inches)	Number of barnacles	Percent of bryozoan coverage
351B	12.0	8.4	0.062	0	0
351C	6.0	4.38	0.062	0	0
351D	4.0	2.9	0.062	0	1
351E	2.0	1.5	0.062	0	0
351F	1.0	0.75	0.062	0	0
351G	0.5	0.37	0.062	0	1
351H	0.25	0.18	0.062	500+	35
351I	0.125	0.09	0.062	500+	100
351J	0.06	0.045	0.062	500+	100
351B	12.0	8.4	0.125	0	0
351D	4.0	2.9	0.125	0	0
351F	1.0	0.75	0.125	0	10
351H	0.25	0.18	0.125	40	17
351J	0.06	0.045	0.125	500+	100
351C	6.0	4.38	0.032	0	0
351E	2.0	1.5	0.032	0	0
35 G	0.5	0.37	0.032	2	5

TABLE 2–11[70]

Panel Thickness Relation to Fouling Initiation (Data to 42nd Immersion Month)

Code	Thickness (inches)	TBTO® (phr[a])	Months to first appearance: Algae	Barnacles	100% fouled (months)
351B	1/32	12	17	24	33
351B	1/16	12	29	42+	–
351B	1/8	12	42+	42+	–
351D	1/32	4	10	17	18
351D	1/16	4	18	30	–
351D	1/8	4	42+	42+	–
351F	1/32	1	7	13	19
351F	1/16	1	6	12	17
351F	1/8	1	10	18	23
351H	1/32	0.25	2	3	8
351H	1/16	0.25	2	4	15
351H	1/8	0.25	6	2	20

[a]Parts per hundred parts rubber.

Table 2–12[62]

Fouling Threshold Values

Category	TBTO® dissolution rate, μg/cm²-day
Algae	5.5
Barnacles	0.80
Bryozoa	0.55
Bugula	0.25
Tubeworms	0.20
Tunicates	0.20 (estimated)

B151 Recipe[70]

Ingredient	Formulation (parts)
Neoprene WRT	100
HAF black	29.6
PBNA	2
Zinc oxide	5
Stearic acid	0.5
Ethylene thiourea	0.5
Magnesium oxide	2

Comparing these data with those of the 351 series, it is seen that the higher carbon black loading leads to earlier fouling at lower TBTO content.

Likewise, the type of carbon black used is an important factor in fouling onset. Using neoprene Formula 351 various carbon blacks (at 14.5 parts) were added and panels were processed and immersed. Table 2–14 shows 5-month immersion results.

Vulcanization time and temperature also effect the antifouling longevity of a given compound. Using Formulation 351 with 14.5 parts of ISAF black and 0.25 parts of TBTO, a number of test panels were prepared. Cure temperature was held constant at 149°C while the cure time was varied as shown in Table 2–15. The compound in question has an optimum cure of about 30 min; 45 through 120 min constitutes an overcure which is essentially greater cross-link density. Antifouling natural rubber compounds were prepared from the master recipe shown below:

Ingredient	Formulation (parts)
Natural rubber	100
HAF carbon black	40
PBNA	1
Zinc oxide	5
Stearic acid	3
MBTS	0.6
Sulfur	2.5

TBTO and TBTS were added at various concentrations. Cure conditions were 15 min at 151°C. Table 2–16 shows the test results after 35 months' immersion at Miami Beach.

The superiority of TBTS in a natural rubber led

to its study in several other elastomers. Table 2–17 shows recipes used, and Table 2–18 shows fouling observations to the 35th tropical immersion month.

Other organotin compounds have also been evaluated in the 351 neoprene recipe. TBTF (tributyltin fluoride) at eight parts loading in 0.0625-in.-thick panels remained foul free after 35 months. TBTAc (tributyltin acetate) at eight parts loading in 0.075-in.-thick panels showed no fouling at 21 months, while TPTC (triphenyltin chloride) at five parts loading in 0.032-in. panels showed fouling onset at the 20th month.

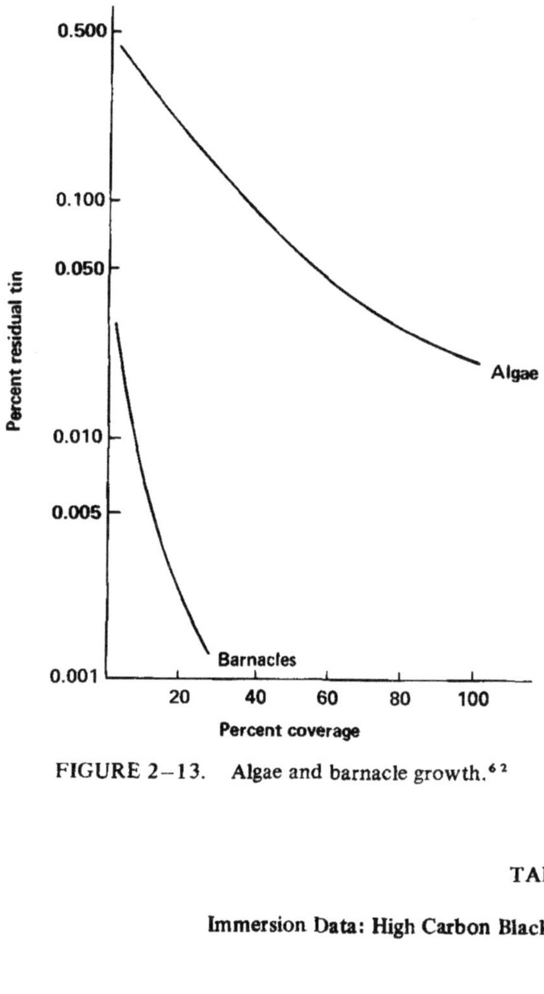

FIGURE 2–13. Algae and barnacle growth.[62]

The value of antifouling rubber (Nofoul) can best be illustrated in comparison to conventional paint coatings (Table 2–19).

Rubber and plastic materials exposed to the marine environment show rapid deterioration due to both chemical and biological attack. The mechanisms involved are not well understood although attempts have been made to enhance our understanding of the phenomena involved.[74]

The effective foul-free lifetime of B. F. Goodrich antifouling rubber can be extended by the use of a membrane-reservoir concept. Many materials that have high organotin solubilities also lose the toxicant much too rapidly; acrylonitrile polymers of high nitrile content and polyurethanes are examples. Also, sponge-type materials such as foamed natural rubber or neoprene will mechanically hold large amounts of TBTO or TBTS, but once immersed the toxicant is completely depleted within a year or less.

In order to overcome this rapid loss rate it is possible to saturate the rubbery material so that it serves as an organotin reservoir, and then to laminate onto that reservoir layer an elastomeric material through which the toxicant will diffuse slowly.[69] This technique permits the incorporation of as much as 25 parts/100 parts base elastomer of the toxicant, whereas the solubility limit for long-term simple Nofoul formulations is about 10 parts.

TABLE 2–13[70]

Immersion Data: High Carbon Black Content Test Panels (Miami, Fla. Immersion)

			Months to (average)			
Code	Toxicant	Loading (parts by weight)	1% algae	1 barnacle	50 barnacles	Completely fouled
B151	0 (control)	0	1	1	1	1
B152	TBTO®	1	7	18	19	21
B153	TBTO	3	24	26	27	29
B154	TBTO	5	26	28	–	–
B155	TBTO	8	30	31	–	–
B156	TBTS	2	20	26	28	–
B157	TBTS	4	22	27	–	–
B158	TBTS	6	21	–	–	–
B159	TBTS	10	–	–	–	–

Another, perhaps more practical, method of increasing the toxicant loading is to establish a three-phase system. The organotin is first absorbed or adsorbed on an inert material such as phenol/formaldehyde microspheres, and then the now-loaded microspheres are added to the rubber mix as a filler.[75] It is possible to achieve loadings as high as 100 parts of the organotin agent to 100 parts of the base elastomer as compared to the normal 10- to 12-part organotin solubility maximum of first generation antifouling rubber. Exceeding the solubility limit in a two-phase organotin/rubber system leads to drastic loss of pertinent physical properties and often destroys processibility.

TABLE 2–14[70]

Influence of Type of Carbon Black on Fouling Onset (0.72% TBTO® Content)

Carbon black type	Particle size range (millimicrons)	5-month fouling performance
SAF	11–19	No fouling
ISAF	20–25	No fouling
EPC	26–30	No fouling
FF	31–39	No fouling
FEF	40–48	Fouling in Month 4
HMF	49–60	Fouling in Month 3
SRF	61–100	Fouling in Month 2
FT	101–200	Fouling in Month 2
MT	201–500	Fouling in Month 2

TABLE 2–15

Effect of Cure Time on Antifouling Performance (Average)

Cure time (minutes)	Months to first barnacle	6-month observation; number of barnacles
20	–	None
30	–	None
45	6	2
60	3	2
90	2	20
120	2	5

In the case of various plastics matrices, TBTO and other organotins are essentially insoluble. Polyvinyl chloride, for instance, will accept and hold a few percent of TBTO through mechanical occlusion, but it will not release upon immersion, so that no antifouling capability is observed. Again, the use of a three-phase system will allow the incorporation and release of sufficient quantities of TBTO to prevent fouling for 8 to 12 months. TBTO is partially soluble in several plasticizers. About 5% TBTO can be added to dioctylphthalate (DOP) and this material incorporated at 30% or more into polyvinyl chloride. Upon immersion, DOP migrates slowly towards the plastic surface, carrying with it TBTO molecules which in return are released into the sur rounding waters to repel fouling.

Other Antifouling/Organotin Rubber Systems

Scientists at the Defense Standards Laboratory in Australia have been developing and evaluating organotin/elastomer systems for some time. Woodford has reported on the incorporation of TBTO, TBTF, and TBTAc in natural rubber, nitrile rubber, and polychloroprene.[76] TBTO and

TABLE 2–16[70]

Antifouling Evaluation of Natural Rubber

Code	Toxicant	Toxicant concentration (parts)	Months to: 1% algae	1 barnacle	50 barnacles	Completely fouled
B1	0 (control)	0	1	1	1	1
B2	TBTO®	1	11	13	15	18
B3	TBTO	3	18	16	20	21
B4	TBTO	5	26	22	24	28
B5	TBTO	8	26	26	35+	35+
B6	TBTS	2	8	11	19	19
B7	TBTS	4	20	18	21	23
B8	TBTS	6	20	22	23	23
B9	TBTS	8	29	35+	35+	35+

TABLE 2–17[70]

TBTS Formulations

	CB[a]	Hycar®[b]	SBR[c]	Butyl[d]	EPDM[e]
Rubber	100	100	100	100	100
ISAF black	30	–	–	–	85
SRF black	–	40	–	–	–
HAF black	–	–	40	–	–
EPC black	–	–	–	50	–
Zinc oxide	3.6	3.6	5	5	5
Stearic acid	–	–	1.5	3	0.5
Lauric acid	3.0	3.0	–	–	–
MBTS	–	–	0.8	0.5	0.5
TMTD	–	0.4	0.15	1.0	1.5
Santocure®[f]	1.7	–	–	–	–
Sulfur	2.5	1.5	1.75	–	–
TBTS					
Cure time	15 min	15 min	30 min	30 min	60 min
Cure temperature (°F)	307	307	307	307	307

[a]Ameripol® CB, cis-1,4-polybutadiene rubber.
[b]Hycar 1043, butadiene/acrylonitrile with approximately 34% acrylonitrile; Hycar 1001, similar but containing 38 to 40% acrylonitrile; Hycar 1014, similar but containing about 18% acrylonitrile.
[c]Styrene-butadiene elastomers No. 1001, 1013, and 1015.
[d]NBS No. 388.
[e]Nordel® 1070 ethylene-propylene polymer.
[f]*N*-cyclohexyl-2-benzothiazylsulfenamide.

TABLE 2–18[70]

Immersion Evaluation of TBTS-containing Elastomers (35-month Test, Miami Beach)

	Months to	
Elastomer base	1 barnacle	Complete fouling
Hycar® 1043	35+	35+
Hycar 1014	30	35+
Natural rubber	29	35+
SBR 1013	29	35+
CB	25	35+
SBR 1015	24	35+
Hycar 1001	23	35+
Butyl	22	35+
SBR 1001	22	35+
EPDM 21	21	25

TBTAc were found to be soluble at varying levels in the given elastomers while TBTF was found to be nonsoluble through examination of scanning electron micrographs of cryogenically fractured surfaces. Woodford lists the controlling parameters for the release of organotin from elastomers as:

Solubility of the organotin in the rubber
Solubility of the organotin in sea water
Water absorption properties of the elastomer.

Woodford also found that elastomers with higher water absorption made the use of TBTF more favorable. It is suggested that TBTF is the superior antifouling agent for incorporation in moderately water-resistant rubber formulations. Thin solution coatings of organotin/rubber failed as antifouling surfaces whereas vulcanized sheet rubber provided at least 12 months foul-free life.[76,77]

This research effort was continued by deForest, Pettis, and Phillip, who provided extensive data in a 2-year summary report.[78] Immersion effects on the mechanical properties are present (10 to 15% decrease in ultimate elongation, tensile strength, and tear strength) but do not significantly affect the durability of the materials. Table 2–20 presents the compositions and fouling observed after extended immersion periods.

Tin loss rates were also measured, and it was

TABLE 2–19[73]

Comparison Between Nofoul® and Antifouling Paint

	Nofoul	Paint
Loss mechanism	Diffusion-dissolution	Leaching, exfoliation
Predictability (longevity)	Yes, based on conventional diffusion equations	No
Toxicant	TBTO®, TBTS, TBTA, TBTF, or mixtures	Generally copper, few organotin
Toxicant content	2–10%	85–95% for copper, 30–55% for organotin
Electrical character	Nonionic	Cuprous oxide types are ionic
Application state	Solid	Liquid
Form	Sheet, molded item	Paint, putty
Thickness	0.06–0.125 in.	0.002–0.012 in.
Longevity (foul-free life)	Over 90 months in tropical waters	Generally 4–8 months, exceptional 12 months in tropical waters
Peel strength	100+ lb	5–15 lb
Tensile strength	2000 psi	25–100 psi
Toxicity	Low (except when hot)	Moderate
Shelf life	Over 5 years	12–24 months
Appliability	Depends on object; buoys easy, hulls somewhat difficult	Usually easy
Universality	Repels all fouling forms, prevents worm and limnoria penetration	Repels all fouling forms, susceptible in many instances to worm attack
Cavitation resistance	Excellent	Poor
Economics	$2–$3/ft² applied	About $0.25–$0.40/ft² applied
Abrasion resistance	Excellent	Poor
Mechanical penetration resistance	Good	Poor

shown that the natural rubber composition lost at the most rapid rate, nitrile at the lowest rate, and neoprene (polychloroprene) at an intermediate rate.

The Goodyear Tire and Rubber Company has developed a sheet coating material, designated MA852, which is an organotin/neoprene system similar to Nofoul, at least in short-term performance.[79]

It has long been known that certain organotin materials can serve as monomers and be polymerized to macromolecules by various means, such as the following compound synthesized in 1965 by B. F. Goodrich scientists:

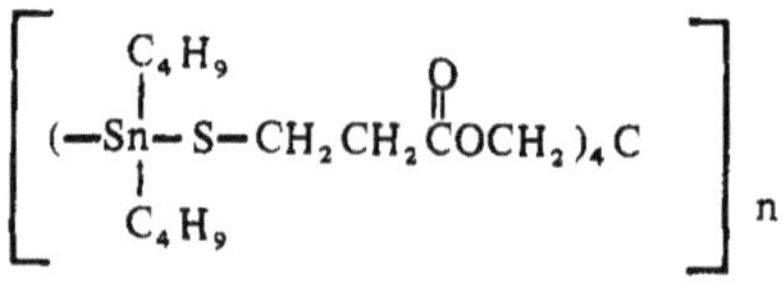

Dunn and Oldfield have developed a new concept in antifouling elastomers by adding an unsaturated organotin monomer to an elastomer mix and curing with a peroxide.[80] Three reactions can occur: cross-linking of the elastomer, homopolymerization of the monomer, or covulcanization of the elastomer and the organotin

TABLE 2–20[78]

Evaluation of Organotin/Elastomers After Extended Immersion Periods

Code	Elastomer	Toxicant	Toxicant concentration	Immersion period (days)	Attached epibiota[a]	Fouling rating[b]
WP1A	NRX	TBTO®	2%	547	S+, T, B	2
WP1B	NRX	TBTO	5%	547	S–, B–	7
WP1C	NRX	TBTO	10%	547	S	8
WP2A	NRX	TBTF	2%	547	S–, T–, B	5
WP2B	NRX	TBTF	5%	547	S–, B–	7
WP2C	NRX	TBTF	10%	547	S–, B–	9
WP3A	NRX	TBTAc	2%	547	S, T–, B+	3
WP3B	NRX	TBTAc	5%	547	S–, B	5
WP3C	NRX	TBTAc	10%	547	S–, B	5
WP4	NRX	0	0	547	S+, T+, B+	0
WP5	Nitrile	0	0	636	S+, G+, W+, B+	0
WP6A	Nitrile	TBTO	2%	636	S, B, W–	4
WP6B	Nitrile	TBTO	5%	636	S–, B–	8
WP6C	Nitrile	TBTO	10%	636	S–	9
WP7A	Nitrile	TBTF	2%	636	B+, G, T, W	0
WP7B	Nitrile	TBTF	5%	636	S, B–	6
WP7C	Nitrile	TBTF	10%	636	S, B–	8
WP8A	Nitrile	TBTAc	2%	636	B+, T, Z	0
WP8B	Nitrile	TBTAc	5%	636	S, B–	8
WP8C	Nitrile	TBTAc	10%	636	S–, B–	9
WP9	Neoprene	0	0	730	B+, T, Z	0
WP10A	Neoprene	TBTO	2%	730	S+, B, W–	4
WP10B	Neoprene	TBTO	5%	730	S, B–	6
WP10C	Neoprene	TBTO	10%	730	S	9
WP11A	Neoprene	TBTF	2%	730	S, B, Z–	4
WP11B	Neoprene	TBTF	5%	730	S	9
WP11C	Neoprene	TBTF	10%	730	S–	9
WP12A	Neoprene	TBTAc	2%	730	S+, B+, Z–	2
WP12B	Neoprene	TBTAc	5%	730	S, B–	6
WP12C	Neoprene	TBTAc	10%	730	S–	8

[a]S = slime, T = tube worms, Z = bryozoa, B = barnacles, G = bugula, W = watersipora, – = slight or few, + = heavy or numerous.

[b]Fouling rating is subjective covering the range from 0 (completely fouled) to 10 (completely clean).

compound. In covulcanization the polymer radical can react with tributyltin acrylate to form an intermediate which can further react to form a cross-linked network. In this case the organotin compound is firmly bound into the polymer network. Up to 50% TBT acrylate can be so incorporated into different types of elastomers. The following base elastomers (and one plastic) were used in this study: chlorosulfonated polyethylene (CSM), ethylene-propylene terpolymer (EPT), ethylene-propylene copolymer (EPR), natural rubber (NR), SBR, styrene-butadiene black copolymer (SBS), polybutadiene (BR), nitrile-poly (vinyl chloride) blend (NBR/PVC), polychloroprene (CR), butyl rubber (IIR), nitrile rubber (NBR) and polyethylene plastic (PE).

Dicumyl peroxide was used to cross-link TBT acrylate and all elastomers save butyl rubber and polychloroprene, where zinc oxide and sulfur and lead oxide were used, respectively. Sea water immersion for 2 years caused slight to moderate changes in mechanical properties. Organotin release rates from covulcanizates were materially slower than those of simple compositions where TBTO is only an additive. In some cases the loss rate was retarded to the point where fouling occurred, but in most cases the peroxide cure led to increased antifouling life. Table 2–21 sum-

TABLE 2-21[80]

Residual Organotin Content Of Covulcanizates

Compound	Base polymer	Organotin content	Original organotin content	1-year immersion residual organotin content (%)	2-year immersion residual organotin content (%)
RT1	CSM	15% TBT Acr	15%	13	10
RT2	EPT	15% TBT Acr	15	13	12
RT3	EPR	15% TBT Acr	15	13	9
RT4	NR	15% TBT Acr	15	11	11
RT5	SBR	15% TBT Acr	15	13	11
RT6	SBS	15% TBT Acr	15	13	14
RT7	BR	15% TBT Acr	15	13	12
RT8	NBR/PVC	15% TBT Acr	15	15	14
RT9[a]	CR		15	–	–
RT10	PE	15% TBT Acr	15	17	15
RT11	IIR	15% Poly (TBT Acr)	15	12	10
RT12	NBR	15% TBT Acr	15	14	11
RT13	NBR	7.6% TBT Acr	7.6	4	2
RT14[b]	NBR	15% TBT Acr	15	16	13
RT15	NBR	7.5% TBT Acr	7.5	7	7
RT16	NBR	30% TBT Acr	30	26	19
RT17	NBR	50% TBT Acr	50	29	35
RT18	NBR	15% Poly (TBT Acr)	15	13	13
RT19	NBR	5% TBTO®, 10% TBT Acr	15	14	14
RT20	NBR	5% TBTAc, 10% TBT Acr	15	11	10
RT21	NBR	5% TBTF, 10% TBT Acr	15	13	13
RT22	NBR	15% TBT Acr	15	11	8
RT23	NBR	15% TBT Acr	15	13	12
RT24	NBR	15% TBT Acr	15	12	14

[a]RT9 could not be analyzed due to the lead curing agent.
[b]Compound RT14 uses 30 parts silica rather than 60 parts carbon black (RT12) as the reinforcing agent. Likewise, RT22 uses calcium carbonate, RT23 uses clay, and RT24 uses sodium fluoride for reinforcement.

marizes compounds and organotin loss over a 2-year period, while fouling conditions after 2 years' immersion are given in Tables 2–22.

Since slime films are easily washed off on moving objects, they do not in general disturb stream lines, and, since the added weight is negligible, they are relatively unimportant of themselves. Consequently, panels showing only slime and/or only periodic slight algae fouling can be considered good. RT9 and RT13 appear to be superior materials.

Scientists at the U.S. Naval Research and Development Center have synthesized organometallic polymers and applied them directly or in a volatile solvent to immersion test panels.[81] Organometallic addition products of acrylic, vinyl, and cellulose polymers were synthesized. These included organotin esters of carboxymethyl cellulose, organometallic polyacrylates and polymethacrylate, and organotin esters of poly (methyl vinyl ether/maleic acid). Materials showed good antifouling performance after 18 months' immersion in the very heavily fouled waters at Pearl Harbor, Hawaii. Polymers containing tributyltin and/or tripropyltin groups showed the best antifouling character. It is reported that "an optimal antifouling performance material exhibiting a minimal amount of leaching can be produced by varying the ratio of organometallic monomer to inert comonomer along the copolymer backbone."[81] Since effective antifouling properties are observed with an order of magnitude (or better) less organotin loss, one advantage would be less environmental impact.

Organometallic polymers may be added directly to glass fiber/plastic hulls at time of manufacture, thus, conceivably, eliminating the need for periodic antifouling paint application.[82]

TABLE 2-22[80]

Marine Fouling After 12 and 14 Months' Immersion

Compound	Fouling[a] Year 1	Month 14
RT1	B–, A–, S+	B–, Z–, A, S
RT2	A–, S–	S+
RT3	A–, S	A–, S+
RT4	B–, Z–, A– S+	A–, S=
RT5	B–, A, S	B–, S+
RT6	B–, Z, A, S	B–, Z+, A, S
RT7	A, S+	B–, S
RT8	B–, S=	B–, S+
RT9	A–, S–	S
RT10	A–, S+, B–	B–, S+
RT11	A, S	S+
RT12	S+	B–, S
RT13	S	S
RT14	S+	S+
RT15	A–, S+	B–, S+
RT16	A+, S–	S+
RT18	B–, A–, S+	B–, S+
RT19	A–, S	B–, S+
RT20	A, S–	S+
RT21	A–, S	S+
RT22	A–, S	S+
RT23	A–, S=	S+

Note: Panels RT17 and RT24 were lost in a storm.

[a]A: algae; B: barnacles; C: slime; Z: bryozoa; –: slight fouling, 1 to 9% panel coverage; =: moderate fouling, 10 to 30% panel coverage; +: heavy fouling, over 30% panel coverage

Toxicant release from organometallic polymers is hypothesized as being through a hydrolysis reaction.[83]

Organometallic polyesters and epoxy resins have also been prepared and have demonstrated 100% antifouling performance after 17 and 7 months, respectively, at Pearl Harbor.[84]

Organolead Systems

Organolead antifoulants have been under evaluation for several years. Preparation methods have been discussed elsewhere.[85,86] Various organolead materials impregnated in wooden poles have been shown to be superior to creosote in protecting against marine borers.[87] Organolead antifouling paints have been previously discussed; however, if we assume that the conventional paint releases through a leaching mechanism and the organotin/elastomer systems function through a diffusion-dissolution mechanism, then the organolead materials appear to be substantially different. It has been hypothesized that the antifouling agent released is not the organolead material added, but water-soluble polymeric substances that react with the organolead during preparation of the paint.[56] Analytical data indicate that the polymers act as a vehicle for the release of at least triphenyllead acetate. The polymeric bases used are non-water soluble; however, the reaction products with the organolead do enter the water at a slow, continuous rate.[88]

Diffusion-dissolution Mechanism

The leaching mechanism common to most antifouling paints of the conventional type is very difficult to handle in a mathematical sense. Critical parameters such as porosity growth rate and tortuosity are extremely difficult, if not impossible, to determine for a given material. Cuprous oxide loss from such surfaces *may* be continuous, but it is not at a constant rate. Various areas of a given paint film probably release at different rates.[89] A partially depleted antifouling paint section can be microtomed without difficulty and each layer subjected to copper analysis. X-Radiation techniques show a steep toxic gradient.[71] The near surface layers show complete depletion while the lowest layers have the same copper content as the nonimmersed paint film.

As depletion occurs, porosity increases and the cuprous oxide loss rate decreases, probably due to a water stagnation layer within the pore structure, i.e., the movement of water within the paint film is considerably slower than flow at the film-water interface.

As a rule, fouling attachment initiates on paint films *prior* to the total depletion of toxicant. Again, fouled paint can be sectioned and examined by various techniques, and the author has found that paints, both organotin bearing and cuprous oxide types, retain considerable toxicant in the lower layers while perhaps only the first 2 mil of a 4-mil film, for instance, are totally depleted. It is also noted that, generally, when a paint film fails fouling accumulation is very rapid, a duration of 6 to 8 weeks elapsing from incipient fouling to 100% coverage under tropical marine conditions.

In contrast to paint films, antifouling vulcanized rubber in the partially toxicant-depleted state shows no measurable toxicant gradient. The loss rate decreases as the organotin concentration falls with immersion time, but sectioning and analytical examination by both wet chemical methods and spectrophotometry failed to detect any concentration gradient.

Comparing leaching-type paints with antifouling rubber indicated that the mechanisms, as predicted, were vastly different. An organotin such as TBTO added to neoprene or other elastomeric polymers is soluble therein to a specific limit, as previously mentioned. Once incorporated at a concentration below or at the solubility limit a solution equilibrium results. Upon exposure to water TBTO molecules on or near the surface pass into water through a dissolution mechanism. The rate of dissolution is not clearly established for various materials, but it seems to vary by about 1% for each 4°C rise in temperature and probably varies with salinity. Tests performed on an underwater rotating device showed no significant increase in loss rate with velocity.[71] Also, deep sea evaluation indicated that hydrostatic pressure had very little influence on loss rate.

As surface and near-surface TBTO molecules pass into the watery interface a localized solution disequilibrium occurs. Under solution pressure internal solute molecules migrate towards the depleted area so that equilibrium is reestablished. Since the process is continuous, the net result is a continuous loss of toxicant at the surface. *No* pore structure is established.

The diffusion-dissolution phenomenon, where dissolution is rate controlling, can be mathematically described with a fair degree of accuracy. One important consideration lies in determining the lifetime of antifouling rubber in relation to two controllable factors: toxicant concentration and sheet thickness. For release of a toxic agent from one side of a flat surface,[62,65]

$$\frac{dQ}{dt} = -KC = -K\frac{Q}{a} \tag{1}$$

where C is the concentration of toxicant in the matrix, K is a proportionality constant, Q is the quantity of toxicant in the matrix per unit area, and a is the sheet rubber thickness.

Solving the above equation,

$$Q = Q_o e^{-\frac{Kt}{a}} \tag{2}$$

where Q_o is the initial toxicant concentration (at time 0) and t is the elapsed time.

Q_R, the total amount of toxicant released, is given by:

$$Q_R = Q_o\left(1 - e^{-\frac{Kt}{a}}\right) \tag{3}$$

The foul-free lifetime, T, of the antifouling elastomer is the time that it takes for the toxicant concentration in the matrix to reach the threshold level for fouling attachment, C_T. Solving Equation 2 for T results in the following expression that predicts lifetime:

$$T = \frac{a}{k} \ln \frac{Q_o}{Q_t} = \frac{a}{k} \ln \frac{C_o}{C_t} \tag{4}$$

Thus, the lifetime is directly proportional to rubber thickness. Predicted values of Nofoul sheet are given in Table 2–23.

Tables 2–24 and 2–25 illustrate barnacle fouling of antifouling rubber in relation to both TBTO concentration and sheet thickness. Note the slow onset of fouling compared to the very rapid fouling accumulation rate observed with paint films. This series consisted of replicated 7 X 9 in. sheets *bolted* to the substrate so that TBTO loss is from *both* sides.

Figure 2–14 illustrates the distinction between leaching and diffusion-dissolution. The diffusion-dissolution mechanism can have diffusion or dissolution as the rate-controlling step. With commercial antifouling rubber the migration of organotin through the matrix is more rapid than the loss at the surface; thus, dissolution is controlling. However, by the proper use of additives diffusion can be sufficiently retarded so that it becomes the controlling phenomenon. In such a case toxicant loss is proportional to the square root of the elapsed time. Harris has suggested that the Higachi expression, shown below, describes the release mechanism.[90,91]

$$Q = \sqrt{Dt(2A-C_s)C_s}$$

where Q is the amount of toxicant released after time, t, per unit exposed area, D is the diffusion coefficient of the agent within the matrix, A is the total amount of toxicant per unit volume, and C_s is the solubility of the agent in the binding matrix.

Organotin/elastomer formulations are chemically complex. A reactive material such as TBTO

TABLE 2–23[71]

Predicted Foul-free Life of Nofoul® Sheet (for Calcareous Fouling)

Thickness (inches)	Four parts (years)	TBTO® concentration; six parts (years)	Eight parts (years)
0.040	1.5	2.1	2.4
0.050	2.0	2.9	3.3
0.060	2.3	3.5	3.9
0.070	2.5	4.0	4.4
0.080	3.0	4.2	4.9
0.090	3.9	5.1	5.6
0.100	4.1	5.4	6.3
0.120	4.6	6.8	7.8

TABLE 2–24[71]

Barnacle Accumulation – Miami Beach (1/16-in.-thick Toxic Neoprene Sheet)

	TBTO® concentration (parts per hundred parts rubber)									
	0	1/16	1/8	1/4	1/2	1	2	4	6	12
Month	Number of barnacles									
1	11	5	3	0	0	0	0	0	0	0
2	49	8	5	0	0	0	0	0	0	0
3	100+	7	4	0	0	0	0	0	0	0
4	100+	7	3	0	0	0	0	0	0	0
5	100+	7	4	5	6	0	0	0	0	0
6	500+	8	6	6	4	0	0	0	0	0
7	500+	7	8	13	2	0	0	0	0	0
8	500+	9	7	15	3	0	0	0	0	0
9	500+	18	27	24	6	0	0	0	0	0
10	500+	31	29	21	8	0	0	0	0	0
11	500+	36	24	16	6	0	0	0	0	0
12	500+	39	37	27	5	1	0	0	0	0
13	500+	59	54	49	8	3	1	0	0	0
14	500+	60	59	45	6	2	0	0	0	0
15	500+	53	55	47	7	6	0	0	0	0
16[a]										
17	500+	68	b	b	5	5	0	0	0	0
18	500+	80	b	b	6	7	2	0	0	0
19	500+	95	b	b	15	10	5	0	0	0
20	500+	100+	100+	b	95	80	19	0	0	0
21	500+	100+	100+	b	96	85	25	0	0	0
22	500+	100+	100+	b	b	94	29	0	0	0
23	500+	100+	100+	100+	100+	100+	34	0	0	0
24[a]										
25	500+	100+	100+	100+	100+	100+	40	23	0	0
26	500+	100+	100+	100+	100+	100+	70	26	0	0
27	500+	100+	100+	100+	100+	100+	50	22	3	0
28	500+	100+	100+	100+	100+	100+	60	19	3	0
29	500+	100+	100+	100+	100+	100+	62	22	6	0
30	500+	100+	100+	100+	100+	100+	70	20	8	0
31	500+	100+	100+	100+	100+	100+	67	12	7	0
32	500+	100+	100+	100+	100+	100+	53	12	7	0
33	500+	100+	100+	100+	100+	100+	70	5	5	0

[a]Data not gathered this month.
[b]Heavy bryozoan encrustation made barnacle count impossible.

TABLE 2–25[71]

Compound 105B[a] – Florida Immersion

Immersion month	Sheet thickness (inches) 0.025	0.032	0.050	0.0625	0.075	0.100	0.125
	Number of barnacles						
2	0	0	0	0	0	0	0
4	0	0	0	0	0	0	0
6	0	0	0	0	0	0	0
8	0	0	0	0	0	0	0
10	1	0	0	0	0	0	0
12	0	0	0	0	0	0	0
14	0	0	0	0	0	0	0
16	0	0	0	0	0	0	0
18	0	0	0	0	0	0	0
20	19	0	0	0	0	0	0
22	28	18	4	0	0	0	0
24	32	55	4	0	0	0	0
26	36	58	2	0	0	0	0
28	45	56	2	0	0	0	0
29	55	65	4	0	0	0	0

[a]6.5 parts TBTO®.

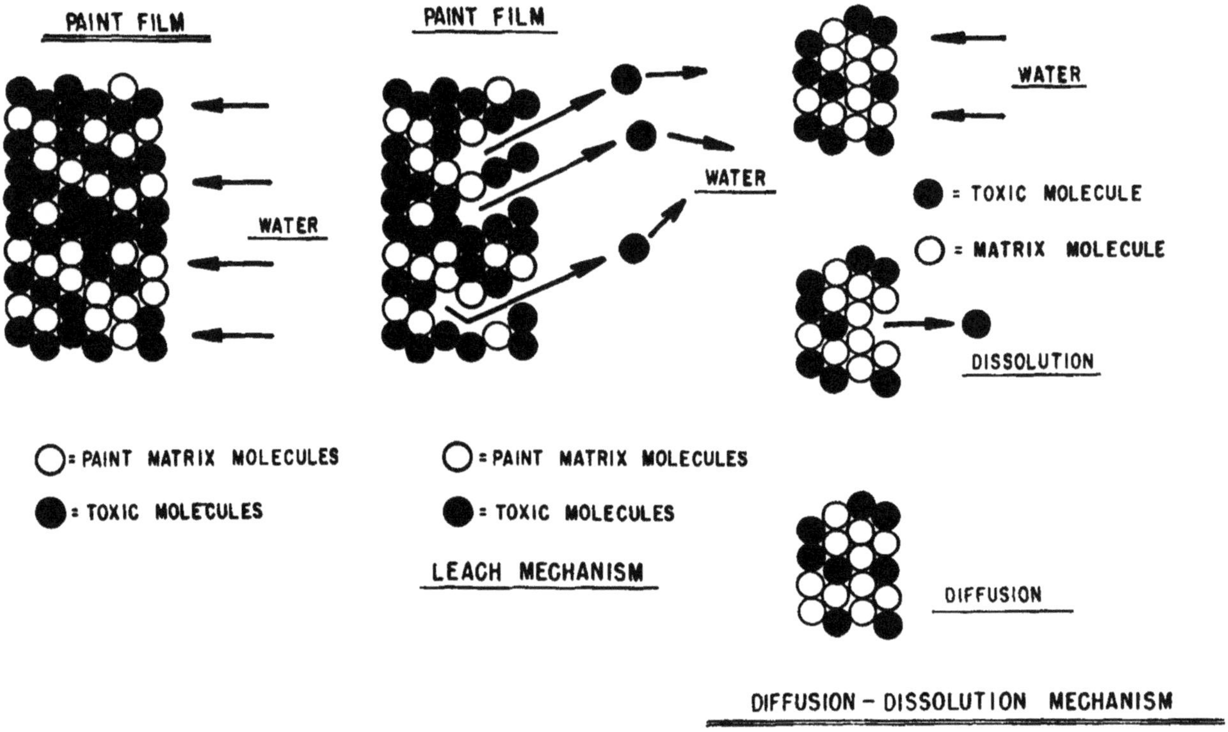

FIGURE 2–14. Distinction between a leaching mechanism and diffusion-dissolution.[62]

when added to rubber can and will combine with various compounding ingredients, especially during vulcanization. It is well known that TBTO will react with sulfur-bearing compounds, amino acids, and fatty acids. When TBTO is added to standard chloroprene recipes,[70,75] for instance, the materials released upon contact with water are unreacted TBTO, TBTO-dilaurate (from lauric acid), TBTS, dialkyl tins, and other unidentified organotins. Furthermore, Kanakkanatt has discovered that TBTO will actually combine with the polychloroprene matrix and provide cross-linking.[92]

In most antifouling elastomers the rate of diffusion is relatively unimportant since dissolution is rate controlling. However, one can decrease diffusion to the point where dissolution is no longer rate regulating. Diffusion can be varied by several means. Probably the most practical method is based on increasing the diffusion path length by the use of absorptive and/or adsorptive additives. Carbon black is the logical choice although it is likely that certain waxes and perhaps fine-structured clays will act in a similar manner.

Diffusion rate can be measured by holding a toxicant-loaded rubber sheet in integral contact with a blank (nonloaded) of the same formulation. Pressure of several thousand psi is necessary to assure adequate contact. Solute molecules then migrate through the interface, in much the same manner as sulfur or sulfur-bearing additives will migrate from the higher to the lower loaded sheet.[93] With TBTO/neoprene WRT 0.025-in.-thick sheets, an equilibrium concentration is reached in about 40 days.

The addition of carbon black to a rubber compound does not alter free volume nor greatly affect segmental motion, as polymer segments adsorb to the carbon black surface.[94] Thus, it is unlikely that carbon black retards diffusion by altering free volume. It is well known that the absorptive power of carbon black depends upon particle size and structure. The carbon black particle possesses a discrete porosity,[95] although internal area arising from such pores does not appear to influence performance effects of the elastomer.[96]

Fine-structured blacks, having an increased porosity and thus a greater surface area for adsorption and/or absorption, retard TBTO diffusion rate. Both concentration of the fine-structured black and the degree of black-elastomer interaction (chemisorption)[97] are basic parameters to be considered in investigating diffusion rate. Unfortunately, these parameters are difficult to define for a given system. For instance, the adsorption of rubber on carbon black increases with increasing molecular weight of the elastomer,[98] while the molecular weight and degree of chemisorption depend upon the vulcanization system used.[99] As a further complication, organotins can influence vulcanization rate.[76] We shall return to the nature of diffusion and the role of elastomers and additives as rate determinants in the next chapter.

Factors influencing dissolution rate of essentially insoluble materials (TBTO water solubility is no greater than 30 ppm at 25°C) are not well understood. The structure of liquid water is controversial.[100] Electrolytes may "break" the water structure (a combination of H-bonded macromolecules and monomeric molecules), and nonpolar solutes at the interface tend to enhance structure. The dissolved ions and gases present in marine waters vary in amount and type with geographic location, depth, and temperature. The nature and concentration of marine electrolytes are fairly well known;[101] however, the processes occurring at an elastomer surface releasing nonelectrolytes into an electrolytic milieu where the medium itself varies between a polymeric and monomeric structure are complex indeed, and a theoretical model does not exist as yet.

Applications

Nofoul rubber is applied to objects such as ship hulls, buoys, and various structures by use of an adhesive system. The invention of a long-term marine adhesive by Ronald Senderling is a critical factor in the efficacy and, hence, commercial prospects of Nofoul.[102] The procedure, though somewhat complicated, has led to excellent rubber-to-metal, -wood, -fiberglass, and -concrete adhesion over time periods in excess of 6 years. To the author's knowledge there has never been a case of adhesive failure in the many (both large- and small-scale) applications made to date. In many cases involving bond strength test panels the mode of failure during peel is cohesive in the elastomer (or the substrate) and not adhesive.

Essentially, application is as follows: The substrate is properly prepared by sandblasting and/or degreasing. Two coats of HLN-11, a two-step epoxy primer, are applied, followed by one coat of

HLN-21, a two-step tie coat, and then a coat of HLN-30, the two-step neoprene adhesive. Each coat is cross-linked by a curing agent. Shortly after HLN-30 has dried, it is wetted by swabbing with a solvent, and then the Nofoul sheet stock is pressed into place using a flat-face steel roller. Tack is very strong, and once the sheet adheres even immediate removal is extremely difficult. No problems are evidenced through slippage, voids, wrinkles, etc. even when the stock is stretched over three-dimensional sharp curves.

The recommended temperature for applying Nofoul lies between 10 and 32°C. Humidity can be a problem, and to avoid it temperature should be at least five units higher than the relative humidity.

Seams are handled by skiving and buffing so that a smooth surface results. Seam discontinuities increase drag (and thus fuel consumption) on vessels.

Nofoul rubber can be molded to any size or shape normal to the rubber industry. For instance, rubber hose or tubing or large molded structures such as sonar bow domes can be prepared. Also, the material can be vulcanized in place when the object covered is small enough to fit in an autoclave.

Chapter 3

MOLLUSCICIDES AND SCHISTOSOMACIDES

MOLLUSCICIDES

Public Health Problem

Molluscicides are materials that destroy molluscs such as snails and slugs. Various species of snails serve as hosts for parasites that affect man, his domestic animals, and wildlife of economic or ecological importance to man. The following parasitic diseases constitute serious human health problems.

Name	Infected population (number of persons)	Direct and indirect annual mortality (estimated)
Schistosomiasis	240–300 million	2–4 million
Fascioliasis	2+ million	Few hundred
Angiostrongyliasis	7+ million	Very low
Clonorchiasis	45+ million	Very low
Paragonimiasis	Unknown	Low
Fasciolopsiasis	10+ million	Unknown

Angiostrongyliasis

Angiostrongyliasis is a disease of the central nervous system arising from infection with *Angiostrongylus cantonensis,* a nematode. It is endemic in Hawaii, various other Pacific islands, Australia, Thailand, Indonesia, Taiwan, Southeast Asia, and the Philippines. The rat is the normal definitive host of the parasite, and various species of snails and slugs are intermediate hosts. Transmission to humans is through ingestion of snails or slugs, or vegetables contaminated with infective larvae by these molluscs. Susceptibility to infection is general.

The disease may last from several days to several months with periodic recurrence of symptoms. Infection in humans is characterized by severe headache, stiffness of the neck and back, and temporary paralysis. Deaths are rare.

Clonorchiasis

Clonorchiasis is highly endemic in southern China and throughout Japan, Taiwan, South Korea, and Vietnam. Man is infected by eating raw or improperly cooked fish. During digestion larvae of the trematode *Clonorchis sinensis* are liberated from cysts and move through the bile duct to the biliary system. The worm matures in the biliary radicles and deposits copious eggs which reach the external environment with feces. Eggs, containing a fully developed miracidium form of the parasite, are ingested by susceptible snails. Hatching occurs within the snail and another larval form of the trematode, the cercaria, develops and when mature emerges into water. Upon contact with any of about 40 piscine species, cercaria penetrate this second intermediate host and encyst.

Man, dogs, domestic cats, swine, and other animals serve as reservoirs for the adult fluke. Clinical symptoms of infection result from inflammation of the bile ducts, toxemia, and sometimes secondary pathogenic bacteria. Obstruction of the biliary system often leads to cirrhosis, enlargement of the liver, ascites (abdominal fluid), and edema. The disease is chronic and may last for 30 to 40 years in one individual, but is usually not fatal. Susceptibility is universal and there is no evidence of acquired immunity. Specific medical treatment is lacking, although certain medicinals can be used to relieve symptoms.

Fascioliasis

Fascioliasis is generally considered an infection of ruminants, and will be treated as such in the next section. It does affect man, however, in localized endemic areas of South America and the Caribbean. Relatively few cases have been diagnosed in the United States and Europe.

Fasciolopsiasis

Fasciolopsiasis is a trematode infection of the small intestine, particularly in the duodenal area. Massive infections produce acute intestinal

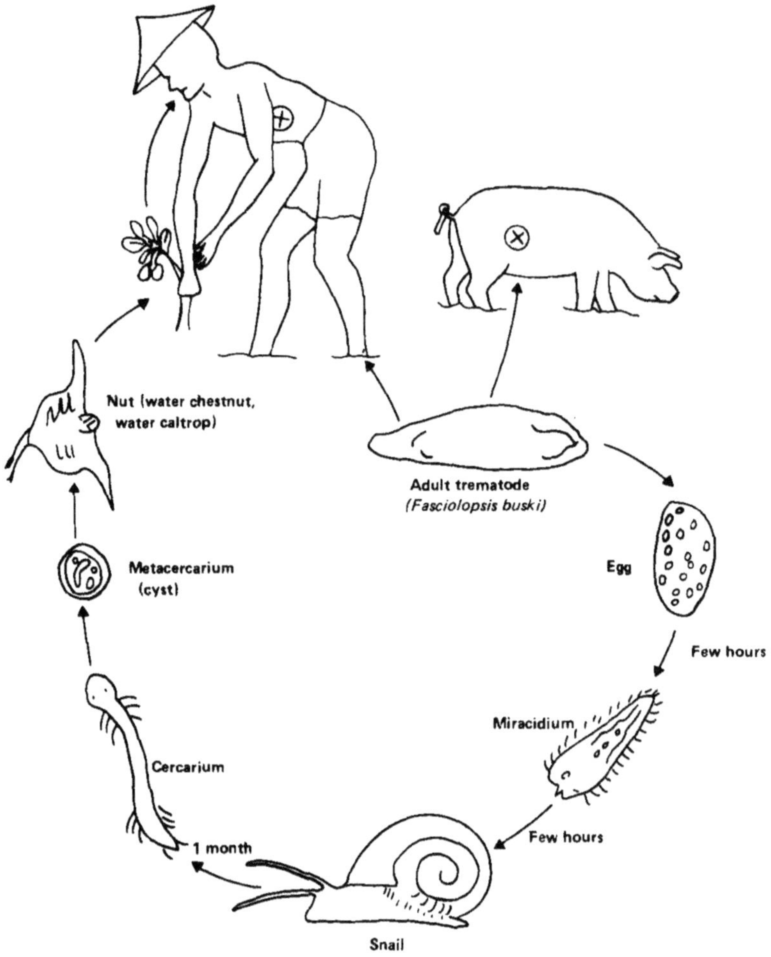

FIGURE 3–1. Life cycle of *Fasciolopsis.*

obstruction, edema of the abdominal wall and legs occurs, and ascites and sometimes secondary anemia is likely. Death is unusual. The disease is widespread in central and southern China, Taiwan, Vietnam, Pakistan, Thailand, Malaya, and neighboring areas.[103] Man, dog, and swine are definitive reservoir hosts of *Fasciolopsis buski,* a large (2- to 7.5-cm) trematode.

The mode of transmission (Figure 3–1) is as follows: Eggs are passed with feces from the definitive host. They hatch in water into miracidia which locate and penetrate various snail species, which serve as intermediate hosts. Cercariae develop in the snail; when mature they migrate into the external environment and encyst on various water plants. Man infects himself by ingesting such plants without cooking. The water chestnut is a chief form of human infection.

Paragonimiasis

This trematode infection occurs extensively in Korea, Taiwan, and Japan, with scattered foci in China, the Philippines, Southeast Asia, Africa, and South America. Man, dog, cat, swine, and a number of wild carnivores are definitive hosts. Transmission occurs when man consumes various species of fresh-water crabs and crayfish that contain infected larvae. The larvae emerge in the duodenum, penetrate the intestinal wall, and migrate through the tissues, encysting in the lungs or other areas and developing into mature egg-laying adults. Eggs leave the definitive host through sputum and/or feces, reach fresh water, and hatch into miracidia. The miracidium in turn seeks and enters a susceptible snail. The cercarian larval form emerges from the snail in approximately 3 months to find and penetrate fresh-water crustacea.

Clinical manifestations of the disease in man depend upon the organ system affected. Commonly, the lungs are involved and cystic lesions develop. Other organs as well may be involved, e.g., lymph nodes, genitourinary tract, and brain. Infection lasts many years.

Preventive measures involve sanitation and chemical control of snails. Drugs are available that alleviate the clinical manifestations and occasionally afford a cure.

Schistosomiasis

Schistosomiasis is by far the major snail-borne parasitic disease, and at this time the world's Number One malady arising from parasite infections. There are two basic forms of schistosomiasis: one involving a snail-man-snail cycle, and a second, relatively trivial type, in which the parasite involved has a bird-snail-bird transmission cycle where man intervenes. This minor version, termed schistosome dermatitis, arises from parasite penetration of the skin, usually triggering a localized rash that subsides in a few days. However, an allergic reaction can occur, leading to serious illness and occasionally death.

Schistosomiasis is transmitted to man when he contacts cercariae. Larva are aquatic and infectious for perhaps 8 to 12 hr. Human contact may be through drinking, bathing, washing of clothes or domestic animals, swimming, or merely traversing infected water. Cercariae penetrate the dermis in 8 to 20 sec, usually, enter the bloodstream, and eventually migrate to the capillaries around and in the liver or spleen. Here they mature rapidly into adult worms, reproduce sexually, and lay countless eggs. Oviposition often occurs on the exterior urinary tract and gut. Natural processes move the eggs through the intervening tissue and into the intestines and urinary system where they pass into the exterior environment during excretion.

The eggs, which may lay dormant for several months, hatch within a few hours' contact with nonsaline water. The free-swimming miracidial larval form emerges and seeks, contacts, and enters specific snail species. Of the over 5000 known snail species probably no more than 80 or so are implicated in schistosomiasis transmission. Within the host mollusc the miracidium forms a sporocyst, and many cercariae, all of the same sex, are formed by nonsexual reproduction.

When the sporocyst is mature cercaria emerge, leave the snail, and infest the surrounding waters seeking (or perhaps happening upon) a human. Thus the transmission cycle repeats itself.

The worm burden within the human body is not a major determinant of the etiology of the disease. Rather, the egg masses that do not reach the exterior world gradually block blood vessels, leading to the gradual malfunctioning of vital organs. Cells die as the oxygen and nutrient flow is curtailed by lodging of the egg mass in the capillaries. The usual course is for the liver, spleen, and bladder to be first affected, then the upper intestinal tract and stomach, followed by the lungs; in rare cases even the brain may be involved.

The female trematode (or fluke) may oviposit thousands of eggs at one laying and repeat this every 20 min over a 5- to 7-year lifespan. The presence of 2000 or more adult worms in the human body is usually needed for the manifestation of pathological symptoms. Each human entrance into cercariae-infested water may lead to one or more penetrations. Thus, in endemic areas the victim is often infected as an infant, and as he grows older the worm burden increases, peaking at 15 to 18 years of age.

The outward symptoms of the disease are commonly observed as a general weakness and malaise, as liver and spleen dysfunction increases. Splenomegaly, hepatomegaly, hydronephrosis, impaired growth, hematuria, and greatly lessened resistance to bacterial infections are clinical manifestations. Mental impairment is not rare. Mortality usually arises from secondary pathogenic infections such as hepatitis, or neoplastic growths within the bladder.

The physical strength and endurance of the schistosomiasis victim decreases with the increasing trematode load, and in the nonfatal cases anywhere from a 10 to 80% debility may exist. The high endemicity in agricultural regions results in poor crop production due to the inability of the farmer to work at a normal rate. This in turn leads to malnutrition and impoverishment. The worldwide economic burden, although not assessed, must be enormous.

Infectivity levels may be near 100% for a given locale, and overall infection as high as 70% of the total populace in some nations. Severely infected countries include Egypt, Iraq, the Sudan, South Africa, Malagasy, Tanzania, Kenya, Brazil, and China. Moderately infected nations include Taiwan, the Philippines, Rhodesia, Yemen, Saudi Arabia, Ghana, Niger, Zaire, Nigeria, Chad, Dahomey, Ethiopia, Mozambique, parts of Indonesia, and Venezuela. Other nations in which schistosomiasis is a public health problem include Syria, Iran, Cambodia, Lebanon, Upper Volta,

Senegal, Somalia, Gabon, Tunisia, Libya, Togo, Uganda, Malawi, Central African Republic, Angola, Cameroon, and various Caribbean islands, including Puerto Rico. The disease has been eradicated in Israel and Cyprus, and is near eradication in Japan and Portugal.

Schistosomiasis is increasing throughout vast areas of the world, often resulting from the harnessing of water for irrigation and hydroelectric plants, which tends to spread the snail hosts into hitherto schistosomiasis-free areas. The Aswan Dam in Egypt, for instance, has given rise to over one million additional cases of schistosomiasis.[104]

In order to control or eradicate schistosomiasis it is necessary to interrupt the transmission cycle. Basically this problem can be and has been historically approached by three basic methods: medical therapy, sanitary engineering, and snail control. Engineering technology can be used to provide safe drinking water, sewage disposal schemes, concrete-lined irrigation canals, safe swimming, bathing, and washing facilities, and so on. On a small scale such measures will indeed reduce human contact and hence infectivity.[105] However, the cost of such approaches is not realistic for most of the populations at risk.

Schistosomiasis can be medically arrested through a therapeutic regimen. Fairly safe vermifuges are available, although careful measure of dosage is required and often unpleasant side effects result.[106,107] One medicinal, niridazole, destroys the functioning of the trematode's reproductive system without killing the worm.[108] Immunity cannot, as yet, be induced,[109] although some degree of natural immunity may exist in indigenous populations.[110] Thus, medical treatment is of little value if the patient will in a few days be again in contact with infested water.

The primary conventional method of intervention in the transmission cycle lies in the chemical control of snails using snail-destroying chemicals termed "molluscicides." Such materials suffer a number of disadvantages, but their use is almost always more cost effective than medical treatment or sanitation.

Three forms of schistosomiasis are known: *Schistosoma mansoni*, found in the Caribbean area, South America, and much of Africa; *Schistosoma hematobium*, spread throughout Africa and the Middle East; and *Schistosoma japonicum*, found in Asia. The etiology of each disease varies in terms of the organ systems infected.

Extensive work has been done in the study of the nature of schistosomiasis, populaces affected, development of diagnostic and therapeutic medical techniques, identification, biology, and ecology of the various snail hosts, snail rearing methods, and so on. Warren and Newill in their monumental bibliography list 10,286 references in the scientific literature from 1852 to 1962.[111] Certainly an additional 5000 or 6000 reports are now available. Yet, in spite of all the research endeavor applied to this dread malady, man is slowly losing the struggle!

Cercarial dermatitis or "swimmer's itch" is a skin eruption arising from penetration of human skin by cercaria of various species of nonhuman schistosomes. It is globally distributed with both fresh water and salt water types. In the United States it is spread throughout the Great Lakes region, Iowa, Nebraska, Minnesota, and the Dakotas. It also occurs in most Canadian provinces, Mexico, Japan, Malaya, Australia, New Zealand, France, Switzerland, and areas of the United Kingdom. Marine forms (so-called "clam digger's itch") are found along the Atlantic and Pacific coasts of the United States.

Schistosome dermatitis is seldom fatal, although humans have died from allergic reactions. However, it does disable rice paddy farmers in Asia, thus reducing crop output, and in Michigan and other areas use of recreational water facilities is restricted during several months of the year unless chemically treated.

The transmission cycle is similar to that of schistosomiasis except that birds, fish, or other animals are the primary hosts. Infected waterfowl, muskrats, and mice pass schistosome eggs into a watercourse through the feces. The free-swimming miracidial form hatches in a few minutes and seeks snails of the Lymnaeidae or Physidae families. If a suitable host cannot be located the miracidium will succumb within 12 hr or so. If a target snail is found penetration occurs and the miracidia mature as sporocysts. After 5 weeks, give or take a few days, cercaria emerge and seek a definite host; once located, attachment occurs, histolytic substances are secreted, partially dissolving the epidermis, and penetration takes place. The cercaria seek a bird, rodent, fish, etc. to serve as the primary host. Unfortunately, man sometimes intervenes in the cycle.

Humans are accidental hosts. Cercaria penetrate but usually die just below the epidermis. Cercarial proteins elicit a hypersensitive response. The dermatological reaction in man is variable and depends upon the degree of hypersensitivity induced by previous exposure.[112]

Agricultural Problem

In that human schistosome infection reduces the physical (and mental) ability of the farmer to produce foodstuffs, it presents an agricultural problem. Since this aspect was discussed in the previous section our concern at this point is snail and slug damage to growing crops, and schistosome infection in domestic animals.

Fascioliasis is a widespread sheep and cattle disease arising from infection with *Fasciola hepatica*, a trematode parasite. The adult worm inhabits the bile ducts and sometimes other organs within the infected animal, where it produces eggs almost continuously. Eggs are voided with the feces and remain viable for months, even under unfavorable environmental circumstances. When temperature is above 50°F and eggs are in contact with water, hatching occurs in approximately 9 days.

Free-swimming ciliated miracidia emerge and seek a susceptible snail: *Lymnea bulinoides, L. cubensis, L. columella, L. natalensis,* and others. Once contact is made the miracidium burrows through the soft tissue of its host, entering the pulmonary chamber or lymphatic system. Here the cilia are lost, and the organism enlarges and becomes a sporocyst. Each germ cell in the sporocyst develops through parthenogenesis into a second larval stage, termed the redia. Redia production varies from three to eight (or occasionally more) in each sporocyst. In 8 days or so redia burst through the cyst wall, migrate towards the liver, and reproduce there by parthenogenesis for several generations. Finally, each original or daughter redia yield several cercaria, which burrow out of the host snail and swim about, settling within an hour or two on vegetation. The tail is lost as this larval form changes into the metacercaria, which encysts. Encysted larvae remain viable for weeks or months if not subjected to high temperatures. Infested vegetation is ingested by sheep or various bovines. The cyst wall is digested and larvae liberated which burrow through the intestine, reach the liver, and enter the biliary system. In several weeks mature flukes appear.

The lifespan of *Fasciola hepatica* is unknown although estimates range from several months to several years. The adult may produce half a million or more eggs, while each egg may yield through parthenogenetic multiplication up to 300 larvae.

Sheep and cattle contact is made during browsing on wet pasture lands, marshes, or shoreline vegetation around ponds and other watercourses.

The economic losses arising from fascioliasis must be enormous on a world-wide scale, although exact figures are not available. In the U.S. and other nations the livers from infected animals are condemned. Losses from this source are $0.5 million in Puerto Rico alone.[113] Much more importantly, fascioliasis substantially reduces milk and meat production from cattle, and meat and wool output from sheep.[114]

Vegetable crops are subject to damage by snails and sheep. Total economic loss is unknown, but it is large enough to support several bait-type product lines. Except in isolated areas such as parts of California, the problem appears to lie essentially with home garden plots. World-wide crop losses from amphibious snail infestations may be very large.[115] Unfortunately, agricultural pest snails have been moved by man's interference with nature to hitherto uninfested areas.[116]

Marine snails such as *Thais haemastoma,* the "oyster drill," are predators on commercially valuable shellfish. Economic losses along the Atlantic and Gulf coasts of the United States may exceed $5 million per annum.[117] This problem will be examined in a later section on bait molluscicides.

Conventional Technology

Due to the ubiquitousness of host snails, conventional methods of intervening in the transmission cycle are widely varied in accordance with the nature of the infested environment and molluscan ecology (Figure 3–2). The weakest link in the cycle has been assumed to be the snail; thus, chemical control of the molluscan host using a toxic substance introduced into the watercourse is the favored approach. Choice of molluscicide and dispersal method will depend upon environmental factors and economy. Destruction of the host snail benefits the populace involved by preventing new infections and stabilizing the worm burden in those with the malady. Since egg masses lodged in the victim's body are the major causal factor, it is necessary to destroy the ability of the adult trematodes to manufacture eggs. Thus, a useful approach would be to combine medical therapy with chemical control.

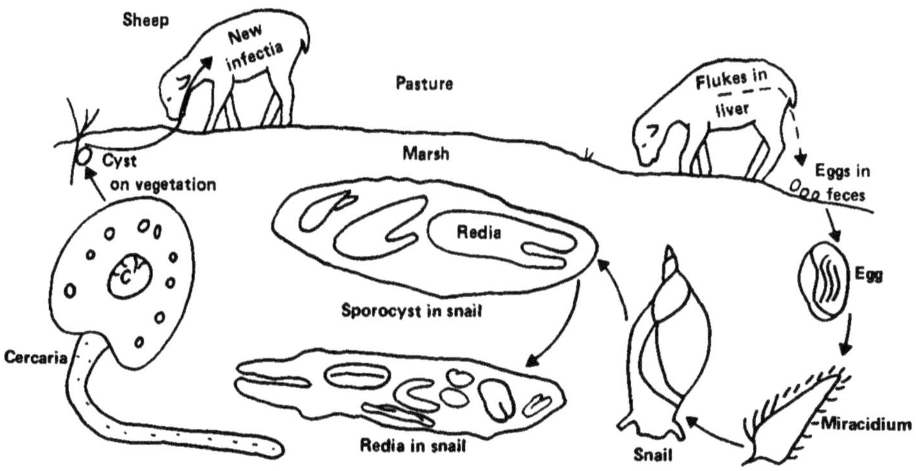

FIGURE 3-2. Fascioliasis transmission cycle.

Chemical treatment has never completely halted transmission save in relatively few instances where an entire isolated endemic area could be treated. Engineering techniques such as provision of safe drinking water and sewage disposal systems can materially assist in separating the human victim from cercariae and the snail host from contact with miracidia.

Medical treatment is limited in usefulness. Trematodes can be destroyed using drugs of various potency, although care must be taken to avoid serious, and even fatal, side effects.[106-108,118-121] Since immunity does not arise from therapy, nor are vaccines available, the "cured" individual must be prevented from recontact with infested water. The best approach would seem to be to first interrupt transmission with molluscicides, and then begin a drug regimen.[122]

Hemofiltration to remove adult worms is perhaps a theoretically sound approach, but does not appear to work with infected humans.[123]

Prophylaxis has been attempted through the use of topically applied lotions, which offer varying degrees of protection. Applications of cercariacides such as copper oleate in alcoholic solutions have been found effective against non-human schistosomes for 6 or more hours.[124,125] Use of cercariacides in ointment bases provided only a few hours' protection against *Schistosoma mansoni.*[126-128] Recently it has been demonstrated that inert film-forming elastomers can be topically applied with good dermal adherence and sufficient longevity for value in preventing Rhus dermatitis.[129] Refinement of these preparations has led to the creation of lotions offering nearly complete cercariae penetration protection for several days.[130,131] Such protection is of little value to exposed indigenous populations, and should be considered useful only as a prophylactic measure for short-term exposures such as might be encountered by chemical treatment personnel, snail survey teams, or military units traversing an infested watercourse, swamp, etc.

The selection of the proper molluscicide from a number of readily available materials is one of the most important aspects of chemical control. Both environmental and economic factors must be considered. Choice is usually based upon previous experience, although a model mathematical approach involving chemical costs, labor costs, necessary concentration, stream velocity, etc. is available.[132,133]

Copper Sulfate

Probably the oldest efficient molluscicide used was copper sulfate pentahydrate. This highly water soluble material can be added to an infected watercourse as powder, crystalline lumps, solution, etc. Dispersal can be through a simple regulated drip system with water solutions, or by hand or manual dispersion of the solid form. One notable method is to place copper sulfate in cloth bags and hang them in the infested water, where solvation and permeation into the surrounding water require several hours.

Copper sulfate as a molluscicide suffers several disadvantages. Downstream carriage is poor due to rapid detoxification. Amin reports *Biomphalaria* mortality at 18% and *Bulinus* mortality at 8.7% only 3 km downstream from the point of drip application.[134] Copper ion has a relatively short half-life in water, depending upon water quality factors and bottom soil. It is readily absorbed by muds, clay, and humic acid.[135] Living organisms, especially algae, absorb and chemically react with copper,[136] thus purging it from the watercourse. Cupric ion undergoes chemical reactions with dissolved ions such as calcium, magnesium, carbonate, and others. Rate of loss through chemical mechanisms increases with increasing alkalinity.

The net result of rapid detoxification means that abnormally high copper sulfate dosages must be used under field conditions (see Table 3–5 later in this chapter). Although laboratory evaluation of snails against copper ion indicates that molluscicidal dosages are not piscicidal, field dosage requirements are 20 or more times greater and fish readily succumb.[137]

Copper sulfate use in quantity is now limited to Egypt, the Sudan, and the Middle East. Its basic advantage is low expense. In many areas it has been found too ineffective for major use.[138]

Other soluble copper compounds have been evaluated as potential molluscicides, but appear to suffer the same drawbacks as copper sulfate.[137,138]

Exposure to copper ions at concentrations lethal to the snail provokes various defense mechanisms. The target may leave its water habitat, reentering later after the toxic concentration has fallen to levels below the mortality threshold. Thus, transmission of schistosomiasis is not interrupted. The exposed mollusc often will withdraw into its shell, expel a protective air bubble, and secrete a mucous plug to decrease copper ion contact with soft body tissue.[139]

In order to avoid adverse physiological response and prolong copper life in natural waters by overcoming deactivation factors, various new experimental compounds such as copper-bearing macrocyclic polyamines and polyethers, and various copper organic chelates have been synthesized.[140] Bioassay has indicated that the molluscicidal activity depends on the degree of stereochemical exposure of the copper moeity.[141] It is possible that complex copper compounds exert a more toxic effect on snails than that seen with the ionic form of copper.[142]

Insoluble copper compounds have been evaluated. Long-term residues with copper carbonate[143] and cuprous oxide[144] have been reported. Copper pentachlorophenate has been used extensively in Venezuela with efficacious results.[145]

Sodium Pentachlorophenate

Sodium pentachlorophenate (NaPCP) has been in use as a molluscicide since 1948. For years it was the molluscicide of choice in many areas, gradually supplanting copper sulfate. Cost effectiveness is better than that shown with copper compounds and comparable to niclosamide. It suppresses egg laying at relatively low concentrations[146] and can be used as briquets good for about 60 hr of continuous release.[138,147]

NaPCP has been effectively used against amphibious hosts of *Schistosoma japonicum* in the Philippines[148] and Japan,[149,150] as well as against *S. mansoni* and *S. haematobium*.

Several severe disadvantages are present that render use of NaPCP unattractive. Compared to the more recent niclosamide and trifenmorph, relatively large NaPCP dosages must be used to achieve adequate snail control, so fish and other nontarget biota suffer accordingly. Perhaps more importantly, NaPCP presents a high risk of intoxication to those handling it.[152]

Niclosamide

Niclosamide, 5,2′-dichloro-4′-nitrosalicylanilide, is at least ten times more active than sodium pentachlorophenate.[153] It was first formulated in a 70% wettable powder as the ethanolamine salt, and later a 25% emulsifiable concentrate became available.

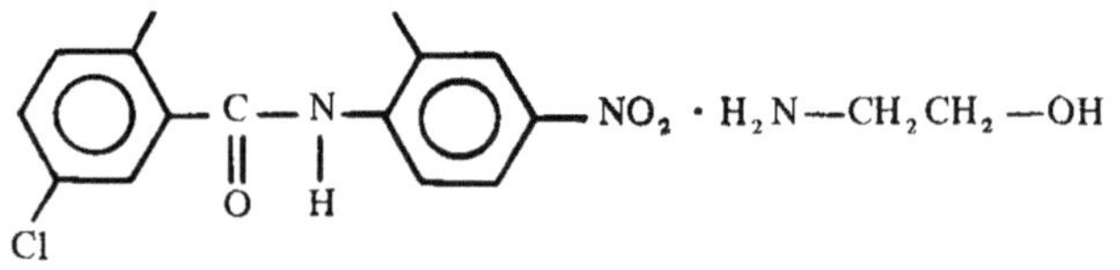

Structural formula; 2-amino ethanol salt of niclosamide (Bayluscide®)

Bayluscide has been extensively evaluated against host snails for the three major forms of schistosomiasis, under laboratory and field conditions.[154-161]

Niclosamide formulations have several advantages over trifenmorph and NaPCP. All stages of the mollusc are affected and the ovicidal quality is high.[155]

Snail control is achieved at relatively low toxicant concentrations. Bayluscide can be readily dispersed by constant flow devices,[156] meter and pump technique,[157] power spraying,[157-159] and even by air. Downstream transmission is good.[156] Bayluscide has been used in combination with copper sulfate against *S. dermatitis* host snails.[161] Unlike copper sulfate, NaPCP, and other materials, Bayluscide is not repellent to snails at practical concentration levels.[162] Residual activity may be prolonged for weeks[163] (Figures 3–3 and 3–4).*

Other salicylanilides have been evaluated and a number appear to be biological efficacious in laboratory studies.[164,165]

Bayluscide suffers a number of disadvantages. It is attacked and decomposed by various microorganisms,[166] and effectiveness is a function of pH,[167] with increasing loss in toxicity as the pH drops below 7.2.[168] Adsorption occurs on organic matter in water,[169] which may or may not affect snail control depending upon the browsing habits of the snail. Dispersion may be difficult in certain snail habitats.[152] Bayluscide is very toxic to fish[170] (it is used as a piscicide in the United States[170-172]) and other organisms.[172,173]

Trifenmorph

Trifenmorph (*N*-tritylmorpholine, Frescon®, WL8008) is a commercially available product showing considerable promise in snail control. It has undergone rigorous laboratory and field evaluation.[174-180] Molluscicidal quality under alkaline water conditions is higher than that shown by niclosamide salts, and trifenmorph is less hazardous to fish and other nontarget biota than is niclosamide. Dispersal of available formulations is relatively easy. The granular form provides a few days of continuous release upon exposure to natural waters.

Frescon has been used with success in the control of *Fasciola hepatica.*[181,182] Residual toxic effects have been noted.[179] It is essentially nontoxic to plants, mammals, nematodes, birds, and other animals.[182]

Disadvantages to the use of trifenmorph exist. Dosage requirements are too high for competitiveness with NaPCP and yurimin in amphibious snail control.[177] Frescon hydrolyzes in water to the inactive carbinal, the rate increasing with decreasing pH.[183] *N*-tritylmorpholine is neither ovacidal[184] nor cercariacidal.[182] Fish are destroyed, although Frescon is far safer in this respect than niclosamide.[138,180,185] Downstream carriage may not be as good as that observed with other molluscicides.[186] In many instances intoxication of the infested watercourse is short lived and residual toxicity of any duration absent.[187] Probably the major drawback is the rapid inactivation in acidic water.[152]

Other Molluscicides

A number of molluscicides of vegetable origin have been discovered and investigated.[154,188,193] Herbicides and larvicides have been evaluated as potential molluscicides.[154] These include various carbamates,[138,148] acrolein (handling and dispersal problems render it an unlikely candidate),[189] paraquat (1,1-dimethyl-4,4′-dipyridilium dichloride),[189,190] diquat (1,1-ethylene-2,2′-dipyridilium dibromide),[148,190] 2,4-D amine (amine salt of 2,4-dichlorophenoxyacetic acid),[189] Lindane (1,2,3,4,5,6-hexachlorocyclohexane),[191,192] fenthion (*O-O*-dimethyl, 0-3-methyl-4-methylthiophenyl phosphorothionate),[192] DDT (dichlorodiphenyl-trichloroethane),[192] and malathion (*O,O*-dimethyl-*S*-(1,2,-dicarboxyethyl)phosphorodithioate).[148,192] Organophosphorus and chlorinated hydrocarbon-type insecticides showed little efficacy compared with that of niclosamide.

Yurimin (3,5-dibromo-4-hydroxy-4-nitrobenzene) compares favorably with NaPCP for *Oncomelania* control and has been used successfully in Japan. Unlike NaPCP it is relatively stable against ultraviolet radiation. Yurimin is piscicidal at molluscicidal dosages. It shows no phytotoxicity, low mammalian toxicity, and is judged to be safe to handle.

Experimental Molluscicides

A number of experimental molluscicides which are incorporable in controlled release elastomeric

*For Figures 3–3 and 3–4 see page 89.

matrices have also been evaluated in the conventional sense.

Triphenyllead acetate (TPLA) – Hopf and his co-workers demonstrated that TPLA was molluscicidal at practical dosage levels under laboratory conditions.[194] Hira and Webbe investigated the effects of sublethal concentrations of TPLA on *Biomphalaria glabrata* and on the development of *Schistosoma mansoni* in the snail.[195] Shiff et al. conducted field tests of TPLA in a river habitat and an irrigation storage reservoir.[196] It was found that the 6-hr LD_{99} for TPLA against *B. pfeifferi* was 3.0 ppm and the 24-hr LD_{99} 0.62 ppm. Shiff et al. further showed that TPLA at 0.25 ppm had negligible toxicity to plankton.[196] Field evaluation at approximately 0.5 ppm showed good snail control. No evidence of dead fish or amphibia, or signs of phytotoxicity were observed in the field tests. TPLA results against *Oncomelania* at a 10-ppm concentration were not as encouraging, showing a 24-hr mortality of only 91%.[148]

DS®-2787 – DS-2787, manufactured by the Diamond Shamrock Company and commercialized as a fungicide,[197] has shown considerable potency as a molluscicide. DS-2787 is a thermally stable noncorrosive powder melting at 250°C. The structural formula is

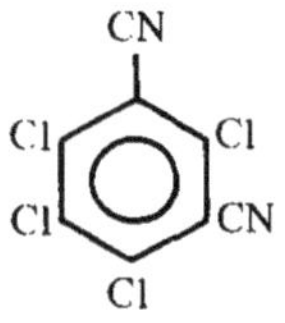

2,4,5,6-tetrachloroisophthalonitrile

Adult *Biomphalaria glabrata* are rapidly destroyed at dosages of 10 to 20 ppm, immature snails at 5 to 10 ppm, and ova at 0.5 to 1 ppm.[198]

Organotins – Organotins as candidate molluscicides were initially studied by Hopf and Muller.[199] A number of materials were examined in the lab and lethal dosages determined over the 1962–1969 period, as illustrated in Table 3–1. A number of obscure organotins were also evaluated against *Helisoma* species by Freiberg.[210]

By 1966 it was recognized that organotin compounds have high toxicity for snails and eggs, provided exposure times are 6 to 24 hr.[211] Deschiens indicated that snail control in small ponds could be achieved at 0.015 ppm TBTO in 3 to 6 days' exposure with no effect on fish.[212]

A limited number of field tests have been performed. In Cameroon, Deschiens et al. applied TBTO to ponds at 0.015 ppm and 0.03 ppm, which provided effective snail control without damage to fish or plankton.[209] Field plot tests against *Oncomelania quadrasi* required 8 g/m^2 of TPTA for effective control.[148] In downstream carriage test Berrios-Duran et al. showed that TBTO applied at 1 ppm for 6 hr was almost as effective as niclosamide against *Biomphalaria glabrata* (Table 3–2).[186]

Biological Control

The use of biological control agents against schistosome-bearing snails has been under consideration for some time. The leech *Helobdella fusca* has been observed to feed on very young *Biomphalaria glabrata* snails.[213] Nematode parasites of snails are known, but their capacity to induce mortality under natural conditions remains unknown.[214] Various species of protozoa parasitize snails.[215] Michelson discovered that *Plistophora husseyi*, a microsporidian, parasitized snails, causing extensive tissue damage and mortality.[216] This parasite appears to be specific for members of the physidae.

Mycobacteria infections have been noted in two snail species.[217,218] Infections are transferable to several species of human-schistosome-bearing snails.[218]

Probably the only biological control approach showing limited usefulness in field studies arises from the use of the snail *Marisa cornuarietis* (Gray) as a predator on schistosome-bearing molluscs.[219,220] Small-scale field tests over a 9-year period have indicated that *Marisa* will displace *B. glabrata* in natural ponds.[220]

Chemical Control Methodologies

Chemical treatment regimens vary with environmental factors. Water quality and flow rate is of primary importance in the selection of an appropriate molluscicide and its dosage. Other considerations involve the presence of food fish, growing crops, and use of the treated waters by domestic animals and humans. Chemical "control," as opposed to eradiction, implies that the program is ongoing and that the same area will be retreated at periodic intervals. The duration between treatments will depend upon the rate of snail repopulation, detection of infected snails, etc. as established by surveillance. Eradication is

TABLE 3–1

Laboratory Screening of Potential Organotin Molluscicides

Compound[a]		Mollusc	Results (24-hr bioassay)	Ref.
1.	TBTO®	*Bulinus tropicus*	LC_{50} = 0.01 ppm	200
	TPTA	*B. tropicus*	LC_{50} = 0.075 ppm	
2.	TBTA	*Biomphalaria glabrata*	LC_{90} = 0.30 ppm	201
	TPTA	*B. glabrata*	LC_{90} = 0.58 ppm	
	TPTO	*B. glabrata*	LC_{90} = 0.66 ppm	
	TPTA	*B. glabrata*	LC_{90} = 1.24 ppm	
	TBTA	*B. glabrata* (ova)	LC_{90} = 0.046–0.072 ppm	
	TPTA	*B. glabrata* (ova)	LC_{90} = 0.040–0.075 ppm	
3.	TBTA	*Oncomelania quadrasi*	LD_{50} = 8.835 ppm	202
	TPTO	*O. quadrasi*	LD_{50} = 41.9 ppm	
4.	TBTA	*O. quadrasi*	LD_{90} = 2.70 ppm	148
5.	TBTA	*B. glabrata*	LC_{95} = 0.57 ppm	203
	TBTO	*B. glabrata*	LC_{95} = 0.84 ppm	
6.	TBTA	*B. glabrata*	LC_{50} = 0.1–0.3 ppm	204
	TBTO	*B. glabrata*	LC_{50} = 0.05–0.1 ppm	
	TBTPC	*B. glabrata*	LC_{50} = 0.1–0.4 ppm	
7.	TBTA	*B. glabrata*	LC_{90} = 0.115 ppm	205
	TBTO	*B. glabrata*	LC_{90} = 0.105 ppm	
8.	TBTA	*Bulinus contortus*	LC_{100} = 0.30 ppm	206
	TBTO	*Bulinus contortus*	LC_{100} = 0.075 ppm	
9.	TBTA	*Biomphalaria glabrata*	LC_{90} = 0.115 ppm	207
	TBTA	*B. glabrata* (ova)	LC_{90} = 0.45 ppm	
10.	TBTA	*Bulinus sudanica*	LC_{100} = 1.137 ppm	208
	TBTA	*Bulinus nasutus*	LC_{100} = 0.954 ppm	
11.	TBTO	*B. camerounensis*	LC_{100} = 0.015 ppm	209

[a]Compound codes:

TBTO – bis(tri-*n*-butyltin) oxide.
TBTA – tri-*n*-butyltin acetate.
TPTO – tri-*n*-propyltin oxide.
TPTA – triphenyltin acetate.
TBTPC – tributyltin pentachlorophenate.

not considered to be an achievable goal, save in very isolated areas, in relation to available methods and economics.[221]

Two general methods of snail control are presently practiced: area control and focal control. The former is used where transmission is widespread and the snail habitat fairly indigenous to that of the populace. Focal control is probably limited to situations where the snail habitat is small and well defined, or where chemicals can be applied to transmission points, i.e., where the human at risk contacts the water, such as along a lake shore.

Chemical treatment methods will vary greatly with the nature of habitat. A very general classification can be considered where habitats are categorized as man-made lakes, natural lakes, irrigation schemes, drainage schemes, ponds,

TABLE 3–2

Results of Short-distance Field Screening Tests of Molluscicides

Molluscicide	Application time × concentration (hours × ppm)	Mortality[a] of laboratory-reared snails; exposure distance (meters)				
		–5+	30	1050	1375	1540
Niclosamide	1 × 1	0	77	77	12	–
	1 × 2	5	100	92	70	–
	1 × 3	0	100	100	100	39
	2 × 3	0	100	100	100	100
Frescon®	1 × 0.5	0	100	78	15	–
	1 × 1	0	90	100	0	–
	1 × 2	0	100	100	95	–
	2 × 1	0	100	100	90	–
TBTO®	6 × 0.25	0	100	43	0	–
	6 × 0.5	0	95	80	70	–
	6 × 1	0	100	100	100	100

[a]5 m upstream from application point.

From Berrios-Duran, L. A., Ritchie, L. S., and Wessel, H. B., *Bull. WHO*, 39, 316, 1968. With permission.

reservoirs and marshes, slow-moving watercourses such as small streams, fast-moving watercourses such as rivers and mountain streams, and diffuse, periodically water-saturated areas such as bogs and pastures.

Man-made Lakes – The damming of sizable rivers in order to harness water power for hydroelectricity and irrigation has led to the recent creation of a number of lakes. Slow-moving water and plentiful vegetation attract snails. Presence of food fish draws people, and if the humans are infected soon the lake becomes a source for schistosomiasis contact and transmission. Lake Volta (Ghana) will serve as an example. The construction of the Akosomba Dam across the Volta River led to the flooding of a considerable area as the Volta and Afram Rivers backed up. Water quality was monitored during the 1963 to 1969 filling period and thereafter, along with studies of lacustrinization, both primary (phytoplankton and higher plants) and secondary (zooplankton, invertebrates, and fish).[222] Lake area is approximately 3200 mi^2.

Within several years plants favorable to snails appeared in Lake Volta. Snail populations increased thereafter. Humans from other areas of Ghana moved to the lake shore to take advantage of the fish life. By 1969 the catch of *Tilapia* and other fish was around 60,000 metric tons per annum. Undoubtedly, humans infected the waters for by 1969 7.6% of the *Bulinus truncatus* rohlfsi snails surveyed showed schistosome infection.[223] The World Health Organization, aware of the near-epidemic increase in schistosomiasis among man-made lake shore populaces, established a study program to evaluate various means of transmission intervention.[224] It is important to note that "a comprehensive control strategy should be directed not only at the man and snail sub-systems, but also at the two aquatic sub-systems."[224] The only apparently feasible method of destroying the larval forms of the schistosome lies in the use of controlled release materials, as will be discussed in the next chapter.

Other man-made lakes have shown a rising incidence of schistosomiasis, e.g., Lake Kainji (Nigeria), where schistosomiasis infection rates in the surveyed villages vary from 18 to 62%,[225,226] and Lake Nasser (Egypt).[227]

Molluscicidal treatment of an entire lake of any size would be prohibitively expensive. The present approach lies in only treating the human/snail contact areas along the lake margin. Conventional treatment would not last more than a day or two if there is any water movement at all. Once the treated area is detoxified by dilution or natural processes the snails from surrounding areas will migrate to the contact area and transmission will again initiate.

Natural lakes – Many large water bodies such as Lake Victoria are infected. Treatment of lakes has usually been considered a useless task since economics has always dictated that only contact areas be treated. Consequently, very little has been done to control snails in lakes.

Irrigation schemes – The need for snail control was most evident in irrigation schemes, and, consequently, chemical control methods were first developed for this area. Mollusciciding of a system where channels are straight and water flow rate controllable appears deceptively simple. Application of the chemical agent to the head of the main canal ought to be carried to secondary and tertiary channels by water flow; thus, the entire scheme should be easily treated.

Unfortunately, practice is far more difficult than theory. Schemes such as the Gezira in the Sudan where 400,000 feddans (1 feddan = 1.038 acres) are under control are far too large for simple one-point application.[228] In this instance the chemical agent, copper sulfate, detoxifies before reaching very far downstream.[229] Thus, a number of mechanical feed ("drip") apparatus must be maintained and properly set to administer the correct dose to a given canal. Where irrigation schemes are small, e.g., village size, then focal control is possible.[230] Modest irrigation schemes can be handled by area control techniques with proper planning.[231]

Irrigation reservoirs and ponds – Irrigation reservoirs in endemic areas often serve as a snail habitat and foci of repopulation after the irrigation channels are treated. Snails can be controlled to some degree by water management such as rapid drawdown, stranding snails and thus permitting mortality through dehydration.[232]

Normally, irrigation reservoirs and ponds or impoundments have a small enough water volume to permit a complete treatment with chemical agents. Such agents can be scattered by hand or blower (granules or powders) or sprayed from back-pack sprayers (emulsifiable concentrates). If water is static, agent dispersal may be slow and the center portions of the water body will remain as sanctuaries for snail populaces. This is especially true if treatment is from the shore areas. Focal control at human contact sites is possible in farm ponds, pools, and impoundments,[233] but it may not be a practical control method where the selected agent has little residual action.

Drainage schemes – By their nature drainage ditches seldom have continuously flowing water. The usual effect is flow during and after rain, then gradually drying puddles. Automatic feed systems are useless, but one acceptable control technique is to spray the entire ditch with a suitable control agent.[159]

Streams and rivers – Mollusciciding of fast-flowing streams and rivers is considered impractical. However, snails avoid rapidly flowing water and thus are found in fairly calm areas near the shore line. Focal control can be used to eliminate such populations. Slow-moving streams are prime sources of snail transmission from untreated to treated areas. Blanket area treatment has little lasting value if the rivers and streams feeding that area remain snail-ridden. Conceptually, the best method of attacking snails in a stream is to use a continuous chemical feed system at the headwaters, but, the streams usually originate in many small trickles of water which tend to be geographically or economically inaccessible. Focal treatment at human contact points may be the best approach.

Area treatment is not realistic if the infested zone is of great size and comprises many water bodies. Here the focal point treatment approach (Figures 3–5, 3–6, and 3–7)* can provide significant control if continuous snail surveillance is made and retreatment begun when repopulation commences.[234-237]

Controlled Release Systems

Development; Historical Survey

Controlled release molluscicides were a logical extension of the antifouling rubber program. The concept of long term slow release of a biocide from an elastomer was proven through the evaluation and later commercialization of Nofoul rubber. During the evaluation of organotin-releasing elastomers against insect larvae it was called to the author's attention by Dr. Donald B. McMullen of the Walter Reed Research Institute that such materials might be of value in snail control. A program was established at the University of Akron in 1965 to evaulate the molluscicidal potency of various candidate antifouling elastomers.

Formulations releasing TBTO, TBTS, and TBTA were evaluated against *Helisoma trivolvis*,

*For Figures 3–5 thru 3–7 see page 89.

TABLE 3–3[70,245]

Controlled Release Molluscicides Distributed to Investigators, 1966–1968

Code	Base matrix	Agent	Agent loading (%)	Calculated longevity[a] (months)	Loss rate
443A	Chloroprene	TBTO®	5.8	90	30 $\mu g/cm^2$-day
351B	Chloroprene	TBTO	8.2	36	45 $\mu g/cm^2$-day
402A	Chloroprene	TBTF	5.8	24	Unknown
633A	Natural rubber	Bayluscide®	5.8	30	Unknown
378E	Acrylonitrile	TBTO	3.0	12+	55 $\mu g/cm^2$-day
419A	Chloroprene	TBTAd[b]	5.8	28	Unknown
419B	Chloroprene	TBTR	5.8	24	Unknown
300J	Natural rubber	TBTO	4.4	40	Unknown
300G	Natural rubber	TBTO	8.2	18	Unknown
300K	Natural rubber	TBTO	8.2	21	Unknown
324C	Chloroprene	TBTO	11.7	66	Unknown
389B	Chloroprene	TBTO	4.9	28	Unknown
B59	Acrylonitrile	TBTS	7.5	42	Unknown
633B	Chloroprene	{Bayluscide TBTO}	8.0 3.0	–	50 $\mu g/cm^2$-day
105B	Chloroprene	TBTO	6.5	60+	Unknown

[a]Calculated from antifouling life and/or measurement of residual tin content.
[b]TBTAd = tributyltin adipate.

and excellent mortality rates noted. Materials thought superior were sent to various interested colleagues. Berrios-Duran and Ritchie examined antifouling rubber pellets and found that they remained molluscicidal after daily washings over an 18-week period.[238] Long-term laboratory bioassays on controlled release TBTO, TBTR (tributyltin resinate), TBTS, and TBTO/Bayluscide mixes were performed by Hopf and Gall.[239] It was noted that several formulations remained active over a 26-month period. Also, Hopf and Gall observed that formulation and vulcanization conditions influenced longevity.

DaSouza and Paulini evaluated five TBTO formulations and one slow release Bayluscide formulation in laboratory and field tests.[240] Both pools and drainage ditches were used as test environments over a 4-month exposure period. Pool tests showed 100% snail mortality. Drainage ditch evaluations, even with flowing water, showed a continuous reduction in living snail populations.

In Tanzania, Fenwick treated several pools with formulation 633B and observed a reduction of snail populations over a 9-week period.[241] It was observed that *Biomphalaria pfeifferi* mortality in running water varied with dosage. At four pellets/l 100% mortality was achieved in 72 to 96 hr, while at one pellet per 8 l 192 to 216 hr was required to achieve the same mortality (Table 3–3).[242]

Lack of data concerning the environmental impact and mammalian toxicity of organotins led to the incorporation of a conventional control agent, Bayluscide, in chloroprene and other matrices.[243] Controlled release is well confirmed and longevity half-lives in excess of 2 years are predictable. Mammalian toxicity and phytotoxicity are low. The human allergic reaction to powdered Bayluscide is not evident after the material is incorporated in rubber. Fish toxicity is high whether Bayluscide is used in the conventional sense or via controlled release.

The basic slow release Bayluscide materials are shown in Table 3–4, with the loss rates given.

All 1121 series are chloroprene based of the following master recipe:

Ingredient	Formulation (parts)
Neoprene WRT	100
Zinc oxide	5
Magnesium oxide	4
FEF carbon black	15
PBNA[a]	2
MBT[b]	1
Lauric acid	3

[a]Phenyl-β-naphthylamine.
[b]Mercaptobenzothiazole.

Loss rate curves were plotted over a prolonged laboratory immersion test with periodic sampling and Bayluscide determination by the Struffe method.[246] Basically, the active nitro group on the ethanolamine salt of niclosamide is reduced to the corresponding amine with a sodium-hydroxide-in-methanol solution. The amine concentration is then measured spectrophotometrically at 335 nm. 1121F, considered superior, was monitored over a 3274-hr period.

It is noted in the Table 3–4 that an abrupt break occurs in loss rate between 1121F and 1121G: although the increase in agent loading is only 1/5, the loss rate increases by almost a factor of 4! This is hypothesized as arising from the fact that the Bayluscide-in-chloroprene solubility is about 29%; below this, loss is through a diffusion-dissolution mechanism, whereas when this value is exceeded, the excess (over the solubility threshold) is rapidly washed out by a leaching process.

TABLE 3–4

Slow Release Bayluscide® Compounds

Code	Bayluscide concentration (%)	Loss rate (average)	
		Loss/in.^{2}-hr	Loss/g-hr
1121A	5.8	–	–
1121B	7.1	3.6 μg	1.44 μg
1121C	10.3	–	–
1121D	13.4	7.0 μg	2.80 μg
1121E	18.9	8.9 μg	3.56 μg
1121F	27.8	10.5 μg	4.20 μg
1121G	35.0	39.0 μg	15.60 μg
1121H	43.5	51.2 μg	20.48 μg

From Cardarelli, N. F., in *Molluscicides in Schistosomiasis Control*, Cheng, T. C., Ed., Academic Press, New York, 1974, 177. With permission.

The ethanolamine salt of niclosamide, isolated from the 70% wettable powder (Bayluscide) via repeated dimethylformamide extraction, was examined for water stability. At pH 7.0 the solubility of the purified material was about 50 ppm. A 1150-hr stability test in distilled water indicated a half-life of over 6000 hr. Once incorporated in rubber, the half-life is greatly extended, as shown in Figure 3–8.

Little work has been done with controlled release niclosamide. Laboratory tests of the 1121 series at the Technical Development Laboratories of the National Communicable Disease Center (Savannah, Ga.) indicated that target snails could be continuously destroyed over a 22-week period, using 6-hr exposure periods.

Figure 3–8 illustrates the increase in water stability achieved through incorporation of the agent in an elastomeric base.

Attempts at the incorporation of trifenmorph (Frescon) in a long-term controlled release system have not been overly successful to date.[247] The best achievement has been with a 20% active

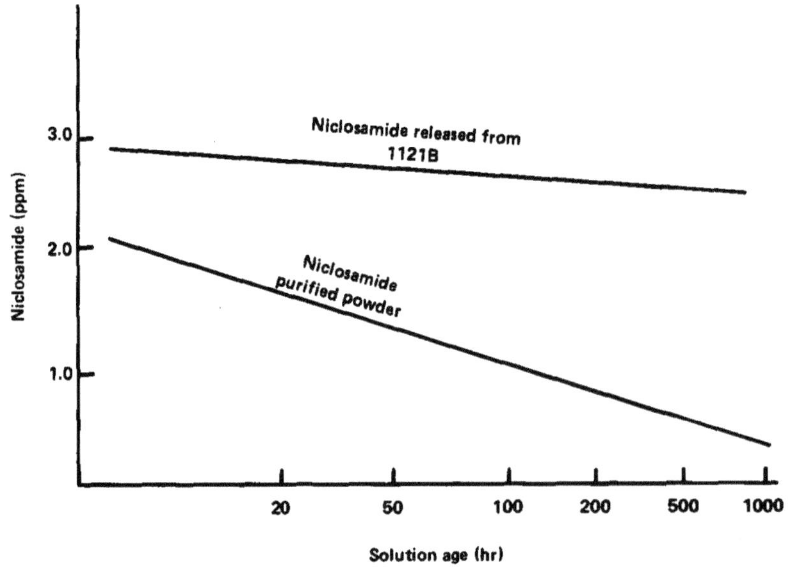

FIGURE 3–8. Deterioration of niclosamide in water.

loading in a synthetic natural rubber base. This material provided a 97% mortality in a 10-day bioassay after an initial 215 days' water soaking (with periodic water changes) period.

Over the past several years the incorporation of copper sulfate in an elastomeric leaching type matrix has been undergoing evaluation. Since inorganic copper salts show little, if any, solubility in all rubber matrices examined to date, establishment of a diffusion-dissolution type system was not possible.[248]

In the preliminary formulation work it was ascertained that copper sulfate pentahydrate loadings of up to 86% by weight could be incorporated in an ethylene-propylene-type matrix. Epcar® 5465, an oil-extended polymer manufactured by the B. F. Goodrich Company, was found superior in this respect; 80% loadings in natural rubber was the maximum achievable. Highly loaded materials, once exposed to water, lose their agent content rapidly due to leaching and degradation of the very porous matrix. A longer biologically active life is achieved at agent concentrations of 60% or less in the binding matrix.

Copper sulfate in the anhydrous, monohydrate, and pentahydrate forms was investigated, as well as copper carbonate, a low water solubility chemical. Due to its deliquescent nature, anhydrous copper sulfate presented processing problems that would defeat its use under typical manufacturing conditions. The pentahydrate form loses water of hydration as steam during vulcanization. In a low-pressure curing system this leads to bubble formation in the stock and processing difficulties. In a high-pressure cure, the steam, unable to escape, appears to combine with the sulfate radical-forming sulfuric acid which destroys the stock. Copper carbonate releases at too slow a rate for efficacy. Consequently, the monohydrate form was found to be the most acceptable.[244,249]

Vulcanization conditions have little effect on loss rate per se, although proper curing retards matrix degradation and enhances the strength properties of the binding element.

Controlled release copper sulfate formulations exhibit typical leach-type loss curves. Loss is proportional to exposed surface area.

Several hundred formulations were prepared showing varying degrees of biological activity against target snails. Loss rate cannot be varied directly by the use of additives such as carbon black. Two methods of adjusting the loss of copper ions were discovered. Pellet geometry is a major factor. Since increasing or decreasing the surface-to-volume ratio the loss rate is varied accordingly, the processed pellet or granule size is significant. It was also observed that the water quality, in terms of pH, mineral content, and dissolved gases, especially carbon dioxide, affected loss rate. Copper ions in the elastomer/water interface tend to combine with dissolved ions and form a film of copper carbonate and other insoluble materials on the test specimen surface, which in turn retards, and may completely halt, further copper ion movement from the matrix into the interface. This problem was overcome by the addition of ammonium sulfate, a leachate, to the matrix. The slow loss of ammonia ion prevents film build-up.[244,248,249]

Superior copper sulfate systems have provided 4- to 6-month biological lifetimes. Over this period, 100% snail mortality has been noted in periodic testing of ammonium sulfate/Epcar 5465/copper sulfate monohydrate systems.

Toxicology

Controlled release molluscicide formulations have been based almost exclusively upon the use of organotin and inorganic copper toxicants. A few materials bearing niclosamide as the active agent have been prepared. Present evaluation of organolead materials is under way. Relatively little is known regarding the chronic effects of any of the materials, only copper intoxication being somewhat documented in the medical literature. Acute effects of copper materials and niclosamide on various life forms are known, although ecological impact from continuous application has received little attention. Organotin and organolead compounds of interest as molluscicides have not been thoroughly investigated in terms of acute or toxic effects, mechanism of intoxication, effects on nontarget biota, detoxification phenomena, environmental half-life, and so on. A body of data has been slowly accumulating and is surveyed in this section. Table 3–5 is a general overview of the molluscicidal properties of the aforementioned materials with information helpful to field evaluation personnel. Data are strictly in terms of *conventional* use and do not include values relative to controlled release formulations. TBTF is added to the table due to increasing evidence of its superiority over other organotins.

Molluscicide effects on nontarget biota have

TABLE 3–5[250]

Molluscicides and Their Properties

Property	Molluscicide						
Common name	Niclosamide	Trifenmorph	Copper sulfate	TPTA	TBTO®	TBTF	TPLA
Physical form	Crystalline solid	Crystalline solid	Crystalline solid	Crystalline solid	Liquid	Solid	Crystalline solid
Active ingredient	Ethanolamine salt of 2'5-dichloro-4'-nitrosalicyl-anilide	*N*-trityl-morpholine	Copper ion	Triphenyltin acetate	bis(Tri-*n*-butyltin oxide)	Tributyltin sulfide	Triphenyllead acetate
Solubility in water	230 ppm (pH dependent)	<0.1 ppm	32%	>500 ppm	10–30 ppm		1.55%
Toxicity							
Snail, LC_{90} (ppm × hr)	3–8	0.5–4	20–100	1.9	0.9–1.3		>3.6
Snail egg, LC_{90} (ppm × hr)	2–4	240	50–100	0.6	0.55		
Cercaria, LC_{90} (ppm)	0.3			<0.05	<0.001		<0.07
Fish, LC_{90} (ppm)	0.05–0.3			2.4–4.8	1–3		
Rat, acute oral LD_{50} (mg/kg)	5000	1400		<200	<250		12.3
Herbicidal activity	No	No	Yes	Yes	(Slight)		
Stability (affected by)							
UV radiation	Yes		No	Slight	No	(No)	Slight
Mud, turbidity	Yes	No	Yes	Yes	Yes		Yes
pH	Optimum 6–8	Yes	Yes	No	No	(No)	
Algae, plants	No	No	Yes	Yes	Yes		Slight
Storage	No	No	No	No	No	(No)	(No)
Handling qualities							
Safe	Yes	Yes	Yes	Yes	Yes	Yes	No
Simple	Yes	Yes	Yes	No	Yes	Yes	No
Formulations	70% WP[a] 25% EC[b]	16.5% EC[b]	100% granules	100% powder	95% liquid	100% powder	100% powder
Field dosage							
Aquatic snails (ppm × hr)	4–8	1–2	20+	4–10 5–8	4–20		
Amphibious snails on moist soil (g/m^2)	0.2						

[a]Wettable powder.
[b]Emulsifiable concentrate.

not been investigated in detail. Shiff and Garnett examined the effects of niclosamide, sodium pentachlorophenate, copper sulfate, and trifenmorph on cladocera, copepods, diatoms, and insect larvae.[173] Only trifenmorph showed no noticeable effect. The above four materials have been in use for years with relatively little concern over their environmental impact. Unfortunately, newer materials such as the organoleads and organotins are viewed with considerable "alarm," and those concerned with environmental quality have asked for far more detailed evidence concerning ecological impact and safety than ever required for the conventional molluscicides. For instance, long-term environmental effects of niclosamide are essentially unknown, and short-term effects have only been subjected to cursory study.

Organotins

Organotin compounds are defined as materials possessing at least one tin-carbon bond. Four series of compounds are known and commercially produced: R_4Sn, R_3SnX, R_2SnX_2, and $RSnX_3$, where R is a butyl, octyl, or phenyl group and X is a halide, oxide, hydroxide, carboxylate, or thiolate.

R_4Sn compounds are mainly used as intermediates in the preparation of other organotins. However, tetrabutyltin finds limited use as a corrosion-inhibiting additive for mineral-based lubricants. Tetraphenyltin is an acid scavenger useful in chlorinated dielectric fluids.

R_3SnX compounds find wide use as biocides. Tributyltin compounds are used as wood preservatives, bactericides, the active ingredient in antifouling paints, mildewcides, and pesticides. Some have an antifeedant action against insects, and others are chemosterilants.[251] Tributyltin acetate, sulfide, fluoride, and oxide are used in antifouling preparations. Tributyltin benzoate is a germicide. Tributyltin chloride has been used as a rodent repellent. Triphenyltin acetate is an effective pesticide against *phytopthora infestans*, potato blight, while triphenyltin hydroxide is an antifeedant. Probably the excellent wood preservative features of triphenyltin chloride will lead to its commercialization in the marine area.

Tricyclohexyltin hydroxide is registered and used as a residual nonsystemic miticide. Hexamethylditin is marketed as an insecticide and antifeedant formulation by the Pennsalt Chemical Corporation, Cornwells Heights, Pa.

R_2SnX_2 compounds are used as stabilizers in polyvinyl chloride (PVC) and as catalysts in polyurethane and silicone rubber production. Dibutyltin di(2-ethylhexoate) and dibutyltin dilaurate are used as catalysts. The latter also finds use as an intestinal parasiticide for treating poultry. Dibutyltin maleate, di(*n*-octyl)tin maleate polymer, and di(*n*-octyl)tin-S,S′-bis(iso-octylmercaptoacetate) are used as PVC stabilizers, the latter two having FDA approval for use in plastic containers for food products.[252]

Organotin-stabilized PVC may slowly lose organotin upon exposure to various environments. Tissue reactions to PVC catheters have been reported and investigations have indicated that organotin leachates are responsible.[253]

It has been established that tin is a trace element necessary for life, although its specific function in the mammalian body is unknown.[254] Schwarz and his co-workers have demonstrated that tin is a trace element essential for growth and well being in rats.[255,256] Addition of tin as inorganic salts or organic derivatives to tin-free rat diets results in a constant growth effect of 30 to 50% of the initial growth rate. The effect is seen with trimethyltin hydroxide, dibutyltin maleate, stannic sulfate, potassium stannate, and stannous tartrate. Optimum results were noted at 1 ppm tin in the diet.

It has long been known that various organs of the human body contain trace amounts of tin, although it was felt that it existed solely as an impurity of no benefit to man.

Tetraalkyl tins are toxic to man, and exposure can lead to reversible paralysis and encephalopathy. Di and tri alkyltins are, in some cases, vesicants and most show a lachrymatory property. Triethyltin is very active and its ingestion can lead to a chronic poisoning of the central nervous system and inhibition of glycose oxidation.[257,258] Tripropyltin compounds are less reactive and the tributyltin materials are far less toxic, indicating that increasing the molecular size of the alkyl radical leads to greater safety.[259]

Bis(tri-*n*-butyltin) oxide, TBTO, applied directly to rat skin at high dosages (33 to 100%), led to dermatitis ulcerans.[260] Very low dermal concentrations showed no effect. TBTO administered orally by a stomach tube to rats at high dosages led to an acute intoxication with gross symptoms such as preening depression, squinting eyes, labored respiration, diarrhea, and ataxia.[261]

Gross autopsies showed hyperemic lungs, irritation of the gastrointestinal tract and peritoneum, congested kidneys, and damage to the adrenal glands.

TBTO, even at small dosages, shows severe eye effects and rapid corneal perforation in some instances.[262]

A considerable body of literature exists concerning the toxic effects of tin compounds on mammals and other animals, and postulated mechanisms of toxicity.[263] Unfortunately, investigators are interested in the more active materials such as the triethyltins and not the compounds of value to controlled release, TBTO, TBTAc, and TBTF for instance.

In so far as one can search the literature there exist no references as to acute or chronic organotin poisoning in man. In the author's experience in the factory production of antifouling rubber and controlled release molluscicides the following observations were made:

1. Skin contact with undiluted TBTO, TBTAc, and TBTF lead to no dermatological reaction if removed immediately with a suitable solvent such as hexane or ethanol. Water is of no value. If contact is prolonged, especially with TBTO, a mild dermatitis is evidenced lasting a few days with no sequelae. In some instances a burning sensation, similar to exposure to diluted acid, is felt.

2. TBTO vapors irritate the mucous membranes, with varied response in different individuals. Hot TBTO vapors observed during vulcanization, and more importantly in the buffing of Nofoul rubber seams, can cause swelling of the membranes of the nose and throat and the symptoms of a severe sore throat. Removal from the contaminated atmosphere leads to an immediate decrease in symptoms and full recovery within a few hours. Adequate ventilation and/or the use of a face mask are helpful. Lachrymatory response of some individuals is noted as well.

3. In the only case of injury known to the author an oiler carrying buckets of TBTO from the drum to the millmen succeeded in splashing a few drops in one eye, and did not seek immediate medical assistance. Once the situation was known the eye was washed and treated. Inflammation with some pain was evidenced for several days. Healing was rapid and after 4 days no gross symptoms remained. Repeated checks indicated no sequelae. The results would probably have been similar with a solvent splash and far less serious than that experienced with acids or various additives in common use in the rubber industry.

There is no history of human injury arising from contact with or ingestion of controlled release organotin elastomers. This would be expected in that the amount of organotin readily available would be a monolayer of the active molecule on the surface, an extremely small amount. The author has worn TBTO-containing elastomer sheet in tight contact with the forearm for extended time periods without dermatitis or any gross symptoms.

TBTO is toxic to bacteria; however, its effect on soil microbes at levels up to 100 ppm is negligible.[264] Triphenyltin acetate has a 3- to 10-day life span in soil, triphenyltin hydroxide 3 days or less.[265] The products of decomposition are of greater stability, however, lasting from 12 to 25 days.

Others have hypothesized that the triaryl organotins are not decomposed but tightly bound to soil particles.[266] Katsumura indicated that TBTO is strongly bound to soil particles.[267]

Phytotoxicity of organotin compounds is largely unknown. The most comprehensive studies have been performed with TPTA and tricyclohexyltin hydroxide, with general results indicating that these materials do not penetrate the test vegetables and fruit trees and thus are not translocated; residues tend to be inorganic tin.[268–270] The compendium of phytotoxicity studies in Table 3–6 was furnished in 1972 by the M & T Chemicals Company of Rahway, N. J.

Trialkyltins are probably more phytotoxic than the triphenyltins.[271]

Evaluation of TBTO and TBTF against the banana plant has indicated that these materials are not translocated from the surrounding soil to the fruit.[272]

Sheldon has presented toxicity test results of in-use antifouling organotin materials: TBTO, TBTF, and TPTF. General conclusions are shown below.[273]

1. Acute toxicity studies

	Acute oral LD_{50}, rat	Acute dermal LD_{50}, rabbit	Primary skin irritation, rabbit	Primary eye irritation, rabbit
TBTO	234 mg/kg	11,700 mg/kg	Severe	Extreme
TBTO (in paint)	–	–	Moderate	Severe
TBTF	200	680	Minimal	Extreme
TBTF (in paint)	4,556	10,250	Extreme	Extreme
TPTF	11,700	1000–3000	Minimal	Extreme
TPTF (in paint)	–	–	Moderate	Extreme

2. In 90-day subacute dermal toxicity evaluation of TBTF against albino rabbits a dose level of 68 mg/kg is toxic and 14 mg/kg is a no-effect level.

3. TBTF, in a 6-month study, showed no carcinogenicity potential.

The fate of organotins in mammals is not well known. Ethyltin trichloride is probably not metabolized but excreted unchanged.[274] Molluscicidal organotin reactions in mammals are unknown. Organotins tend to be species specific in toxicity; for instance, the dibutyltin oxide oral LD_{50} in rats is 100 mg/kg and 2000 mg/kg in fowl.[271] Thus, each important nontarget organism would have to be assessed for each compound.

TABLE 3–6

Phytotoxicity of Triorganotin Compounds

Pre-emergence Rating[a,b] (50 lb/acre)

Compound	Pea	Radish	Sorghum	Japanese millet	Wild oats
TBTF	0	0	0	0	0
TPTH[c]	0	0	0	0	0
TCHH[d]	0	0	0	0	0

Post-emergence Rating[a,e] (4,000 ppm)

Compound	Tomato	Crabgrass	Corn	Beans	Wild oats
TBTF	10	10	0	8	0
TPTH	7	0	3	0	10
TCHH	0	0	0	0	0

Pre-emergence Rating[a,b] (10 lb/acre)

Compound	Morning glory	Zinnia	Mustard	French millet	Japanese millet	Crabgrass
TBTO	0	0	0	0	0	0

Post-emergence Rating[a,c] (10 lb/acre)

Compound	Morning glory	Zinnia	Mustard	French millet	Japanese millet	Crabgrass
TBTO	10	10	10	2	5	9

[a]Rating system 0–10: 0 = no damage, 10 = complete kill.
[b]Seeds planted, soil surface sprayed with test compound, ratings at 2–3 weeks.
[c]TPTH = triphenyltin hydroxide
[d]TCHH = tricyclohexyltin hydroxide
[e]Plants grown to 4 in., sprayed to run off, ratings made 10 to 14 days after application.

TABLE 3–7

Toxicity of Tributyltin Compounds to Fish

Compound	Species	Exposure time	LC_{50} (ppm)	LC_{100} (ppm)	Ref.
TBTAc	*Carassius auratus* (goldfish)	24 hr	–	0.075	206
	Lebistes reticulatus	24 hr	–	0.075	206
	(guppy)	48 hr	–	0.03	
TBTO	*Tilapia nilotica*	5–6 days	0.03	–	275
	Heterotis hemichromis	5–6 days	0.03	–	275
	Crassius auratus	24 hr	–	0.075	276
	Lebistes reticulatus	24 hr	–	0.075	276
	Rainbow trout	24 hr 48 hr		0.028 (lethal) 0.02 (lethal)	277
	Minnows, bass, perch	2 months	Bass and perch unaffected at 1 ppm; 100% minnow mortality at 1 ppm		73

Environmental detoxification effects are largely unknown. Triphenyltins are decomposed by ultraviolet radiation as follows:[271]

$$\underset{\text{(toxic)}}{(C_6H_5)_3Sn-} \longrightarrow \underset{\text{(slightly toxic)}}{(C_6H_5)_2Sn} \longrightarrow (C_6H_5)Sn- \longrightarrow \underset{\text{(nontoxic)}}{-Sn-}$$

TBTO and TBTF may hydrolyze to a carbonate form, at least in sea water.[273]

Organotins are piscicidal, Table 3–7 being a summary of pertinent fish toxicity data.

Toxicity analysis of single-dose TBTO evaluated against *Lebistes reticulatus* provided the following LC_5 data:[278]

30-day LC_5 (TBTO) = 0.007 ppm
60-day LC_5 (TBTO) = 0.006 ppm
90-day LC_5 (TBTO) = 0.004 ppm

Multiple daily dosage effects on guppies were as follows:[279]

30-day LC_5 (TBTO) = 0.0022 ppm/day
60-day LC_5 (TBTO) = 0.0014 ppm/day

Matthiessen investigated both the acute and chronic effects of TBTO on *Tilapia mossambica*, an African food fish. In this thorough study the lethal time (LT) values of TBTO at various concentrations (Table 3–8) and, importantly, sublethal effects were noted, the latter being carried out at 0.008 ppm. It is postulated that acute TBTO intoxication leads to death through asphyxiation.[280] The following sublethal effects were noted: a reversible opacity of the eye, sluggishness, difficulty in balancing, discoloration of the skin, and changes in behaviorally related display patterns. Effects diminished after a few weeks.

Nofoul rubber was evaluated by the U.S. Army in mouse inhalation studies and drinking of effluent water. No changes in body weight, histology, or activity were noted.[281]

The toxic mechanism underlying snail destruction through ingestion of TBTO or other organotins is largely unknown. Gross histological examination of snail organs exposed to various dosages of organotin indicated a selectivity of effect. Attack is focused upon mucous membranes, the gastrointestinal tract, and connective tissue.[282] No pathology was evident in muscle or nerve tissue. By 60 min, postexposure cells in connective tissue and membranes began to collapse. Internal hemorrhage was apparent as impermeable membranes became permeable to body fluids. Mortality arises as hemolymph from the internal sinuses floods the internal organs. It is

TABLE 3–8

Acute Toxicity of TBTO® to *Tilapia mossambica* (ten fish per run)

TBTO concentration (ppm)	LT_5[a] (minutes)	LT_{50} (minutes)	LT_{90} (minutes)
0.119	64	88	115
0.095	33	47	63
0.089	170	277	405
0.066	355	470	580
0.043	330	433	540
0.030	460	630	820
0.021	2,600	3,800	5,100
0.018	3,100	4,600	6,300
0.009	>140,000	>140,000	>140,000
0.004	>140,000	>140,000	>140,000

[a]LT: lethal time value.

hypothesized that TBTO reacts with various, though not all, amino acids, thus depolymerizing specific proteinaceous tissue leading to the collapse of cell walls, destruction of membrane properties, and profound breakdown of liver, heart, and other organ function due to massive hemorrhage.

Niclosamide

The toxicology of niclosamide has been investigated in some detail by Hecht and Gloxhuber.[283] The LC_{50} values for the ethanolamine salt given in a single dose was 2.5 g/kg (rats) and 1.0 g/kg (mice). Repeated oral doses for up to a year did not cause gross histopathological changes in dogs or rats. Dermal LD's are high. Humans given single oral doses at 0.3 g/kg showed no overt symptoms of intoxication. The compound is rapidly excreted after absorption, with no evidence of accumulation in rat tissue.[284]

The snail intoxication mechanism is unknown although it is said that niclosamide blocks oxygen uptake.[138]

Chronic effects, carcinogenic potential, and the fate of niclosamide in the environment remain largely unknown. It is a powerful piscicide, destroying fish at concentrations sublethal to most snails; in fact, the major U.S. usage is keyed to destruction of trash fish and the lamprey eel. Degradation, other than bacterial,[39] and detoxification factors are largely unknown.

The author has observed allergic reactions in those handling the 70% wettable powder in industry, although this varies from individual to individual. Mouse paralysis and mortality at concentrations below 1 ppm with very short-term exposure has also been observed. The general consensus that niclosamide is harmless to human handlers and the environment remains to be proven to the satisfaction of this author.

Organoleads

Several organolead materials have shown considerable promise as antifouling agents and molluscicides.[285] Since organoleads are soluble in elastomers, potential exists for the creation of long-lasting controlled release pesticides. Table 3–9 depicts chemical and toxicological values of organolead materials of interest.

In antifouling compositions triphenyl- and tributyllead acetate and triphenyl- and tributyllead laurate appear to be the most promising.[85] Dibutyllead diacetate and tributyllead acetate show wood preserving properties superior to that of creosote.[87] Other organolead compounds may be useful as cotton preservatives.[289]

Tributyl- and triphenyllead adjuncts show considerable phycotoxicity, comparable to that of organomercury and organotin compounds.[290] Use as controlled release algicides and fungicides is suggested. Dibutyllead diacetate appears to be a suitable antihelmintric against tapeworms.[292] A dosage of 5 mg/kg given to chickens destroyed the parasite without harming the host.

Environmental impact parameters, chronic effects on humans and other mammals, and

TABLE 3–9[286-288]

Organolead Materials

Name	Melting point (°F)	Toxicity: Dermal LD_{50} (mg/kg)	Toxicity: Oral LD_{50} (mg/kg)	Water solubility (g/l)
Triphenyllead acetate	421–424	300	228	
Tributyllead acetate	186–191	600+	32	2.6
Tributyllead 2-ethylhexanoate	176–194			
Triphenyllead laurate	243–248	10,300+	800	
Hexaphenyldilead	302–311	10,200	34,600+	Insoluble
Dibutylead dichloride	212+	1,500	300	6.4
Tripropyllead acetate				15
Dibutyllead diacetate		3,720	155	

detoxification mechanisms are unknown. High percutaneous toxicity for some of the organoleads would indicate that caution be used in handling. As with the organotins, once incorporated in an elastomer the toxicity hazard is likely to be significantly reduced.

Copper Ion

Copper is a trace element essential to animals and plants, and is a major constituent in a number of enzymes: ceruloplasmin, vital to iron utilization; lysine oxidase, and tyrosinase in humans; plastocyanin, essential to photosynthesis in plants; and hemocyanin, necessary for oxygen transport in invertebrates. At dosages in excess of trace requirements copper ion is toxic and has been in use for years as an algicide, herbicide, and molluscicide.

It is generally recognized that small amounts of copper ion are nontoxic to mammals; for instance, copper-bearing algicides are cleared for use in potable water, and various copper chemicals are applied to growing crops to control various pests.[292] However, human ingestion of large amounts of copper salts, e.g., copper sulfate, can be fatal. Acute poisoning is known, though rare, resulting in widespread damage to capillaries, kidneys, and liver. Death, if early, results from shock, and if later is due to hepatic or renal failure. The mean lethal dose (human) for copper sulfate is estimated as 15 g.[293] Chronic intoxication in mammals can be induced in the laboratory. Long-term feeding of small doses results in liver cirrhosis and may cause extensive necrotic damage to tissue.

The toxicity hazard of soluble copper compounds is tempered by the fact that rapid detoxification through chemical activity occurs in the use environment. Residual effects are low or absent. Human exposure to toxic levels is usually only a remote possibility as regards the use of copper sulfate as a molluscicide.

The mechanism of copper intoxication in snails is under investigation. Cheng and Sullivan theorize that "copper enters *B. glabrata* no further than the outermost epithelial layer of the body, where its effects may be extracellular, intracellular or in the plasma membrane."[294] Copper accumulations are observed on the epithelial covering of the rectal ridge, perhaps indicating that this is the cidal site.

DS-2787

DS-2787, 2,4,5,6-tetrachloroisophthalonitrile, is a molluscicidal agent of interest in that (1) it can and has been incorporated in controlled release formulations and (2) it is a nonmetallic of very low mammalian toxicity. DS-2787 is fairly inert, being thermally stable (mp 250 to 251°C) and chemically stable in moderately alkaline and acid aqueous media.[197] Water solubility is 0.5 ppm at 25°C.

It is of very low level toxicity to rodents (oral LD_{50} = 10,000+ mg/kg) and canines (5000+ mg/kg). The dermal LD_{50} for rabbits is in excess of 10,000 mg/kg body weight. Repeated applications of 500 to 1000 mg/kg to shaved and abraded rabbit dermis produced a mild to moderate irritation.[197] Antimicrobial activity is high, and DS-2787 is commercialized as a fungicide. Its half-life in soil is dependent upon the type, but ranges from 4 to over 40 days at constant temperature and moisture.

TABLE 3–10[295]

96-hr LC_{50} for Three Species of Fish

Species	LC_{50} for p,p'-DDT	LC_{50} for DS-2787
Rainbow trout	0.48 ppb	250 ppb (est.)
Bluegill sunfish	1.90 ppb (est.)	386 ppb (est.)
Channel catfish	18.0 ppb (est.)	438 ppb (est.)

DS-2787 is piscicidal, although far safer than many other pesticides. Table 3–10 compares DS-2787 and *p,p'*-DDT in that regard.

DS-2787 was used as a fungicide in bait molluscicide formulations tested on a banana plantation in St. Lucia (B.W.I.), and consequently was evaluated against bananas. At the limits of detection, 0.01 ppm, no translocation of DS-2787 or its metabolite from root to fruit occurred.[296]

Compounding Theory

Considering the controlled release of pesticides from elastomers, the only known studies of the underlying theory have been made in regard to molluscicides – and especially in reference to the diffusion-dissolution release mechanism.

A colleague of the author, Dr. S. V. Kanakkanatt, has studied the effects of various parameters on the diffusion-dissolution characteristics of controlled release organotin molluscicides. He notes that the following variables are of importance:[92]

1. Crystallinity of the polymer
2. Carbon black loading
3. Carbon black particle size
4. Other compounding additives
5. Interaction between molluscicide and polymer
6. Glass transition temperature of the polymer
7. Ambient temperature
8. Water flow rate
9. Biomass character.

The first five can be independently determined and their values related to the diffusion coefficient of the molluscicide. Numbers 6 and 7 can be combined into one variable, which is the difference between the test temperature (T) and the glass-transition temperature (T_g). Numbers 8 and 9 influence the dissolution rate and were not considered in this initial study. Numbers 1 through 5 are relatable to the molecular weight between cross links (M_c).

Kanakkanatt derived a number of equations on the basis of known theory as applied to the phenomena of diffusion based upon the above factors.[92,297] The various expressions are summarized below.

Equation to determine the diffusion coefficient at any M_c (not applicable where dc is <1.5):

$$D_{A2} = D_{A1} - \alpha_D(dc_2 - dc_1) \quad (1)$$

where D_{A1} = known diffusion coefficient, D_{A2} = unknown diffusion coefficient, α_D = cross-link density coefficient, and dc_1 and dc_2 are cross-link densities defined as $dc_1 = \frac{10^4}{M_{c1}}$ and $dc_2 = \frac{10^4}{M_{c2}}$.

At a given volume fraction of rubber, αD is obtained by plotting D, the diffusion coefficient, vs. dc for different values of the volume fraction, V_r; αD indicates the decrease in the diffusion coefficient per unit increase in cross-link density.

Relationship between diffusion coefficient and any carbon black content:

$$D_{B2} = D_{B1} - B_D(bc_2 - bc_1) \quad (2)$$

where D_{B1} = known diffusion coefficient, D_{B2} = unknown diffusion coefficient, B_D = carbon black content coefficient, bc_1 = black content for D_{B1}, and bc_2 = black content for D_{B2}.

Combination of the effects of cross linking and the carbon black loading:

$$D_{(AB)_2} = D_{A1} - \alpha_D(dc_2 - dc_1) - B_D(bc_2 - bc_1) \quad (3)$$

An alternate method of showing the effect of the carbon black content is by using the following theoretical equation:

$$D_f = \left(\frac{2}{3-V_r}\right) D_r \quad (4)$$

where D_f = diffusion coefficient of the black loaded elastomer, D_r = diffusion coefficient without a black loading, and V_r = volume fraction of elastomer in the loaded rubber. Substituting the expression for D_{A2} in Equation 2 for D_r in Equation 4 we get:

$$D_{(AF)_2} = \left(\frac{2}{3-V_r}\right)\left[D_{A1} - \alpha_D(dc_2 - dc_1)\right] \quad (5)$$

Determination of the diffusion coefficient for any elastomeric matrix of known glass transition temperature:

$$D_{E2} = D_{E1} + \gamma_{D_{H,L}} \left[(T_2 - T_{g_2}) - (T_1 - T_{g_1})\right] \tag{6}$$

where D_{E1} = known diffusion coefficient, D_{E2} = unknown diffusion coefficient, γ_{D_H} = temperature coefficient of the diffusion coefficient in the high temperature range, γ_{D_L} = temperature coefficient of the diffusion coefficient in the low temperature range, T_1 = temperature at which the diffusion coefficient is known, T_2 = temperature at which the diffusion coefficient must be calculated, Tg_1 = glass transition temperature of the elastomer for which the diffusion coefficient is known, and Tg_2 = glass transition temperature of the rubber for which the diffusion coefficient has to be calculated.

No significant change in the diffusion coefficient is found for values of T-Tg below 65 in the high range and below 85 in the low range. Above these threshold values the change in diffusion coefficient per unit increase of T-Tg is 0.83×10^{-8} cm^2/sec - A° for the high range and 1.22×10^{-8} cm^2/sec - A° in the low range. Combining Equations 5 and 6,

$$D_{(AEF)_2} = \left(\frac{2}{3-V_r}\right) \Big\{ D_{E1} - \gamma_{D_{H,L}} \left[(T_2 - Tg_2) - (T_1 - Tg_1)\right] - \gamma_D(dc_2 - dc_1) \Big\} \tag{7}$$

and combining Equations 3 and 6 we have

$$D_{(ABE)_2} = D_{E1} + \gamma_{D_{H,L}} \left[(T_2 - Tg_2) - (T_1 - Tg_1)\right] - \alpha_D(dc_2 - dc_1) - B_D(bc_2 - bc_1) \tag{8}$$

Concentration dependence of diffusion coefficient: The following equation combines the effect of cross-link density, dc, and the volume fraction of the elastomer, V_s, which is swollen as the diffusing liquid moves through it.

$$D_{C_2} = D_{C_1} - \left[D_1 + \delta_\alpha(V_{s_2} - V_{s_1})\right] (dc_2 - dc_1) \tag{9}$$

where D_{C_2} = the diffusion coefficient under the new conditions of cross-linking density, dc_2 and volume fraction, V_{s_2}, D_{C_1} = the diffusion coefficient under the original conditions of cross-linking density, dc_1, and volume fraction, V_{s_1} and δ_α = slope of the curve α_D vs. V_s.

The final equations are obtained by substituting α_D in Equations 7 and 8 with $\alpha_{D1} + \delta_\alpha (V_{s_2} - V_{s_1})$:

$$D_{(ACEF)_2} = \left(\frac{2}{3-V_r}\right) \Big\{ D_{E1} + \gamma_{D_{H,L}} \left[(T_2 - Tg_2) - (T_1 - Tg_1)\right] - \left[\alpha_{D_1} + \delta_\alpha(V_{s_2} - V_{s_1})\right] (dc_2 - dc_1) \Big\} \tag{10}$$

$$D_{(ABCE)_2} = D_{E_1} + \gamma_{D_{H,L}} \left[(T_2 - Tg_2) - (T_1 - Tg_1)\right] - \left[\alpha_{D_1} + \delta_\alpha(V_{s_2} - V_{s_1})\right] (dc_2 - dc_1) - B_D (bc_2 - bc_1) \tag{11}$$

These equations combine the effects of carbon black content, glass transition temperature, environmental temperature, diffusant (toxicant) concentration, and cross-link density on the diffusion coefficient of a given diffusing agent.[297]

Using simple organic liquids as diffusants the validity of the above equations has been verified.[92,298] Diffusion coefficients predicted from Equations 10 and 11 agree with experimentally measured diffusion coefficients within the limits of error arising from the presence of several other influencing factors, mainly crystallinity and diffusant-polymer interaction. Figure 3–9 illustrates the dependence of the diffusion coefficient on the carbon black loading, while Figure 3–10 shows the dependence of α_D on the volume fraction, V_s, in carbon black filled compounds.[297]

Kanakkanatt determined diffusion coefficients by both sorption and desorption techniques for various molluscicides and other substances. Table 3–11 presents a few values. The high hydrophobicity of the organotins in molluscicidal rubber is probably responsible for the decrease in water absorption.

The diffusion coefficient for niclosamide at varying loadings in neoprene is given in Table 3–12.

TBTO reacts with various compounding additives, especially lauric acid and sulfur-bearing curatives, in neoprene formulations, and also reacts directly with the polymer, giving a degree of cross-linking.[299] Thus, determination of TBTO

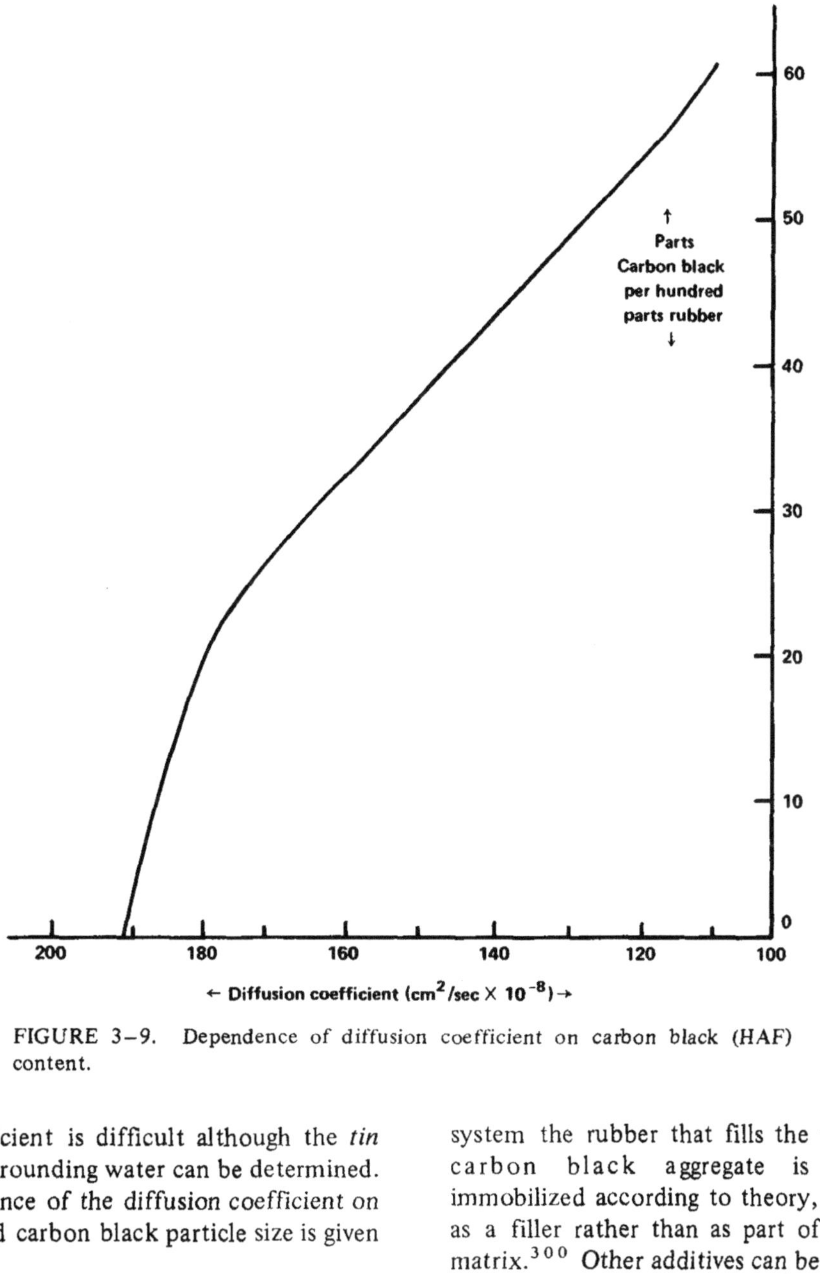

FIGURE 3–9. Dependence of diffusion coefficient on carbon black (HAF) content.

diffusion coefficient is difficult although the *tin* loss into the surrounding water can be determined.

The dependence of the diffusion coefficient on temperature and carbon black particle size is given in Table 3–13.

The diffusion rate is dependent upon polymer diffusant interactions and on the use of additives. Probably the major additive effect on the diffusion coefficient arises from the use of carbon black. Figure 3–11 illustrates the impact of carbon black on niclosamide release.

In an elastomer the influence of carbon black on viscoelastic properties depends upon several factors. degree of dispersion, surface chemistry, and morphological parameters such as size distribution of the particles in a carbon black aggregate and aggregate size. In a rubber-carbon black system the rubber that fills the void spaces in a carbon black aggregate is occluded and immobilized according to theory, and as such acts as a filler rather than as part of the deformable matrix.[300] Other additives can be adsorbed by the polymer or react with it.[301] The resultant "stabilizing" effect tends to increase diffusion path length, thus decreasing diffusion rate.

Solubility of the molluscicide in the elastomer is crucial for long-term controlled release utilizing the diffusion-dissolution method as explained in earlier sections. For instance, TBTO is soluble in the nitrile fraction of acrylonitrile polymers but has very low or no solubility in the acrylic portion. Figure 3–12 illustrates loss of TBTO from several acrylonitriles.

Approximate solubility limits are given in Table

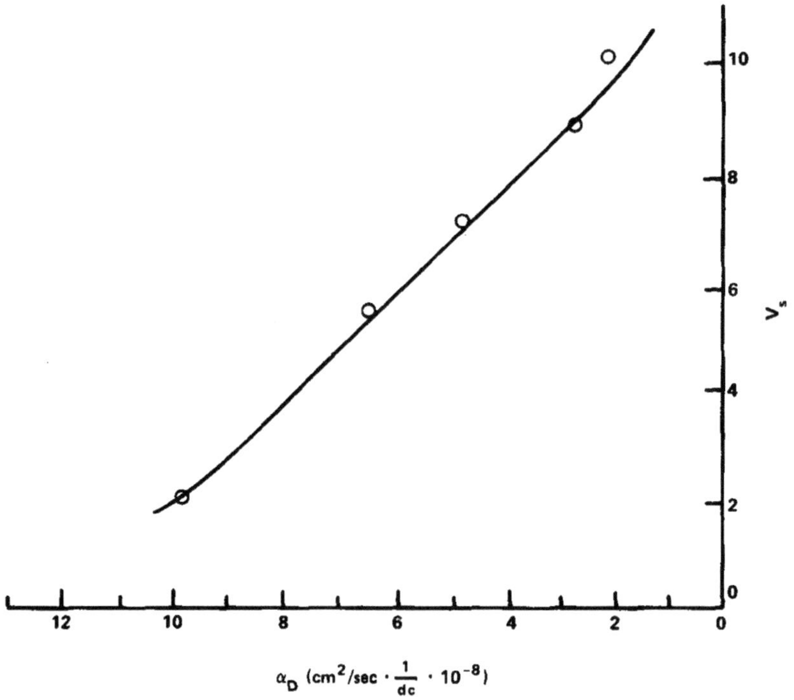

FIGURE 3–10. Dependence of α_D, cross-link density coefficient, on the volume fraction, V_s, in carbon black filled compounds.

TABLE 3–11[92]

Diffusion Coefficient Determination for Water into Molluscicidal Rubber

Code	Molluscicide	Toxicant concentration (%)	Medium	Technique	Diffusion coefficient for water
CBL-4	TBTO®	5.83	Water	Absorption	9.4×10^{-10} cm²/sec
CBL-3A	TBTF	5.84	Water	Absorption	4.9×10^{-10} cm²/sec
PDL-1	None	0	Water	Absorption	5×10^{-9} cm²/sec
CBL-49A	DS-649[a]	7.30	Water	Absorption	2.1×10^{-8} cm²/sec

Recipes

Ingredient	Parts			
	PDL-1	CBL-4	CBL-49	CBL-3A
Neoprene WRT	100	100	100	100
Zinc oxide	5	5	5	5
Magnesium oxide	4	4	4	4
FEF black	14	14	14	–
PBNA	1	2	1	1
MBT	–	1	–	–
Lauric acid	3	3	3	3
TBTO	–	8	–	–
HAF black	–	–	–	15
DS-649	–	–	10	–
TBTF	–	–	–	8

Note: All given formulations were cured at 300°F for 60 min.

[a]3,3,4,4-tetrachlorotetrahydrothiophene pyridine-1,1-dioxide, a potent fungicide of considerable molluscicidal efficacy manufactured by The Diamond Shamrock Company.

3–14. Solubility can be determined by several means of more or less exactness. By microtoming the candidate material solution, equilibrium can be determined by SEM techniques[302] or by X-ray fluorescence.[303] The latter method, immersion pick-up, and migration between sheets of molluscicide-loaded and -unloaded rubber were used to determine the solution limits given.[75,244]

If the solubility limit is exceeded several phenomena may occur. With TBTO at a concentration just in excess of the limit, physical properties of vulcanized materials drop rapidly, as shown in Table 3–15, and the excess, upon immersion, washes out rapidly due to leaching effects. In some instances reversion occurs and the elastomer passes, in part, into solution in the TBTO, creating an impossible processing situation.

TABLE 3–12[92]

Variation in the Diffusion Coefficient with Loading of Niclosamide in Neoprene WRT

Niclosamide loading (%)	Diffusion coefficient (cm^2/sec)
7	2.70×10^{-9}
13	2.09×10^{-9}
28	1.68×10^{-9}
35	1.15×10^{-9}

The solubility limit can be exceeded with only minor sacrifice of physical qualities by the use of a three-phase system. Microballon spheres such as those used in syntactic foams[304] can hold considerable quantities of TBTO. Phenol/formaldehyde microballons are favored in this regard.[75] TBTO diffuses slowly through the sphere walls, filling the interior void, and also is apparently adsorbed on the surface. Thus, by first slurrying microspheres and TBTO for 24 or more hours, and then adding this mix to the elastomer batch, it is possible to hold a TBTO concentration considerably in excess of the solubility limit. Upon immersion, TBTO is slowly desorbed from the

TABLE 3–13[92]

Dependence of Diffusion Coefficient of TBTO® (as Tin) in Natural Rubber on Temperature and Carbon Black Content

Code[a]	Carbon black content (pph)	Carbon black particle size (nm)	Diffusion coefficient $\times 10^{-6}$ (cm^2/sec) 26°C	46°C	70°C	90°C
CB-001	0	–	3.08	4.02	5.42	7.00
EB-001	4	45	2.20	3.10	4.35	5.90
EB-002	8	45	1.91	2.70	3.80	4.96
EB-003	14	45	1.82	2.31	3.11	3.98
EB-004	14	15	1.86	2.25	2.99	3.66
EB-005	14	2,350	1.70	2.10	2.81	3.33

Note: It would appear that the diffusion coefficient increases with rising temperature and decreases with increasing carbon black content. The particle size apparently has little effect.

[a]Recipes

Ingredient	Parts CB-001	EB-001	EB-002	EB-003	EB-004	EB-005
Natural rubber	100	100	100	100	100	100
Dicumyl peroxide	1.6	1.6	1.6	1.6	1.6	1.6
HCCB black	–	–	–	–	14	–
FEF black	–	4	8	14	–	–
MT black	–	–	–	–	–	14

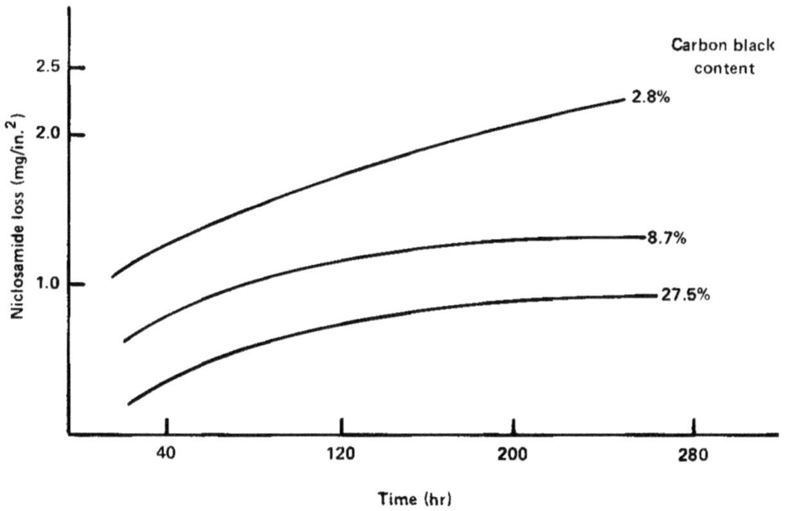

FIGURE 3–11. Relationship between niclosamide release and carbon black concentration. (From Cardarelli, N. F., in *Molluscicides in Schistosomiasis Control*, Cheng, T. C., Ed., Academic Press, New York, 1974, 177. With permission.)

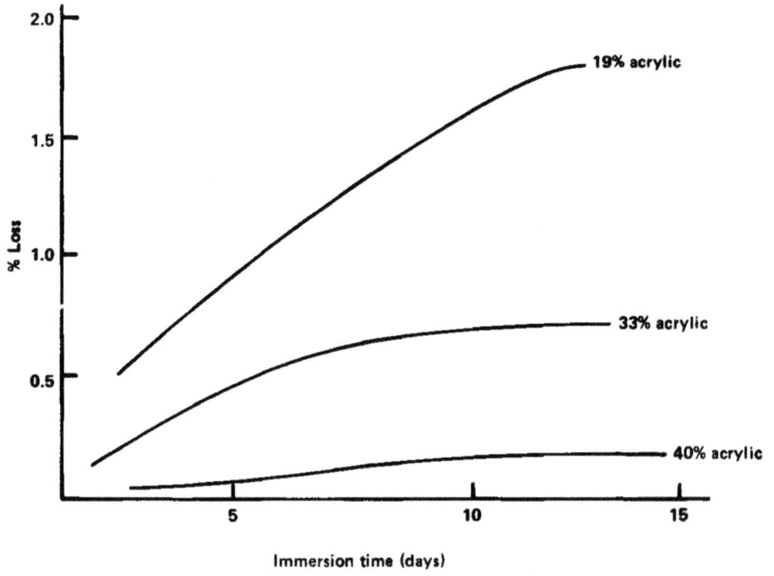

FIGURE 3–12. TBTO® loss from acrylonitrile rubber. (From Cardarelli, N. F., in *Molluscicides in Schistosomiasis Control*, Cheng, T. C., Ed., Academic Press, New York, 1974, 177. With permission.)

TABLE 3–14

Solubility of Molluscicides in Various Elastomers

Agent	Elastomer	Solubility limit (21°C) by weight (%)
TBTO®	Neoprene WRT	9.6
Bayluscide®[a]	Neoprene WRT	29.4
TBTF	Neoprene WRT	10.5
TBTO	Natural rubber	11.2
Bayluscide	Natural rubber*	10.7
TBTO	Ameripol® 1510 (SBR)	7.1
TBTO	Epcar® 5465 (ethylene-propylene dimer)	8.9
TBTO	Acrylonitrile (18% acrylo)	15.8
TBTO	Acrylonitrile (22% acrylo)	12.3
TBTO	Acrylonitrile (26% acrylo)	8.7
TBTO	Acrylonitrile (33% acrylo)	7.9
TBTO	Acrylonitrile (40% acrylo)	4.2
TBTO	Acrylonitrile (52% acrylo)	0.1

[a]Ethanolamine salt of niclosamide (purified).

From Cardarelli, N. F., in *Molluscicides in Schistosomiasis Control*, Cheng, T. C., Ed., Academic Press, New York, 1974, 177. With permission.

TABLE 3–15

Changes in the Physical Properties of Neoprene WRT as the Organotin Solubility Limit is Exceeded

	Physical properties			
Agent	Agent concentration (%)	Tensile strength (psi)	Ultimate elongation (%)	300% modulus (psi)
TBTO®	9.6	2,750	725	1,400
	9.8	375	200	0
TBTF	10.5	2,675	700	1,375
	10.8	1,450	340	800

From Cardarelli, N. F., in *Molluscicides in Schistosomiasis Control*, Cheng, T. C., Ed., Academic Press, New York, 1974, 177. With permission.

microspheres, maintaining maximum solution within the rubber phase. Up to 20% TBTO concentration in neoprene formulations can be achieved by this method.

Also, the Bayluscide loading can be increased by the use of carbon or siliceous glass microspheres. One can thus add up to 20 g of Bayluscide to 100 g of natural rubber where the solution limit is approximately 10.7 g/100 g rubber.

Organotin Materials

An 8-year controlled release molluscicide development program culminated with the creation of MT-IE, a TBTO/natural rubber compound now commercialized by the M & T Chemicals Company as Biomet-SRM®. This material has shown long-term complete control of snails under laboratory conditions at 1 mg formulation per liter of water. This represents a total toxicant reservoir of only 60 ppb TBTO.[305]

In laboratory tests conducted by the author, Biomet-SRM provided 100% biological efficacy against *Biomphalaria glabrata* in periodic 30-day bioassay tests over a 14-month test interval. Also, tests against *Lebistes reticulatus*, the guppy, at 1 mg/l showed an 80% or better survival over 30-day bioassay periods. Biomet-SRM shows a number of advantages besides that attributed to its long-term efficacy: it is equally effective against snail eggs, cost is dramatically less (based upon direct comparison of Biomet-SRM with Bayluscide applications), and there is inherent control of toxicant in treated water, which minimizes human error and handling problems, especially the human toxicity hazard (which would be virtually eliminated).

One disadvantage in the use of TBTO lies in the impossibility of detecting small amounts in water by known techniques. Under ideal laboratory conditions colorimetric determination of tin by the dithiol method has been attempted by the author and a number of his colleagues.[306-309] Realistically, the methodology described will provide quantitative determination of tin in distilled water down to the 10-ppb level or so. Atomic absorption methods will provide quantitative data to perhaps 1 ppb. The problem, however, is *not* in determining tin content in water, but in keeping the tin in the test solution. The great hydrophobicity of TBTO, and presumably other organotins, leads to a rapid migration of tin from water solution into the walls of any vessel whether glass, metal, or plastic (including Teflon®). In recovery experiments at 1 ppm skilled analysts could only recover 20 to 60% of the tin seeded in high purity distilled water. Also, concentration of water through volatilization leads to loss of TBTO. In field waters where a tin content is available in fluctuating concentrations from natural sources, detection of tin from TBTO in the parts-per-trillion range is impossible at this point in time. Likewise, in a field test situation organotin diffusants are rapidly purged from the watercourse by proteinaceous materials, cellulosic substances, suspended soils, and the living biomass – especially algae.

Competing conventionally applied molluscicides such as Bayluscide, trifenmorph, and copper ion can be accurately detected and quantitatively measured in field waters without undue difficulties.[246,310] One might overcome the problem of TBTO detection by using Sn^{113}, a radioisotope, in preparing TBTO for inclusion in test compounds.[311] Controlled release Bayluscide and copper may, of course, present analytical difficulties in the field.

Three methods of evaluating organotin/elastomers in general and Biomet-SRM in particular are in use and are described below.

Laboratory Bioassay

Conventional laboratory methods prescribed over the years are inadequate for controlled release molluscicide evaluation.[190,194,211,250] The Creative Biology Laboratory (Barberton, Oh.) which developed Biomet-SRM for the M & T Chemical Company devised test procedures that would properly distinguish between the acute mollusc intoxication observed with conventional massive dosing and chronic intoxication arising from continuous snail exposure to an ultralow toxicant concentration. *Biomphalaria glabrata* cultured in the conventional manner was used as the test snail.[312] Rearing temperature is critical to growth reproduction and the hatch period.[313,314] At C.B.L. maximum results were achieved using a temperature of 78°F and a pH of 6.4 to 7.2. Leaf lettuce is fed ad libitum. The basic distinction involved in testing controlled release materials against conventional toxicants is that exposure times must be prolonged far beyond the usual 48- to 96-hr bioassay. At least 10, preferably 30, days are required for accurate evaluation of chronic intoxication.

In testing for ovacidal properties, 24-hr egg clutches were exposed to various organotin/elastomer concentrations and development monitored. MT-1 series materials destroyed the gastrula stage after a usually prolonged blastula period. If a viable egg survived beyond the gastrula, the trochophore was generally not affected and death occurred after a greatly prolonged veliger phase.

Santos and his colleagues screened several organotin controlled release materials in the laboratory against *Oncomelania quadrasi*. CBL-9B (floating TBTF formulation) provided an LD_{90} (95% confidence level) of 46 ppm in 48-hr exposures, while 443A (sinking TBTO formulation – B. F. Goodrich Nofoul) gave an LC_{90} (95% confidence level) of 100 mg/l for the same exposure level.[315] Stage-size array susceptibilities were determined in 48-hr exposures; the LC_{50} for 443A was 6.4 mg/l for young snails, 10 mg/l for juveniles, and 22 mgm/l for adults. The LT_{50} for CBL-9B (48-hr exposure) was 154 days.

Mansoori at the Dezful Bilharziasis Research Station (Iran) reported that MT-1E at 1 ppm

TABLE 3–16[317]

Evaluation of Molluscicidal Agents Against *Lebistus reticulatus* and *Biomphalaria glabrata* in Microecological Systems

		Mortality (exposure time, %)							
		30-day		60-day		90-day		120-day	
Toxicant	Dosage (ppm/day)	LR[a]	BG[b]	LR	BG	LR	BG	LR	BG
TBTO®	0.035	98	58	100	100	–	–	–	–
	0.007	2	80	2	96	2	98	2	100
	0.0014	0	4	0	14	2	22	4	36
	0.0007	0	6	2	20	2	24	2	26
	0.00	0	4	0	4	2	4	6	4
TBTF	0.014	96	100	100	–	–	–	–	–
	0.007	22	20	100	100	–	–	–	–
	0.0035	8	6	28	96	52	100	–	–

[a]LR: *L. reticulatus.*
[b]BG: *B. glabrata.*

(pellet, 0.06 ppm TBTO) destroyed *Bulinus truncatus* after 43 days' laboratory exposure.[316]

Microenvironmental Bioassay

A useful intermediate step between laboratory bioassay and field evaluation is a microenvironmental examination of the candidate controlled release molluscicide. For instance, this may consist of small pool tests or the creation of a miniature laboratory microenvironment. The latter is of considerable value in that it permits control of variables such as pH, mineral content, and biomass.

In test microenvironments created by the author a standard soil mixture was used (organic top soil, peat moss, and sand), limestone chips to provide calcium, growing aquatic plants, and *Lebistus reticulatus* as a representative nontarget fish, along with *Biomphalaria glabrata* as the target gastropod.[244,317] The objective was to confirm (or deny) the hypothesis that snail hosts of schistosome trematodes could be destroyed through chronic intoxication while minimizing fish mortality.

As an initial step the LD_5 accumulative dosages for *L. reticulatus* were first determined by microenvironmental bioassay. Values found were

LD_5 (30 days – TBTO) = 0.00205 ppm/day
LD_5 (60 days – TBTO) = 0.00145 ppm/day
LD_5 (30 days – TBTF) = 0.0080 ppm/day.

A similar study was performed combining *L. reticulatus* and *B. glabrata* in one environment. The organotin-induced mortality at various accumulative dosages is shown in Table 3–16. Results indicated that by proper dosage control it should be possible to destroy the target snail while minimizing fish loss.

Evaluation of various experimental controlled release organotin molluscicides were performed at 30-day intervals in microenvironments (Table 3–17). A 10% daily water replacement rate was used to simulate essentially static conditions with a very minor water turnover rate.

Although microenvironmental evaluations indicate that controlled release organotin molluscicides can be used to control snails without untoward piscicidal effects, it must be noted that the test environment changes day by day in chemical and biological character due to the presence and interaction of snails and fish, and the nontarget biota cannot escape the proximity of a toxic pellet. Both conditions would likely be absent under field conditions. Thus, considering this factor and the lack of water movement it is believed that the microenvironmental test procedure is unduly harsh on fish, so results tend to be conservative.

Field Evaluation

The only real test of merit lies in the application of controlled release molluscicides to field

TABLE 3–17

Environmental Bioassays

Code	Toxicant	% toxicant	Pellet wt/l (grams)	Mortality (in 30-day exposure period, %)					
				Month I		Month II		Month III	
				LR[a]	BG[b]	LR	BG	LR	BG
XM-2	TBTO®	14.5	0.1	100	100	100	100	100	100
			0.01	100	100	5	100	0	100
XM-13	TBTF	7.8	0.1	100	100	100	100	100	100
			0.01	60	100	7	83	0	100
XM-123	TBTO	6.0	0.03	100	100	–	–	–	–
			0.003	0	80	–	–	–	–

[a]LR: *Lebistus reticulatus.*
[b]BG: *Biomphalaria glabrata.*

From Cardarelli, N. F., in *Molluscicides in Schistosomiasis Control,* Cheng, T. C., Ed., Academic Press, New York, 1974, 177. With permission.

habitats and monitoring of results. In the typical field test, evaluation is based upon snail mortality as determined by periodic survey of the treated habitat. A control program, in contrast, would be keyed to the decrease in incidence and schistosome egg output among the human populace protected.

MT-1E (Biomet-SRM) has undergone field tests; additional tests are in progress, and at least one area control program is being established at the time of this writing.

The initial small field evaluations were conducted in three pond sites, two drainage canals, one well, one pit, and one tank in Brazil. The largest water volume used was a 160-m^3 pond. Sites showing 100% mortality were periodically recharged with known numbers of viable snails. Results indicated that Biomet-SRM at practical dosage levels of 50 (pellet: 3 ppm TBTO) to 200 ppm (pellet: 12 ppm TBTO) gave control for at least a year in *still* water.[318,319] Mortality in flowing water systems was 100% for several months but partial repopulation eventually was noted.

Pond tests in Iran provided an LT_{75} of 37 days at 50-ppm MT-1E pellet dosage and an LT_{100} of 37 days at 90-ppm pellet dosage.[316]

Upatham and Christie report that Biomet-SRM pellets at 6.25 ppm dosage destroy *B. glabrata* and *Lymnaea cubensis,* the intermediate host of *Fasciola hepatica.*[320]

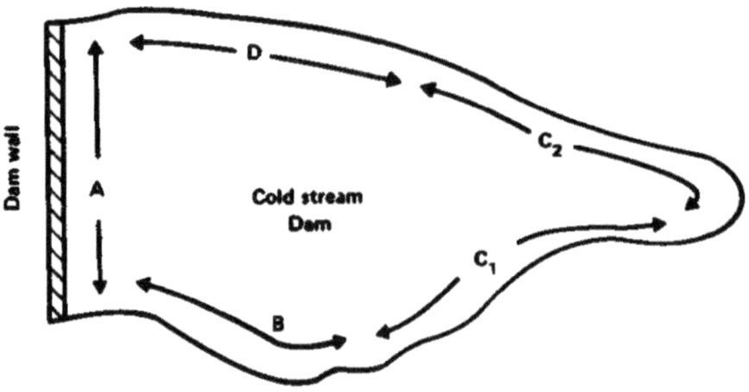

FIGURE 3–13. Schematic of Coldstream Dam (Rhodesia) showing five sectors ("dam" in Rhodesian is best translated as "pond" or "lake"). C_1 : treated area. Not drawn to scale.

The most extensive field tests to date have been performed by Shiff and his colleagues in Rhodesia. Since a number of test sites are in progress each can be conveniently discussed separately.

Middle Sabi irrigation scheme[321] – Some irrigation systems restrict the hazard of schistosomiasis transmission to particular structures. In the Middle Sabi scheme snail breeding is generally found in concrete offtake sumps along the main canal. By painting such sumps with TBTO-loaded antifouling paint, snail control was achieved for 12 months.

Coldstream site[321] – Ninety meters of a 950-m farm reservoir shoreline was treated with Nofoul (compound 443A–5.8% TBTO) (Figure 3–13). Strips 10 cm long were laid parallel through the vegetation near the shoreline. Snail populations fell rapidly in the treated area and slowly in contiguous nontreated areas. No fish or frog mortality was noted.

At the time of treatment (May 1, 1972), sector C_1 had 260 snails per sampling, on June 15, 250, and then only 4 snails per sampling on July 6 and 3 living snails per sampling on July 20. Contiguous areas B and C_2 showed substantial decreases in snail populations.

Belgownie Dams – Several small lakes (dams) were treated with Biomet-SRM pellets. Examination of the application sites after 7 months showed the following:[322] No organotin was detected in water taken from the surface or from the bottom of the treatment site. Bottom water samples contaminated with mud showed 0.18 ppm organotin, while mud scraped off submerged pellets and analyzed for organotin showed 0.09- to 5.07-ppm concentrations. Pellets removed after 7 months' immersion contained 1.28% tin compared to 2.15% tin in unimmersed controls. Thus, since snail control occurred in the treated areas, it is implied that the TBTO released from the pellet remains in the bottom mud, suggesting that the snails succumb through browsing on the muddy bottom.

Rothbury – Rothbury Dam (480 m) was treated with TBTF-containing pellets, and a rapid decline in snail population was noted.[321] To assess the effort on the biotope, samples of bottom mud were taken from treated and untreated sections of Rothbury Dam. No detectable differences in acarina, ostracoda, chironomidae, other insecta, and oligochaeta were noted.

Gilbert and co-workers formulated several TBTO/asphalt and TBTO/SBR rubber compounds containing 17 to 20% TBTO.[323] Molluscicidal activity persisted for over 18 months in several of eight field tests conducted. Application was made to pits, drainage ditches, a concrete tank, an irrigation ditch, and various marsh sites in various areas of Brazil. In general, good snail control was achieved where the water bodies were essentially static, but insufficient control where appreciable flow occurred. It was also reported that immersion in mud and exposure to sunlight did not affect the biological activity of the pelletized formulations, and that TBTO apparently acts accumulatively in snails.

Controlled Release Copper

Inorganic salts of copper are insoluble in known elastomeric matrices, so release through a diffusion-dissolution system is not possible. However, work with baits demonstrated that leach-type release mechanisms are possible. Copper sulfate can be incorporated by mill mixing into chloroprene (neoprene WRT and neoprene AD-10[a]), natural rubber, styrene-butadiene polymers, synthetic natural rubber, cis-polybutadiene, and at least several ethylene-propylene polymers (Nordel® 1070,[a] Epcar® 5465,[b] etc.) In the early formulation work it was ascertained that $CuSO_4 \cdot 5H_2O$ loadings of up to 86% could be incorporated in Epcar 5465 and somewhat over 80% loadings in natural rubber and styrene butadiene materials. Highly loaded materials showed rather rapid leaching and early depletion so that formulations containing 40 to 60% copper sulfate provided a superior biological life. $CuSO_4$ (anhydrous) and $CuSO_4 \cdot H_2O$ were later evaluated as replacements for the pentahydrate in order to increase the leachable copper ion content without increasing the loading.[244]

Vulcanization, carbon black content, or carbon black structure play little, if any, role in loss rate. Loss appears to be a function of loading, base polymer, surface to volume ratio of the granule or pellet, and additives that materially affect the rubber/water interfacial pH.[244,248] Copper ion loss from immersed pellets was spectrophotometrically measured (457 nm) by the cuprethol technique.[324]

Bioassay was performed in the conventional manner, except that exposure time was extended in order to accommodate the chronic intoxication syndrome. Of the nearly 200 experimental formulations prepared, only E-51 provided 100% mortality after 5 months' immersion, and, depending on dosage (0.1 g/l vs. 0.01 g/l) and surface area to volume ratio of the test granule, 95 to 100% mortality after 6 months.[248] Thus, E-51 was commercialized as Incracide E-51®.[c] The recipe is as follows:[248]

[a]Manufactured by E. I. duPont de Nemours and Company, Dover, Del.
[b]Manufactured by B. F. Goodrich Chemical Company, Cleveland, Oh.
[c]International Copper Research Association, New York.

TABLE 3–18[248]

Copper Ion Loss From E-51

Immersion time (weeks)	Average copper ion loss (mg/cm^2-day)	Accumulative copper ion loss, % (0.1-g pellet)
1	0.08	4.0
3	0.06	6.1
5	0.05	7.1
6	0.05	12.2
8	0.05	22.4
10	0.05	26.3
11	0.04	28.3
15	0.05	35.3
19	0.05	47.8
21	0.05	52.3

TABLE 3–19

Lethal Time Values for E-51

Pellet		LT_{99} (first production lot)	
Weight (grams)	Application rate (ppm)	Month 1 (days)	Month 2 (days)
0.1	100	13	16
0.01	10	17	24

Ingredient	Formulation (parts)
Epcar 5465	100
Sulfur	1.0
Mercaptobenzothiazole	0.2
Zinc oxide	3.0
HAF black	50
TMTDS	1.0
Ammonium sulfate	2
Copper sulfate monohydrate	157.2
$CuSO_4 \cdot H_2O$ loading	50%
Copper content (by weight)	17.8%
Hardness (Shore "A")	78
Specific gravity	1.66
Cure conditions (hot air oven)	290°F, 30 min

Copper ion loss from E-51 is shown in Table 3–18, while lethal time (LT) values are in Table 3–19.

Microenvironmental studies have been conducted in accordance with the procedure recently developed.[244,279] Long-term exposure to copper ion and TBTO indicated that proper dosage would permit good snail control with minimal damage to piscine life. In contrast, the ethanolamine salt of niclosamide is piscicidal at all dosage levels, snails being more resistant than the test fish. Copper ion was far less toxic towards *Lebistus reticulatus* than TBTO. Trifenmorph studies showed that this material at a dosage of 0.0098 ppm/day (30 day) was an LD_{99} for *Biomphalaria glabrata* and an LD_5, or less, for *L. reticulatus.*[325]

Table 3–20 presents a compendium of microenvironmental results comparing slow release

TABLE 3–20[247]

Microenvironmental Bioassay (30-day Runs)

					Mortality (%)													
					Month 1		Month 2		Month 3		Month 4		Month 5		Month 6		Month 7	
Code	Toxicant	% toxicant	Base[a]	Pellet dosage (ppm)	LR[b]	BG[c]	LR	BG	LR	BG	LR	BG	LR	BG	LR	BG	LR	BG
XM-2	TBTO®	14.5	SNR	100	100	100	100	100	100	100	100	100	100	100	100	100	100	100
XM-2	TBTO	14.5	SNR	10	100	100	5	100	0	100	0	90	0	100	10	95	3	100
XM-13	TBTF	7.8	CB	100	100	100	100	100	100	100	100	100	100	100	100	100	100	100
XM-13	TBTF	7.8	CB	10	60	100	7	83	0	100	0	77	14	73	10	7	–	–
XM-24	Nicl.[d]	26.3	CB	100	0	7	–	–	–	–	–	–	–	–	–	–	–	–
XM-24	Nicl.	26.3	CB	10	0	10	–	–	–	–	–	–	–	–	–	–	–	–
XM-50	DS-649[e]	29.8	CB	100	100	100	10	7	–	–	–	–	–	–	–	–	–	–
XM-50	DS-649	29.8	CB	10	100	100	3	17	–	–	–	–	–	–	–	–	–	–
XM-111	CS5	78.0	EP	100	100	100	0	10	–	–	–	–	–	–	–	–	–	–
XM-111	CS5	78.0	EP	10	0	97	0	13	–	–	–	–	–	–	–	–	–	–
Biomet®-SRM	TBTO	6.0	NRX	33	100	100	14	100	7	100	0	100	0	100	0	100	3	100
Biomet-SRM	TBTO	6.0	NRX	3	0	80	0	100	0	100	0	97	0	94	3	94	3	97
XM-35	CS5	50.5	SBR	100	94	100	7	67	10	24								
XM-35	CS5	50.5	SBR	10	14	100	0	24	–	–								
XM-38	CS5	63.0	SBR	100	100	100	3	73	10	84	3	87	14	90	3	100	0	94
XM-38	CS5	63.0	SBR	10	7	100	0	17	–	–								
XM-69	CS5	50.0	EP	100	24	100	0	43	–	–								
XM-69	CS5	50.0	EP	10	0	100	14	33	–	–								
XM-114	CS5	80	NRX	100	100	100	3	100	–	–								
XM-114	CS5	80	NRX	10	0	100	0	100	–	–								
XM-18	Nicl.	26.3	SNR	10	21	84	24	84	7	24								
XM-44	DS-649	29.8	SNR	10	100	100	100	100	3	33								
XM-28	Nicl.	41.0	SNR	100	100	100	100	100	57	100	53	100	0	73	–	–	–	–
XM-74	CS5	62.5	EP	100	100	100	3	97	14	97	0	80	0	87	0	97	0	90
XM-74	CS5	62.5	EP	10	3	94	0	73	10	60								

[a]SNR: synthetic natural rubber; CB: cis-polybutadiene; EP: Epcar® 5465; NRX: natural rubber; SBR: styrene-butadiene rubber.

[b]LR: *Lebistus reticulatus.*

[c]BG: *Biomphalaria glabrata.*

[d]Nicl.: niclosamide.

[e]DS-649: 3,3,4,4-tetrachlorotetrahydrothiophene-1,1-dioxide.

TABLE 3–21[326]

Controlled Release Molluscicide Activity Through an Intervening Soil Barrier

Formulation	Toxicant	Toxicant concentration (%)	Base	Month 1 mortality (%)		Month 2 mortality (%)	
				Fish	Snail	Fish	Snail
Control	–	–	–	5.7	13.8	3.8	13.8
Biomet®-SRM	TBTO®	6	NRX	13.0	55.0	11.0	51.0
E-54	CS1[a]	50	EP	10.0	57.0	6.0	53.0
XM-74	CS5[b]	63	EP	10.0	48.0	11.0	56.0
XM-167	Niclosamide	46	CR[c]	52.0	57.0	39.0	51.0

[a]CS1: copper sulfate monohydrate.
[b]CS5: copper sulfate pentahydrate.
[c]CR: chloroprene rubber (neoprene WRT).

copper to other molluscicides. Water changeover is 10%/day with a complete change at the end of each 30-day exposure period.

Microenvironmental studies suggest that it is possible to destroy *B. glabrata,* a host of *Schistosoma mansoni,* without unacceptable fish loss. Release rates vary with the different copper-releasing materials, but generally lie between 0.03 to 0.08 ppm/day with terminal intoxication resulting within a 30-day period. Incracide E-51 pellets release at about 0.03 ppm/day (copper ion) and destroy within 21 days; i.e., 0.63 total ppm is lethal when induced gradually over a continuous exposure period, compared to the need for 40 ppm or more in an actual environmental situation.

The environmental fate of a controlled release pellet is a point of speculation. One important question lies in the determination of biological efficacy if the pellet is buried. Shiff and Yiannakis have indicated that TBTO acts through mud in that snails apparently pick up a lethal dose by browsing on bottom soil.[322] This subject has been examined in a microenvironmental study of controlled release pellet activity through mud. In a series of experiments pellets were buried at a 2-in. depth in the bottom soil of microenvironmental test aquaria. As usual, plants, fish, and snails were introduced; aeration was continuous, temperature a constant 75°F, and a 10%/day water changeover rate maintained. Burial test results, over two consecutive 30-day test periods, are presented in Table 3–21. Ten replicates of 10 fish and 10 snails each were used for each bioassay and 21 control aquaria, each with 10 fish and 10 snails. Biological activity through mud is obvious, and niclosamide activity *on* mud has likewise been observed.[327]

Field trials with copper-ion-releasing materials have just been initiated and little data are available. Shiff and Yiannakis report that E-51 releases copper ion at a rate of 0.3% of the available copper per day, which would predict activity for nearly a year.[322] Pelligrino of the Institute of Biological Sciences (Federal University of Minas Gerais, Belo Horizonte, Brazil) reported in personal correspondence with the author that the granulated E-51 at 1-ppm dosages destroyed snails in both running and still water.

Amin evaluated E-51 in small field trials and noted that *Biomphalaria pfeifferi, Bulinus truncatus,* and *Lymnea* were destroyed.[328] However, he further observed that if the pellet were placed on the bottom of the watercourse only bottom-dwelling snails were affected, and vice versa when placed near the water surface.

Conclusions

Several controlled release molluscicide materials have demonstrated biological efficacy under field test conditions. Both Biomet-SRM and Incracide E-51 are commercially available at competitive prices. The initial sales of Biomet-SRM have occurred and a control program in Rhodesia using this material has begun. In the light of laboratory and field tests to date, both formulations are safer to handle and have much less environmental impact than is seen with copper sulfate applied conventionally, Bayluscide, and the pentachlorophenates. Since snail destruction is achieved through chronic intoxication, only ultralow toxicant concentrations are necessary. Conservative calculation would indicate that approximately one tenth the amount of copper ion is used if the controlled release method is employed. Since TBTO is not utilized as a conventional agent, its ecological advantage cannot be as easily computed. However, compared to Bayluscide, one can compute that the quantity of TBTO necessary is 1/8 to 1/30 that conventionally used.

As a rule, the economic advantage comes from extension of the between-treatment interval, rather than from savings in chemicals. Labor costs, vehicle maintenance, etc. usually account for 95 to 98% of the overall economic burden. Normally, treatments are made or should be made at 3-to-6-month intervals. Slow release copper should provide a 6-month or better release, while the slow release organotins give 2 years or better.

Another factor that cannot be easily assessed is the end result of the use of conventional and controlled release materials. The objective is to interrupt the transmission cycle. Continuous low-level intoxication of the watercourse means that repopulation through newly arrived snails or returnees who escaped intoxication by fleeing from the treated water cannot statistically occur. In a conventional treatment with Bayluscide, NaPCP, etc. intoxication lasts for a few days to perhaps 2 weeks. At the end of this period repopulation commences, and the schistosomiasis transmission cycle is reestablished within 3 months. Biomet-SRM should provide an approximate 27-month transmission interruption, and Incracide E-51 a 10-month interruption per treatment – provided that the pellets, granules, cords, sheets, or whatever forms used remain where placed. Exact results of siltover, current, and external interference with the treatment site cannot be predicted.

Advanced Concepts

Extremely long-term controlled release methods are under investigation. Controlled release elastomers can be produced in any size or shape normal to the rubber manufacturing industry. Concrete irrigation canals, weirs, sumps, and so on can be lined with sheet material that will release the incorporated agent for several years, and through *refillable* systems this can be extended to 20 or more years. Longevity can be extended by the use of a three-phase system, as previously discussed.[75] Likewise, a controlled release membrane laminated to a highly loaded reservoir layer will permit a significant extension of life.[69]

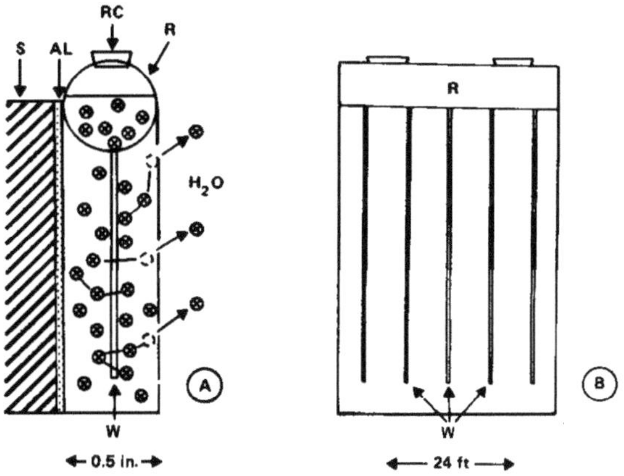

FIGURE 3–14. Increasing molluscicidal longevity by incorporation of a refillable liquid reservoir. A: side view; B: front view. AL: adhesive layer; R: reservoir; RC: refilling cap; S: substrate; W: wick; ⊛: toxicant molecule. (From Cardarelli, N. F., in *Molluscicides in Schistosomiasis Control*, Cheng, T. C., Ed., Academic Press, New York, 1974, 177. With permission.)

Probably the longest biological life can be achieved with the refillable reservoir system depicted in Figure 3–14. This technique is probably limited to the use of liquids such as TBTO, or solids in a transporting liquid solution.[244]

In this system a hollow rubber tube is molded onto a toxic sheet stock containing a wicking fabric; both cotton and rayon work nicely with TBTO. Liquid in the reservoir is dispersed slowly throughout the sheet by passing through the wicks. TBTO movement *upward* in rayon is several inches per day and *downward* in an experimental system appeared to have been about 8 in. per day. Thus, as the sheet material is slowly depleted it is continuously refilled from the incorporated reservoir. The reservoir can be refilled once every 3 to 5 years or so. A lifetime of 20 or more years can be envisioned, provided that the toxicant and the rubber sheet does not undergo chemical or radiation degradation of a nature that detoxifies or substantially alters diffusion-dissolution.

CONTROLLED RELEASE SCHISTOSOMACIDES

At this time the preferred approach to the interruption of the transmission cycle lies in the destruction of the snail host by the use of chemical agents. However, the cycle can be broken

TABLE 3–22

Methods of Intervention into the Schistosome Cycle

Schistosome form	Method of intervention: Conceptual	Practical
Ova (water)	Ovacides	Sanitation, health education
Miracidia (water)	Miracidiacides	Sanitation, environmental control
Sporocyst (snail)	Sporocystacides	None
Snail	Molluscicides	Molluscicides
Cercaria (water)	Cercariacides	Sanitation, water supply engineering
Cercaria penetration	Prophylactic lotions	None in use

at other points. Table 3–22 illustrates the various stages of the cycle and the practical or conceptual method of attack.

Normally, chemical attack is directed toward a target snail(s), and the techniques used depend upon the nature of the infected habitat. With conventional treatment the watercourse remains toxic for a relatively short time – usually several days, perhaps 2 weeks at best under static water conditions. Since all molluscicides undergo rapid detoxification in natural waters, persistence is low and the offending snail must be destroyed rapidly; thus, dosage level must be high enough to induce an acute intoxication. Destruction of the larval forms or eggs of the trematode is of no significance if (1) intoxication lasts but a few days and (2) the host snail is not destroyed. Of necessity, conventional chemical control is aimed at the snail. In contrast, slow release technology allows the long-term, continuous maintenance of watercourse toxicity; thus, it may be conceivable to eliminate the schistosome without destruction of the snail.

Undoubtedly, cercaria and miracidia control will require the use of chemical agents. Purified water supplies and sanitary disposal of wastes is not economically feasible in most of the endemic areas now. Untold billions of dollars would be required in engineering and construction as well as manifest changes in social custom and a herculean educational effort. As in most areas, biological control is touted as the "solution," ignoring the details of implementation. Some work has been done, and introduction of certain fish as cercaria predators may be useful in some areas.[329]

Several investigators have examined the possible use of controlled release materials as schistosomacides. In 1966 at the University of Akron, it was observed that cercaria were quite sensitive to very small amounts of organotin. A little later the use of controlled release materials as cercariacides was proposed.[70,330]

Evaluation of Molluscicides as Cercariacides

A number of experimental and conventional molluscicides have been evaluated against cercariae, with results shown in Tables 3–23 and 3–24.

TBTO is a potent schistosomacide. In studies performed by the author the following generalized picture emerges when *Schistosoma mansoni*-infected snails are exposed to 0.006 ppm/day TBTO.[244]

Exposure day	Effect
2	Number of cercariae released reduced by 90% or more.
4	50%+ snail ova mortality; serious interference with sporocyst development.
6	Cercariae succumb in minutes; number of cercariae released per shed reduced by 95%; 90% ova mortality; 50% mortality among young snails; oviposition frequency per 100 snails reduced.
8	No release of cercariae. Miracidia no longer infectious (if schistosome ova hatch at all).
10	10% adult snail mortality; 100% snail egg mortality; egg laying 5% of normal; 90% juvenile snail mortality.
15	99% juvenile snail mortality; 70% adult snail mortality; no egg laying.
18	99% adult snail mortality

TABLE 3–23[331]

Evaluation of Potential Cercariacides

Agent	Concentration (ppm)	Number of cercariae tested	100% mortality time (LT_{100})
Niclosamide	0.8	108	50–528 sec
Niclosamide	0.4	99	1449–1482 sec
Niclosamide	0.2	140	1800+ sec[a]
Niclosamide	0.1	47	1800+ sec[a]
NaPCP	0.2	45	1800+ sec[a]
TBTCl[b]	0.25	27	1022 sec
DS 649	0.9	17	120 sec
DS 649	0.1	70	1800+ sec[c]
Control	0	60	28 hr[d]

[a]No mortality at 30 min.
[b]Tributyltin chloride.
[c]At the end of 30 min 73% mortality was noted. All swimming motion had ceased by 13 min past exposure.
[d]No mortality within 8 hr past exposure.

TABLE 3–24[244]

Comparative Evaluation of Conventional Molluscicides and Organotins as Cercariacides (Laboratory Exposures)

Property	Niclosamide	Trifenmorph	Copper sulfate pentahydrate	TBTO	TBTF
LC_{90} (1 hr) for snail	3–8 ppm	0.5–4 ppm	0.1–0.3 ppm	0.02–0.08 ppm	0.02–0.05 ppm
LC_{90} (1 hr) for snail ova	2–4 ppm	240+ ppm	50–100 ppm	0.01–? ppm	0.01–? ppm
LC_{90} (1 hr) for cercariae	0.3 ppm	20 ppm	1 ppm	0.001 ppm	0.001 ppm
LC_{50} for fish	0.05–0.3 ppm	2–6 ppm	8–12 ppm	0.03–1 ppm	0.04–? ppm
LD_{50} for rat (oral)	5,000 mg/kg	1,400 mg/kg	5,000 mg/kg	172 mg/kg	200 mg/kg
Effective field dosage (snails)	1–20 ppm	0.05–10 ppm	40–120 ppm	<3 ppm	<2 ppm
Phytotoxicity	Slight	None	Yes	Slight	Slight

From Cardarelli, N. F., in *Molluscicides in Schistosomiasis Control*, Cheng, T. C., Ed., Academic Press, New York, 1974, 177. With permission.

Cercariacide criteria established in 1971 required that 99% of the cercaria population succumb to 0.1 ppm of the agent within 30 min.[332] Criteria were arbitrary, based upon quick kill to prevent cercariae contact with a human, and 0.1 ppm being the highest practical dosage that could be maintained by the use of controlled release materials.

Cercariae mortality from organotin exposure (released from TBTO-containing elastomers) occurs rapidly at levels from 0.001 to 0.1 ppm.[70] However, mortality may not be a useful criterion. Ritchie et al. exposed cercariae to ultralow TBTO concentrations, with results shown in Table 3–25.[333,334]

Suppression of infectivity, which occurs as cercariae lose motility, appears to be the appropriate criterion of effectiveness, rather than mortality.

TABLE 3–25[333]

Immobilization and Suppression of Infectivity of *Schistosoma mansoni* Cercariae by TBTO®

		Motility[a]		Infectivity suppression (percent)	
TBTO concentration	Exposure time (minutes)	Trial I	Trial II	Trial I	Trail II
10^{-7}	5	NP	NP	100	100
10^{-8}	5	NP	P	100	72
10^{-9}	30	P	P	54	6
10^{-10}	30	P	P	40	0
0	30	P	P	0	0

[a]Motility grading: NP = incapable of movement; P = capable of movement.

From Ritchie, L. S., Lopez, V. A., and Cora, J. M., in *Molluscicides in Schistosomiasis Control,* Cheng, T. C., Ed., Academic Press, New York, 1974, 77. With permission.

TABLE 3–26[331]

Cercariae Exposure to Molluscicidal Water

			LT_{90} (hours)		
Code	Toxicant	Dilution factor (percent)	Snail	Cercariae	LT_{100} (hours) – cercariae
443A	6% TBTO®	90	122	1.0	1.3
CBL-9B	20% TBTF	90	60	0.08	0.09
N73	35% Niclosamide	90	600+	0.60	0.73

Evaluation of Slow Release Molluscicides as Cercariacides

In a series of tests, water, in which biocidal rubber was preimmersed, was added at various proportions to containers holding freshly shed cercariae. Next, 0.1 g of the candidate slow release cercariacide was immersed for 390 hr in 3750 cm^3 of distilled water. Organotin concentrations are not directly determinable at ultralow concentrations, so the dosage is unknown. However, Table 3–26 does demonstrate that the slow release materials are lethal to cercariae. Snail LT values are added for comparison. Table 3–27 depicts the results of adding cercariacidal pellets directly to the cercariae-laden water.

Since mortality or lethal time is an inadequate measure of intervention, infectivity of *Schistosoma mansoni* to mice in an environmental situation was examined. Mice tails were suspended in water (3700 cm^3) containing infected shedding snails (cercariae count approximately 3/cm^3) and a CBL-9B (23.7% TBTF) pellet introduced several hours earlier. After a 2- or 3-hr mouse exposure, the animals were recaged and sacrificed 52 days later. Worm burden was noted by perfusion and examination of the liver. Of the 35 mice subjected to the test, only one showed a single worm.[335] Control mice exposed for 2 hr showed elapsed time worm burden/mouse averages of 0 to 2 hr, 7.0 schistosomes; 2 to 4 hr, 12.7; 4 to 7 hr, 15.6; 7 to 9 hr, 3.4; 9 to 11 hr, 2.7; 11 to 14 hr, 4.0; and 14 to 16 hr, 0.0.

A more elaborate screening evaluation was performed next using four controlled release cercariacides. Three groups of albino Swiss mice were exposed to waters laden with *S. mansoni* cercariae. Exposures were performed in large containers filled to a 1-in. depth with warm water.

TABLE 3–27[335]

Direct Addition of Cercariacidal Pellets to Cercariae-laden Water

Code	Toxicant	Toxicant concentration (percent)	Water concentration (ppm)	Pellet dose (ppm)	Cercaria mortality (seconds) LT_{50}	Cercaria mortality (seconds) LT_{100}
443A	TBTO®	6	0.06	100	–	300
443A	TBTO	6	0.05	100	36	540
443A	TBTO	6	0.01	100	–	600
443A	TBTO	6	0.006	100	–	3,600
443A	TBTO	6	0.005	100	72	3,600+
443A	TBTO	6	0.001	100	–	5,400
CBL-9B	TBTF	20	0.01	100	42	390
633A	Niclosamide	6	1.0	100	–	300
633A	Niclosamide	6	0.1	100	–	300
633A	Niclosamide	6	0.001	100	–	3,600
N73	Niclosamide	35	0.1	100	48	2,580
N73	Niclosamide	35	0.01	100	240	2,640
N649	DS649	13	Unknown	100	600	5,400

Nonanesthetized mice were placed in the water without constraint for "whole body" exposure. Maintenance of room and water temperature at 78°F, and having the mice recaged and kept at 78°F for 24 hr prevented presacrifice losses due to exposure. Three heavily infected *Biomphalaria glabrata* were removed from dark isolation and added to each of the five aquaria. The scheme used was as follows:

Group I: Snails introduced at Time 0, mice introduced at Hour 2, cercariacide introduced at Hour 2, mice removed at Hour 3.

Group II: Snails introduced at Time 0, mice introduced at Hour 3.5, cercariacide introduced at Hour 2, mice removed at Hour 4.5.

Group III: Snails introduced at Time 0, mice introduced at Hour 5, cercariacide introduced at Hour 2, mice removed at Hour 6.5.

Pellets used were of 2 cm^2 total surface area, except for CBL-9B, where a 0.5-cm^2 surface area chip was used. Mice were sacrificed at 49 days postexposure and worm burden determined. Table 3–28 depicts the observed results. The loss of seven mice in Group III during the CBL-3 test is presumed due to the ingestion of Bayluscide, which results in partial paralysis and disorientation in mice, thus leading to drowning. Pertinent cercariacide recipes are presented in Table 3–29.

CBL-9B was examined as a candidate cercariacide by Santos, who observed that a pellet dosage of 1 mg/cm^3 reduced cercariae mobility within 5 min and provided a 71-min LT_{50} and 160-min LT_{100}.[315]

Field evaluation has yet to be performed. Cercariae can be recovered in the field with 70 to 80% efficiency by the modified Rowan technique,[336] but various difficulties abound. This probably superior technique has a practical limit of 0.1 cercaria/l.[337] Distinguishing between live and dead cercariae and recognizing human schistosome cercaria as opposed to other larval schistosomes constitute further difficulties. Measurement in turbid water is also quite difficult.

Commentary

The experimental use of controlled release molluscicides as cercariacides has been based upon convenience. A number of cercariacidal agents such as the active and inactive terpenes are known but have not been incorporated in controlled release formulations.[338] Although the physiology and morphology of cercariae has been studied and considerable information made available,[339] to date there are no known reports in the available literature regarding the mechanism of intoxication of cercariae arising from exposure to any toxicant.

Field evaluation is needed in order to adjudge the feasibility of use of present controlled release

TABLE 3-28[335]

Protection of Mice from Cercaria Penetration by the Use of Candidate Slow Release Cercariacides

	Formulation				
	Control	CBL-3	CBL-4	CBL-9B	CBL-49
Group I					
Number of mice exposed	6	6	6	6	6
Number of mice infected	5	0	3	1	0
Average worm burden	6.0	0	3.0	0.7	0
Presacrifice mortality	0	0	0	0	0
Group II					
Number of mice exposed	6	6	6	6	6
Number of mice infected	4	0	0	0	0
Average worm burden	3.5	0	0	0	0
Presacrifice mortality	0	0	0	1	0
Group III					
Number of mice exposed	10	10	10	10	10
Number of mice infected	6	0	0	0	0
Average worm burden	3.6	0	0	0	0
Presacrifice mortality	2	7	1	0	0

TABLE 3-29[335]

Slow Release Cercariacide Formulations

	Recipes (parts)			
Ingredient	CBL-3	CBL-4	CBL-9B	CBL-49
Neoprene WRT	100	100	–	100
Natural rubber	–	–	100	–
Zinc oxide	5	5	2	5
Magnesium oxide	4	4	–	4
PBNA[a]	1	2	1	1
HAF black	15	–	10	15
FEF black	–	15	–	–
MBT[b]	1	1	–	1
BTDS[c]	–	–	2	–
Sulfur	–	–	2.5	–
TBTO®	–	8	–	–
TBTF	–	–	30	–
Bayluscide® (70%)[d]	100	–	–	–
DS649	–	–	–	20
Cure	30 min × 300°F	30 min × 290°F	30 min × 290°F	30 min × 300°F
Agent concentration (percent)	43	5.8	20	13
Agent loss rate ($\mu g/cm^2$-day)	214	143	375	130

[a]Phenyl-β-naphthylamine (an antioxidant).
[b]Mercaptobenzothiazole.
[c]Benzothiazole disulfide.
[d]2-Aminoethanol salt of niclosamide, used as the 70% wettable powder.

materials as cercariacides. It will probably be necessary, if long-term prevalence, incidence, and endemicity studies involving humans are to be avoided, to seed field watercourses with cercariae or infected snails and note infectivity in sentinel animals when cercaricidal dosages used are sub-threshold to the host snail.

Sporocysticides

Exposure of infected snails to minute quantities of various organotins (0.001 ppm and less) suppresses shedding.[335] It is hypothesized that the sporocyst is affected, thus making attack of the sporocyst more convenient than attack of the larval schistosomes. To prevent human contact and infection, cercariacidal action must be rapid. In contrast, challenge to cercariae development within the snail can be prolonged over several weeks, thus perhaps requiring a far lower concentration.

Relatively little work has been done in this area. Upatham and Christie have reported that copper ion at 0.01 to 0.1 ppm and much lower concentrations of TBTO retard sporocyst development.[320]

Miracidiacides

Ritchie et al. have shown that miracidia exposed to 0.01 ppm TBTO for 40 min were incapable of progressive motion and no longer infective.[333] Upatham and Christie indicated that copper ion in the 0.01- to 0.1-ppm range will reduce miracidia infectivity.[320]

Miracidia may be the weakest link in the transmission cycle. Considerable information is available concerning the physiology, morphology, and parasitism of this larval form.[340] Temperature effects,[341] relationship between infection rate and miracidal load,[342] environmental effects on infectivity,[343] water depth effects,[344] range in standing and flowing water,[345] and other factors have been studied.

Since considerable work has been done with miracidia, laboratory techniques are readily available for testing the effects of candidate controlled release miracidiacides.[340,346] Field techniques are generally lacking, although some work has been done with caged sentinel snails.[347]

Another approach lies in the destruction of miracidia ova prior to hatching, which occurs within several hours to several days, depending on various factors.[348] This may provide a longer toxicant exposure period than that which would be normal to miracidia – and, of course, eggs in themselves are not infectious or mobile.

MOLLUSCICIDAL AND SCHISTOSOMACIDAL BAITS

An attractive alternative to the use of conventional or controlled release molluscicides lies in the potential utilization of target-specific biocides or "baits." A bait, in concept, consists of three essential elements: an attractant that is released into the snail habitat and induces movement towards, location of, and ingestion by the snail; a toxic additive, suitably masked by the attractant and present in sufficient quantities to induce lethality; and some sort of binding matrix that provides integrity to the pellet or granule used. The active agent must not be released into the treated area or no advantage accrues over the controlled release approach. Also, the toxicant must act through the route of entry taken, e.g., a stomach poison. Physiological response such as regurgitation must be absent or muted to the extent that a fatal dose has been absorbed prior to the onset of defensive mechanisms: the target must not respond to the agent until it is too late.

The use of baits is ecologically advantageous in several situations. First, it may not be possible to use either conventional or slow release technology. Destruction of the predatory sea snails that devour commercial shellfish such as *Thais haemastoma* would require intoxication of a marine environment, but this would be defeated by the near infinite dilution factor. Target snails may have their foci in potable water supplies, where use of a toxicant in the usual manner may constitute an unacceptable human hazard. The infested habitat may contain piscine life, other animals, or plants (rice, etc.) of economic value that would be destroyed or contaminated through conventional control methodology.

In work keyed to the control of *T. haemastoma*, the "oyster drill," it was found that baits containing niclosamide were ingestible by certain fresh water species that were not hosts for human schistosomes.[349] Unfortunately, most target snails would not accept niclosamide; although attracted to a given bait they failed to ingest it, the probable presence of a slight but sufficient degree of toxicant leakage into the interface having a repulsive effect.

A large number of candidate attractants were

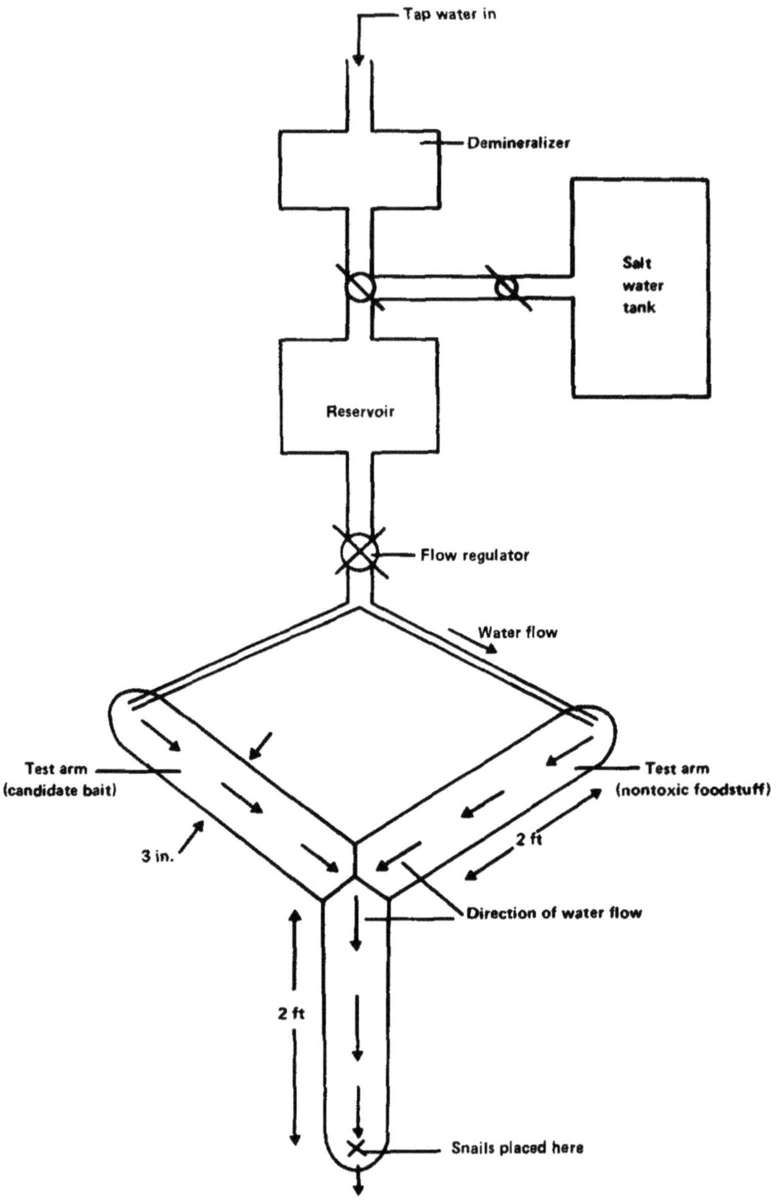

FIGURE 3–15. Y-tube test chamber. Flowing water system adjustable for salt and fresh water.

incorporated in agar and evaluated in a Y-tube apparatus (Figure 3–15) similar to that described by Michelson.[350] Attraction to a given material appeared to be fairly specific. For instance, dextrose is strongly attractive to *Helisoma planorbis* and moderately attractive to *Nassarius obsoletus; Biomphalaria glabrata, Lymnaea emarginata,* and *L. stagnalis* were unaffected by dextrose whereas *Phyra gyrina* and *Polygyra* and *Stagnicola* species were repulsed by it.[351] The variation in attraction from species to species encourages the concept of a bait specific to target snails and unattractive to nontarget molluscs and other members of the biota.

A critical factor in the development of a target-specific bait lays in the selection of a proper binding element. This material must be nonrepulsive to the target, ingestible, and stable under varied environmental conditions. The group of carboxylated acrylic polymers termed Carbosets®, manufactured by the B. F. Goodrich Chemical Company of Cleveland, meet these basic criteria.

TABLE 3–30[351,353]

Marine Snail Attractants

Species	Attractants
Nassarius obsoletus	Fish flour, sodium alginate, dextrose
Luniata heros	Fish flour
Urosalpinx cinerea	Beef heart agar
Ocenebra japonica	None found
O. erinacea	Beef heart agar
Thais haemastoma	Beef heart agar

Carbosets are ingestible by the species tested provided a proper selection is made, and in several instances plasticizers had to be added to soften the bait pellet for ingestion by several species.[351,352] Carboset materials allow the gradual release of attractants such as wheat paste, wheat germ, casein, malt, fish flour, beef heart infusion agar, sodium alginate, and other specific, strong attractants, while binding most conventional molluscicides.[353]

A number of molluscicides were added to the various Carboset/attractant formulations and evaluated against marine, fresh water, and amphibious snails of economic and public health importance. Trifenmorph is the preferred toxicant for fresh water snails and niclosamide for marine snails and amphibious molluscs of the genus *Oncomelania.* Organotins could not be appropriately masked, so their presence in an experimental bait negated ingestion. Copper sulfate baits are ingested at the 1% toxicant content level but with ambiguous results, many snails not succumbing; pentachlorophenates destroyed the attractiveness of a given formulation.[244]

Marine Snail Control

Various snail predators attacking oysters and other shellfish such as mussels and clams lead to sizable economic loss.[354] For instance, the 8000 acres of oyster beds in Alabama gulf waters have an estimated $500,000 annual loss to *Thais haemastoma.*[355]

Control of the oyster drill using toxic baits has been proposed but no method is presently in use.[117,355] Biological control using trematodes has also been suggested.[356] Baits, unless very specific, may also attract crustaceans such as shrimp. Oyster habitats are complex communities of many species and considerable study of the overall ecological effects of baits would likely be necessary prior to use.[357]

Although *T. haemastoma* infestations represent the paramount oyster predator in the U.S. Gulf Coastal region, *Urosalpinx cinerea* (American oyster drill) is the most dangerous and most widely spread predator. Suction dredging is used to remove this menace from oyster beds. Up to 40,000 drills per hour have been so taken from American oyster beds. *Ocenebra erinacea* (Murex) is a major oyster predator along the English coast, while *T. lamellosa* and *Tritonalia japonica* prey on Pacific oyster beds.

Starfish are second to drills in importance of destructiveness of commercial shellfish. Both oysters and clams are subject to attack, and at times this pest may exceed the drill-caused losses along the northeast U.S. coast. Starfish control at present is mechanical, laborious, and expensive. This may well be a candidate area for the development of a specific bait technique.

Attractants for several marine species are given in Table 3–30. Niclosamide appeared to be the superior toxicant of those evaluated.

A typical marine snail bait recipe follows:

MK 39
Carboset 514 – 90%
Beef heart infusion agar – 5%
Niclosamide (ethanolamine salt) – 4.85%
Zinc (curative) – 0.15%

Fresh-water Snails

Over 400 candidate attractants were examined in static and dynamic Y-tube tests against various fresh-water snails.[353,358] Table 3–31 illustrates superior attractants.

Unlike salt-water species, niclosamide was discovered to be a general repellent to fresh-water snails, possibly due to very low-rate leaching. Also, niclosamide apparently interfered with the cross-linking agent used to render the Carboset polyacrylics insoluble under alkaline water conditions.

In a study performed for the National Institutes

TABLE 3–31[353]

Fresh Water Snail Attractants

Species	Problem	Superior attractants
Biomphalaria glabrata	Schistosomiasis host	Wheat paste, wheat flour, wheat germ, casein
Helisoma planorbis	–	Beef gelatin, potato infusion, agar, gelatin
Lymnaea emarginata	Schistosome dermatitis host	Tryptose, celery salt, oregano, wintergreen
L. stagnalis	Schistosome dermatitis host	Glycerol, malt, Tetramin® fish food
Physa gyrina	Schistosome dermatitis host	Potato-dextrose agar, neopeptone, fish flour
Tarebia granifera	Trematode host	None found
Melanoides tubuculatus	Trematode host	None found
H. trivolvis	–	Malt, lettuce/sucrose, dried lettuce

of Health, over 300 experimental baits were prepared and evaluated under laboratory conditions.[358] The ethanolamine salt of niclosamide (0.1 to 10%) and trifenmorph (1 to 5%) were utilized as the bound active agent. Prepared snail food (20 g Tetramin® fish food, 20 g Biorell® fish food, 10 g wheat germ, and 4 g calcium carbonate), wheat paste, wheat germ, fish flour, and casein were evaluated as attractants. Four carboxylated polyacrylates, Carboset 514, Carboset 525, Carboset 526, and Carboset 531, were used as binders. Carbosets are insoluble in acidic water, but soluble from pH 7.1 into the alkaline range. Cross-linking ("cure") is essential or else baits rapidly disintegrate in alkaline water. At low cure levels baits are stable over a pH range of 4 to 8. Calcium ion has since proven superior to zinc as the cross-linking agent.[131,358]

Trifenmorph was found to be the superior toxicant in that its presence in baits is not detected by the snail prior to or during ingestion.

Baits were observed to rapidly deteriorate in air and water from microorganism attack, especially from fungi, thus making a fungicide necessary. Tetrachloroisophthalonitrile (TCPN), TBTO, and TBTA, used as additives at the 0.1% level, provided excellent protection without untoward snail reaction.[358]

Walker et al. evaluated several baits under natural conditions on the island of St. Lucia, with the formulations given in Table 3–32 providing superior field results.[359] Field tests were conducted in partitioned drainage ditches, and the natural biota included snails, algae, vascular plants, frogs, small fish, nematodes, and various insects. Each drain normally contained 6 to 12 in. of water, but up to 24 in. following rain. The pH value was generally alkaline. Prior to testing, indigenous snails were removed and known numbers of collected snails added. Bait pellets were added 4 or more hours after snail introduction. After a given exposure period all snails were removed and placed in recovery aquaria. Live ones were returned for additional exposure.

Snail mortality is noted in Table 3–33. Other than in the single test with Bait 3, no gross disturbance of the biota, other than snail deaths, was observed. Baits 1 and 2 were found to harden rapidly as the calcium ion present (10 to 80 ppm) lead to continuous cross-linking of these formulations. Therefore, polyethylene glycol or Carboset 515 was used to plasticize new formulations.

Baits were not particularly attractive to various piscine species evaluated. Starved fish will feed on baits with no mortality observed under laboratory conditions.

Tests were never completed due to onset of the rainy season and lack of funds. It is reasonable to assume on the basis of short-term field evidence plus laboratory evaluation that a 98% snail mortality can be achieved within 3 weeks in a drainage ditch habitat where competing foodstuffs are fairly scarce.

The use of plasticizing agents, low molecular weight Carboset 515 being preferred, is essential if the ingestibility of a given acrylic-based bait is to be preserved. Likewise, a fungicide such as TCPN

TABLE 3–32[359]

Bait Formulations

Ingredient	Compound recipe (percent dry weight) 1	2	3	5	7	8	10	11
Carboset® 525	69.8	–	–	64.9	–	–	65.0	–
Carboset 526	–	69.8	–	–	64.9	64.9	–	65.0
Carboset 514	–	–	69.8	–	–	–	–	–
Snail food[a]	24.9	24.9	24.9	–	–	–	–	–
Wheat paste	–	–	–	–	–	25.0	24.6	–
Wheat germ	–	–	–	25.0	25.0	–	–	24.6
Trifenmorph[b]	5.3	5.3	5.3	5.3	5.3	5.3	5.3	5.3
TCPN[c]	–	–	–	0.1	0.1	0.1	0.1	0.1
PEG[d]	–	–	–	–	–	–	5.0	5.0
Carboset 515	–	–	–	4.7	4.7	4.7	–	–

[a]Snail food recipe:
10 g Biorell® (fish flour, kelp, insect meal, cod liver oil, and wheat germ mix)
10 g Tetramin® (fish meal, brineshrimp, and plant substances mix)
5 g wheat germ
2 g calcium carbonate.
[b]Frescon® granules, 95% technical grade.
[c]Tetrachloroisophthalonitrile.
[d]Polyethylene glycol.

TABLE 3–33[359]

Snail Mortality: Bait Field Evaluations

Bait	Bait code dosage (grams)[a]	Percent accumulative snail mortality 24 hr	72 hr	120 hr	188 hr
1	22	16.0	21.0	–	24.4
1	100	45.0	–	57.6	–
2	22	52.0	70.1	–	72.8
2	100	41.0	–	72.2	–
3[b]	100	100	–	–	–
5	30	25.0	86.4	–	–
7	30	27.1	55.3	67.8	–
8	30	12.1	59.6	55.8	–
10	30	33.3	48.3	46.5	–
11	30	20.9	43.3	50.7	–
Control	0	–	–	–	1.9
Control	0	–	–	–	1.0
Control	0	–	8.3	14.5	–
Control	0	1.6	–	1.6	–

[a]Amount of bait pellets added per 10 ft ditch section.
[b]Bait decomposed, destroying all snails and killing minnows and tadpoles in the test section as well as a considerable distance downstream.

TABLE 3–34[361]

Oncomelania Baits and Evaluation Results

Ingredient	C525-sfbA(3)	C525-wgbA(3)	C526-sffA(3)	C526-wgfA(3)
	Formulation recipe (percent)			
Carboset® 525	66.0	66.0	–	–
Carboset 526	–	–	66	66
Snail food	25.0	–	25	–
Bayluscide®a	5.0	5	–	–
TBTA	1.0	1	1	1
Zinc complex[b]	3	3	3	3
Wheat germ	–	25.0	–	25
Frescon®c	–	–	5	5
	Accumulative snail mortality (percent)[d]			
Exposure time				
Week 1	38	75	16	16
Week 2	100	100	100	100

[a]Ethanolamine salt of niclosamide, 70% wettable powder.
[b]5.8% zinc ion.
[c]95% trifenmorph.
[d]Dosage rate 0.0014 g bait/cm².

must be used in small amounts to prevent rapid decomposition of the bait structure by micro-organisms. TCPN is molluscicidal and may provide some degree of action against the ingesting target. No synergism between trifenmorph and TCPN was noted. In tests on bananas, the major St. Lucian crop, no TCPN-induced phytotoxicity or translocation to the fruit was found.[360]

Amphibious Snails

Snails that spend a major portion of their lifespan upon moist soil rather than in water present a specially difficult control problem. Conventional molluscicide techniques rely upon the dispersion of the control agent through solution and movement in water. If the target is not in water it is necessary to apply the toxicant to literally every square foot of the habitat by spraying or broadcasting methods. One must take the toxicant to the target for contact to occur. A bait approach might well be meritorious in that the target is attracted to the bait and complete coverage of the habitat is not essential.

Bille et al. prepared finely granulated baits of sufficient smallness and softness to be ingestible by *Oncomelania formosana*, a snail host of *Schistosoma japonicum.*[361] *Oncomelania*, II-1on strain, were reared under laboratory conditions by standard methods using *Chlorella pyrenoidosa* as the foodstuff.[362,363]

Attractant studies with *Oncomelania* were inconclusive. The snail ingested the preferred bait and succumbed, but whether a positive attraction existed or the target randomly discovered the bait could not be ascertained. Both trifenmorph and niclosamide were suitable toxicants although the former seemed to induce somewhat more rapid mortality. TCPN evaluated as a toxicant was ineffectual; TBTA and TCPN were equally effective as preserving agents, but superior kills were seen with the former. Superior bait recipes and laboratory results are shown in Table 3–34. Dosages are applied on moist soil in petri dishes. Field test studies have yet to be performed.

Cercaria and Miracidia Baits

A conceivable weapon against schistosomiasis would lie in the destruction of the larval forms of the parasite by means of a bait. Both miracidia and cercaria will penetrate certain nontarget animals and plants as well as the target. Upatham reported that several nontarget snails, guppies (*Poecilia reticulata* Garmon), and tadpoles (*Bufo marinus* Linn.) serve as decoys to miracidia, thus "pro-

Page 89 →

FIGURE 3–3. Bayluscide® drip feed in a sugar cane irrigation scheme. (Photograph by the author.)

FIGURE 3–4. Large Bayluscide® storage tanks and continuous feed and monitoring system: truck garden irrigation scheme. (Photograph by the author.)

FIGURE 3–5. Hand spraying of a bog (St. Lucia, West Indies). (Photograph by the author.)

FIGURE 3–6. Snail survey in a septic lagoon (Maracay Area, Venezuela). (Photograph by the author.)

FIGURE 3–7. Drainage ditch after Bayluscide® treatment (Cul de Sac Banana Estate, St. Lucia, West Indies). (Photograph by the author.)

FIGURE 4–1. Water hyacinth infestation. (Photograph by the author.)

FIGURE 4–2. Eurasian watermilfoil, Lake Theriot, Louisiana. View directly into water showing sinking 14ACE-B pellets; taken at time of application. (Photograph courtesy of U.S. Army Corps of Engineers, New Orleans District.)

FIGURE 4–7. Ditch covered with water hyacinth prior to application of controlled release herbicide.[481] (Photograph courtesy of U.S. Army Corps of Engineers, New Orleans District.)

A.

B.

FIGURE 4–8. Water hyacinth test area 7 months after controlled release "suspender" application. A: Test area. B: Edge of test area.[481] (Photographs courtesy of U.S. Army Corps of Engineers, New Orleans District.)

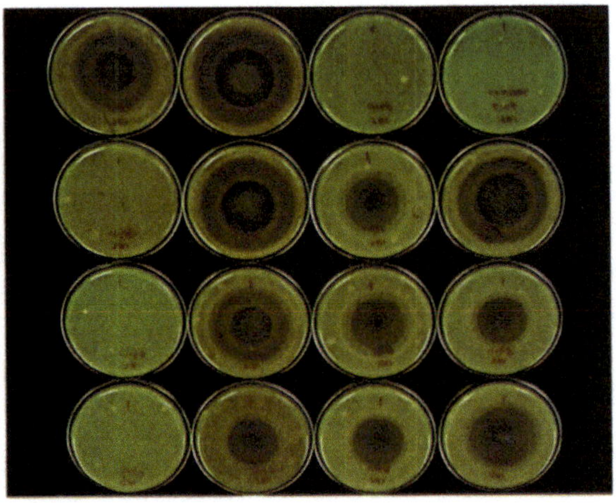

FIGURE 5–1. Fungi inhibition as a function of organotin concentration in controlled release elastomers. In this photograph the zone of inhibition is clearly seen for *Trichoderma* exposed to various controlled release fungicidal materials. The 0.01-g pellet is suspended above the agar growth plate so that an approximate ½-in. air gap exists between it and the inoculum.

tecting" the target mollusc.[364] Miracidia attempts to penetrate nontarget animals may depend upon the presence of a target.[365]

In 1959 Wright suggested that a chemical attraction between the target snail and miracidia existed.[366] MacInnes demonstrated that *Schistosoma mansoni* miracidia react to chemostimulation when in the presence of extracts of host snails, or to agar baits containing butyric, sialic, or glutamic acid.[367] Etges and Decker likewise concluded that miracidia were attracted to host snails.[368] Other investigators have theorized that contact between snail and miracidia is by chance[369] or at least involves no chemical attraction.[370,371] However, a preponderance of the data available indicates that living snails emit an attracting substance that enables miracidia location.[367,368,372]

It is not known whether any serious investigation of the feasibility of miracidia baits has been performed, and available literature is silent on this subject. Certainly the potential exists. Attractants such as butyric, valeric, glutamic, and sialic acid[367] can be incorporated in various penetrable elastomeric matrices and, apparently, released slowly.[351] Various toxic materials such as copper,[320] tissue extracts from cercaria-infected snails,[373] organotins,[333] etc. are known and may be incorporated in bait preparations.

Attractants can be rather specific. For instance, MacInnes reports that butyric acid is strongly attractive to *S. mansoni* miracidia, but considerably less attractive to *Schistosomatium douthitti,* while *Fasciola hepatica* miracidia showed no response at all.[367]

The bionomics of the miracidia habitats are somewhat understood, so baiting techniques could be developed. Miracidia range in static and flowing water is known, and habitat depths have been investigated.[344,345,374] Effects of decoys, illumination, water quality, temperature, dispersion, etc. are at least partially known.[341,343,347,364,370,371,375]

The potential merit of cercariacidal baits was investigated at the University of Akron.[351] In this study pleurolophocercous cercaria were induced into the midpoint of a water-filled glass tube. A candidate "bait" was inserted at one end, but no material at the opposite. Cercaria movement toward or away from the test formulation and penetration was microscopically observed. Nontoxic bait blocks were used. Carboxylated polyacrylics were utilized as the binding element and various attractants incorporated therein. Several acrylic bases containing 5% glutamic acid showed positive cercaria attraction and attempted penetration. Later experiments with *S. mansoni* cercaria were negative towards glutamic acid.

Slug Control

"Slug" is a general term used for certain terrestrial gastropods having no exterior shell. They are distributed among several families in the suborder Stylommatophora. In the United States the following slugs are of agricultural importance:

Limax flavus (tawny garden slug)
L. maximus (spotted garden slug)
Deroceras reticulatum (gray garden slug)
Milax gagetes (greenhouse slug)
D. laeve (gray field slug)
Helix aspersa (brown garden slug)
Arion ater (European black slug)

Conventional control methods rely heavily upon bait preparations. Probably the most widely used pesticide in this respect is metaldehyde, $(OCH\text{-}CH_3)_4$, which is both an attractant and toxicant. Its normal use is in bran-based baits at 1 to 2%. Such baits may also contain 6 to 7% calcium arsenate, which seems to improve toxicity. Tarter emitic, potassium antimonyl tartrate, in a sugar base is also used. A carbamate pesticide, Isolan®, 1-isopropyl-3-methyl-pyrazolyl-(5)*N,N*-dimethyl carbamate, has limited use in bait formulations.[376] Another carbamate material, Zectran®, 4-dimethyl-amino-3,5-xylenol-*N*-methylcarbamate, can be used in grain- or molasses-based baits.[377] Guthion®, an organophosphorus insecticide used in spray applications, will control *H. aspersa* but is ineffective against other species.[378]

In order to evaluate molluscicidal baits against slugs a number of experiments were performed at the University of Akron in 1968–1969. *L. maximus* and *A. ater* were cultured using a method similar to that of Stephenson.[379]

Slugs were attracted to and ingested both niclosamide and trifenmorph baits without any apparent ill effects. Organotin-containing baits were rejected. The most effective attractant appeared to be malt, although bran showed almost comparable results.

Chapter 4

CONTROLLED RELEASE HERBICIDES

NATURE OF THE PROBLEM

Various members of the plant kingdom are nuisances under certain conditions. Higher plants that interfere with man's manipulation of his environment are termed "weeds." For convenience, we can separate these pests into aquatic and terrestrial forms. Lower plants, bacteria, algae, and fungi will be considered in a separate chapter.

Aquatic weeds represent a substantial economic and health burden around the world. Plant growth not only absorbs scarce water in irrigation systems but also blocks the flow of water to growing crops, thus decreasing yield. In many areas aquatic weeds are important contributing factors in malnutrition; rice paddy infestations likewise reduce yield. Blockage of drainage systems can lead to localized inundation of croplands and dwellings. Weed mats such as those produced by the water hyacinth will block navigable waterways, preventing or slowing commerce.

In the United States aquatic weeds constitute a hazard to navigation, reduce or prevent the use of recreational facilities, serve as habitat for harmful insects, can be destructive to fish life, retard water flow in drainage channels, leading to flooding, and, generally speaking, are aesthetically unappealing. Aquatic plant problems are particularly severe in the Gulf States. Alligator weed (*Alternanthera philoxeroides*) infestions will cause complete blockage of navigable waters and are notoriously difficult to control.[380,381] The water hyacinth (*Eichornia crassipes*) is a widespread problem in Florida, Louisiana, Mississippi, and eastern Texas.[381,382] Eurasian watermilfoil (*Myriophyllum spicatum*), which originally entered the United States through the Gulf Coastal area, has spread northward into the Tennessee Valley area, where infestations are increasingly severe, and is rapidly infesting various water bodies in Ohio, Indiana, Michigan, and Illinois.[383,384] Hydrilla species, often considered to be "southern weeds," have moved as far north as Iowa.[385] Southern naiad, (*Najas* spp.), *Cabomba caroliniana*, and coontail (*Ceratophyllum demersum*) are other major aquatic weeds. Even aquatic plants considered to be of minor importance, such as duckweed (*Lemna* spp.), can become a special problem, e.g., by introducing undesirable odor and taste to potable waters.[386]

Aquatic weed problems in other nations can be considered to be of far greater consequence than those observed in the United States. Weed infestations in irrigation systems are directly implicated in low agricultural output in a number of nations. Clearance of weeds from such schemes by hand labor, as in the Gezira system of the Sudan, results in human exposure to water-borne parasites such as larval schistosome trematodes. Comprehensive investigations of present and potential aquatic plant problems have been carried out in various nations (Figures 4–1 and 4–2).[387,388]*

The ubiquitous water hyacinth appears to be the major world-wide undesirable aquatic plant, but plants of a localized geographical area can also create a good many serious problems. For example, floating islands of papyrus, *Cyparus papyrus*, in Lake Victoria, Tanzania, may be up to 1/4 mi^2 in area and weigh 1500 or more tons.[389] Such islands block navigation, foul propellers of ships that inadvertantly collide with them, destroy or obscure marker buoys and navigation lights, damage fishermen's nets, and obstruct port facilities. When they strike shore a swamp-type habitat is established that harbors disease-spreading mosquitoes, schistosome-bearing snails, snakes, rats, and other unwanted biota. Village life is disrupted by obstruction of water intakes and fishing activities and destruction of agricultural land.

The world-wide economic loss arising from aquatic weed infestations is unknown. Aquatic herbicide sales in the United States probably exceed $60 million per annum.

The global necessity of increasing food production means that the *major* detriment to crop yield, terrestrial weeds, must be controlled on a greater geographical scale, one commensurate with the increase in acreage expected, and more effectively than in the past. Weeds reduce crop output through natural competition. In pastures and rangeland weeds depress the more beneficial vegetation. Some plants are poisonous to livestock while others affect the agricultural worker directly through toxic or allergenic reactions.

*For Figures 4–7 and 4–8 see Page 90.

Crop losses through weed infestation is 10 to 15% in developed temperate nations and may be as high as 45% in tropical developing countries.[390]

Terrestrial weeds harbor insects and other animal pests of public health importance, interfere with rights of way, damage roadways and other paved areas, retard forest and orchard growth, and cost the homeowner a great deal (both in money and labor) in his desire to maintain aesthetically pleasing surroundings.

Destruction of weed pests relies heavily upon the use of chemical control agents which in turn can lead to undesirable contamination of the environment. Reduction in the use of such agents in the conventional methodology without a concommitant increase in the use of alternative strategies such as biological control is not a viable solution. Most nations live dangerously close to the thin edge of food shortages. Even in the United States, certainly the most advanced agricultural nation, the food surplus seldom exceeds 6%.[391] A reduction in yield of around 10%, in the United States, which could clearly occur if a significant number of herbicides were no longer available, would lead to a crisis of unthinkable proportions.

Population pressure will lead to expanded acreage, thus increasing the use of herbicides. Accordingly, contamination of the environment with herbicide residues (and those of insecticides, fungicides, etc.) and their toxic breakdown products will likewise increase and could conceivably create a public health crisis as monumental as that foreseen if food production does not keep pace with population growth.

CONVENTIONAL CONTROL METHODS

Aquatic Weeds

Aquatic plant control is practiced through four general techniques or a combination of them. The major methodology involves the use of effective phytotoxic materials. Biological control is in its infancy although rapid progress is being made. Water management can alleviate certain weed problems, but as a rule cannot do a complete control job, and can only be effectively used under very special circumstances. Likewise, mechanical control is a useful adjunct to chemical usage, but is limited in applicability and usually unattractive economically.

The taxonomy and ecology of water weeds have been discussed in great detail in the readily available literature, so no need exists to burden this text with a description of the several hundred aquatic plants of economic importance.[392-394]

Chemical Methods

Copper sulfate and sodium arsenite were the first aquatic herbicides to be utilized (early 20th century) in quantity against algae and water weeds. In 1946 2,4-dichlorophenoxy acid (2,4-D) was placed in use, and a few years later various aromatic solvents appeared on the commercial market. Between 1950 and 1970 several hundred phytotoxic materials were used to some extent in water weed control, but today only 20 or so materials are in widespread use. Many chemical agents have been rejected for economic reasons or due to their unfavorable environmental impact, especially recognized in the destruction of fish. Chemicals such as acrolein, though effective herbicides, have been generally abandoned due to the hazards involved in handling and adverse effects on fish populations. Table 4–1 lists the currently popular herbicides, with pertinent toxicity data.

Alligator weed probably represents the most difficult control problem. It can grow in three habitats: terrestrial, rooted immersed, and free floating. Rooted plants in static water can be controlled with dichlobenil at 5 to 10 lb/acre applied as a granular formulation during the winter months.[395] Silvex® granules at 20 to 30 lb/acre, or the esters used at 8 lb/acre as a foliar spray, were effective in rooted alligator weed control. Foliar application, of 2,4-D agent, although probably the more economical treatment method, only destroys foliage and does not penetrate the mat material, even with prolonged herbicidal contact.[397] Root uptake of 2,4-D is probably greatest when the plants are devoid of leaves, thus indicating a winter treatment. Zur Burg et al. suggest repeated light 2,4-D applications every 6 weeks for effective control.[397] Persistency in various field trials appears to be 4 to 7 weeks with 2,4-D.

Silvex esters have been used against alligator weed. The propylene glycol butyl ether ester at 8 lb/acre acid-equivalent spray and the butoxyethanol ester at 10 to 30 lb/acre granular formulation have been effective.[399] In a tidal area Silvex appears to be superior to 2,4-D.[400]

No satisfactory chemical method of controlling

TABLE 4–1

In-use Aquatic Herbicides

Herbicide[a]	LD_{50}, rat, oral (mg/kg)	LD_{50}, rabbit, dermal (mg/kg)	LC_{50}, fish (ppm)	Ref.
2,4-D acid	370	1,500	100+	418, 419
2,4-D amines	300–1,200			418
2,4-D esters	500–1,200			418
Endothall, dimethylamine salts	51		0.1–7.5	408
Endothall, disodium salt	38–198		80–680	408
Endothall, dipotassium salt	50–125		235–268	408
Dichlobenil	3160	1,350	10–20	415
Silvex® – PGBE	650–1,070		9–22	418, 421
Diquat	400–440	500+		410
Fenac®	1,780	3,160+		418
Dalapon	970			418
Amitrole	25,000	10,000+		418
2,4,5-T	500			418

[a]Chemical names:
Endothall = 7-oxabicyclo (2,2,1) heptane-2,3-dicarboxylic acid.
Dichlobenil = 2,6-dichlorobenzonitrile.
Silvex – PGBE = propyleneglycol butylether ester of [2-(2,4,5-trichlorophenoxy) propionic acid].
Diquat = 6,7-dihydrodipyrido (1,2-a:2′,1′-c) pyrazidiinium dibromide.
Fenac = 2,3,6-trichlorophenyl acetic acid.
Dalapon = 2,2-dichloropropionic acid.
Amitrole = 3-amino-1,2,4-triazole.
2,4,5-T = 2,4,5-trichlorophenoxyacetic acid.

the water hyacinth was available prior to 1948. Initial investigation of available herbicides demonstrated that the isopropyl and butyl esters of 2,4-D and 2,4,5-T were most effective against this weed.[401] Today the dimethylamine salt of 2,4-D is probably the most common herbicide used against the water hyacinth. A spray formulation is usually used, often with a sticking agent, and dosage rates are usually in the neighborhood of 4 lb/acre.[402]

One major disadvantage in having irrigation channels partially blocked with water hyacinth is that evapotranspiration from the plant is greater than that from the water surface alone. Thus, this pest can be a major determinant in water loss prior to reaching the crops. 2,4-D treatment checks such loss.[403] Water hyacinth is also controlled with the butoxyethanol ester of 2,4-D.[399] Nontoxic concentrations of 2,4-D will synergize other herbicides such as ametryne.[404]

Diquat at 1 to 2 lb/acre has been satisfactorily used as a foliar spray.[396]

The potassium salt of Silvex will control water hyacinth, although it is not widely used for this purpose.[398] Silvex has one important advantage over 2,4-D in that it destroys the seeds of this plant at 8 ppm.[405] Amitrol and amitrol-T will also provide reasonable control of the water hyacinth.

Eurasian watermilfoil can be controlled with a number of agents, with 2,4-D possibly the most popular; the butoxyethanol ester at 2 to 3 ppm is effective.[406,407] Various endothall formulations are used at several parts per million.[408,409] Diquat (0.5 to 1.5 ppm), dichlobenil (2 to 10 ppm), Fenac® (3 to 8 ppm), Silvex (2 to 4 ppm), and Silvex-endothall mixtures at 2 to 3 ppm have been found effective against Eurasian watermilfoil. Whereas 2,4-D, diquat and endothall will effectively destroy the immersed portion of this weed, they do not translocate to reproductive structures buried in the hydrosoil. Once the treated watercourse is detoxified through natural processes regrowth is rapid.

Elodea spp. constitute a major pest weed throughout Louisiana, Florida, and scattered locales in other Southern states. Effective control

TABLE 4–2

Selected Aquatic Weeds and Control Agents Utilized

Aquatic plant				
Common name	Botanical name	Herbicide	Dosage (ppm)	Ref.
Bladderwort	*Utricularia* spp.	Diquat	0.25–3	396, 419
		Endothall (amine)	0.5–2.5	419
		Silvex® (potassium)	1–2	394, 398
Coontail	*Ceratophyllum demersum*	Diquat	0.5–1.5	294, 417
		Endothall (sodium)	1–4	396, 416
		2,4-D (isooctyl ester)	2	394, 398, 417
		Dichlobenil		394, 417
		Fenac®		394
		Silvex – PGBE	2	421
Southern naiad	*Najas guadalupensis*	Diquat	1	394
		Dichlobenil	7–10	387, 394, 417
		Endothall (sodium)	1–4	419
Pond weed	*Potamogeton* spp.	Diquat	0.5	394
		Endothall (amine)	1–2	394
		Silvex (potassium)	0.5–2	398–417
		Endothall (potassium)	1	394
		Dichlobenil	1–5	415
Duckweed	*Lemna minor*	Diquat	0.25–1	394, 396, 417
		Endothall (sodium)	4	416
Water lettuce	*Pistia stratiotes*	Diquat	0.5–2	387, 417
		2,4-D (esters)		398
		2,4,5-T (esters)		398
Arrowhead	*Sagittaria* spp.	2,4-D (ester)	1–5	394, 417
		Silvex (ester)	1–2	387, 394
		Endothall/silvex	2–3	394, 416
Parrot feather	*Myriophyllum brasiliense*	2,4-D esters		
		Silvex (potassium)	1–2	396
		Endothall (sodium)		
		Dalapon		398
Smartweed	*Polygonum* spp.	2,4-D esters	1–3	394, 417
		Silvex		394
		2,4,5-T	0.5–2	417
Eelgrass	*Vallisneria americana*	Silvex (sodium)	3–4	398, 422
		Dichlobenil	1–5	415
		Silvex PGBE	2	421
Cabomba	*Cabomba caroliniana*	Fenac		
		Endothall (amine)	0.5–2.5	419
		Dichlobenil		
		Endothall (sodium)	1–4	419
		Silvex – PGBE	2	421

has been achieved with Diquat (0.5 to 2 ppm),[394,396,411] Fenac,[411] sulfuric acid,[412] amine salts of endothall (2 to 4 ppm),[396] acrolein (4 to 7 ppm),[396]. sodium arsenite (3 to 8 ppm),[396] endothall and Silvex mixtures (2 to 3 ppm),[394,416] and the potassium salt of Silvex (1 to 3 ppm),[398] but 2,4-D esters applied conventionally are not particularly effective at economic dosage levels. A diquat and copper sulfate (1 ppm/1 ppm) combination has been successfully used against *Hydrilla verticillata*,[413] and a cutrine and diquat (4:2) mixture is also effective.[414]

Other aquatic weeds are not as ubiquitous or difficult to control as alligator weed, water hyacinth, Eurasian watermilfoil, and *Elodea* spp, but may present severe problems in localized areas. Table 4–2 lists other economically important water weeds and chemicals used in conventional control.

Mechanical Methods

In the early 1900's various mechanical methods were devised to cope with water weed infestations, including gang saws to open clogged channels,

TABLE 4–3

Potential Biological Control Agents

Aquatic weed	Agent	Description	Ref.
Water hyacinth	*Acigona ignitalis*	Flea beetle	387
	Ctenopharyngodon idella	Fish	423, 431
	Neochetina eichhorniae	Insect	424
	Arzama densa	Moth	425
	Trichechus manatus	Sea cow	432
Water lettuce	*Samea multiplicolio*	Moth	387
	Marisa cornuarietis	Snail	387
Vallisneria	*Tilapia zillii*	Fish	387
Alligator weed	*Agasicles hygrophila*	Flea beetle	387, 429
	Amynothrips andersoni	Thrip	425
	Vogtia malloi	Moth	425
Hydrilla verticillata	*C. idella*	Fish	387, 431
Eurasian watermilfoil	*Parapoynx stratiotata*	Moth	428
	Accentropus niveus	Moth	428
	C. idella	Fish	431
Southern naiad	*C. idella*	Fish	431

special barges utilizing conveyor belts to pick up plants, and crusher boats to grind hyacinth mats into pulp. Such methods only offer temporary relief, since breaking up the plants tends to disperse rhizomes which resprout and cause infestation elsewhere. Also, complete plant destruction returns nutrients to the water, contributing to more luxuriant growth of other aquatic forms. Draglines that pick up vegetation from canals and ditches, depositing debris on the banks for later truck pick up and destroyer ships that cut and shred vegetation using a bank of saws are used. The main use of mechanical destroyers and harvesters is to cut a path through weed mats, allowing access for spray boats (Figure 4–3).

Mechanical harvesting does not appear to be as economical as using conventional chemicals unless a means of disposal of the weed residues, such as use in cattle feed, can be developed to offset a portion of the costs. Weed debris left to decompose on the banks of dredged channels furnishes harborage for various obnoxious insects and is aesthetically unappealing; also, breakdown products with substantial nutritional value eventually wash back into the cleared watercourse, enhancing weed growth.

One dramatic new control method lies in the destruction of floating plants, especially water hyacinth, through the use of a laser beam. Exposure levels of 5 to 100 J/cm^2 leads to plasmolysis and eventual death. Laboratory results using a CO_2 laser against water hyacinth and duckweed were gratifying,[420] but this method is not effective against alligator weed.

Biological Control

Noteworthy advances have been made in the control of aquatic weeds through the use of predatory animals. Herbivorous fish, various snail species, and a number of insects have been examined as potential biological controls, and experimental investigations have been made or are underway with the potential biocontrol agents given in Table 4–3.

Biological control of water hyacinth is speculative rather than actual. Native insect predators are not particularly effective and mass release programs would be needed.[424,425] Several pathogens are known to exist, one of which, the fungus *Cephalosporium zonatum*, appears to have the highest potential as a biocontrol.[430,433] The use of the manatee, a voracious consumer of aquatic vegetation, has been attempted, but, unfortunately, poachers have trapped and eaten this

A.

B.

C.

D.

FIGURE 4–3. A: Typical aquatic weed obstruction of navigation occurring in Louisiana streams during 1940s. B: Conveyor used to remove aquatic vegetation from streams for deposit and drying on bank. C: Crusher-type boat used to clear waterways by picking up and crushing aquatic vegetation. D: Sawboat type of destroyer cutting aquatic vegetation for in-place destruction. (Photographs courtesy of U.S. Army Corps of Engineers, New Orleans District.)

animal so that large-scale release is probably not practical.

The most notable success in biological control of aquatic weeds has resulted from the release of the *Agasicles* flea beetle in various locations. Populations have been established in two areas in Texas, and effective control has been achieved in various release sites.[426] However, use of the flea beetle requires modification of the aquatic ecosystem by increasing the density of plant life normally competitive with the alligator weed.[427] Thus, the decrease of one problem plant gives rise to increasing populations of water hyacinth, Eurasian watermilfoil, cattails, and giant cutgrass (this last not being particularly susceptible to herbicidal control).

Alligator weed is subject to fungi and viral disease, although their use as control agents would necessitate overcoming severe dispersal problems.[430]

Hydrilla is subject to several phytopathogens, and their potential for biocontrol is under evaluation.[430] The white amur, *Ctenopharygodon idella*, consumes *Hydrilla* in preference to other higher water weeds tested.[431]

Integrated Control Methods

Integrated control implies the use of several methods simultaneously. Combining herbicidal treatment with release of the *Agasicles* beetle has led to a level of alligator weed control unattainable by the use of either method alone.[380] It has also been observed that insect injury to the water hyacinth accelerates attack by phytopathogens.[433] Thus, one might integrate two distinct biological entities into a single attack on the target weed. Two different chemicals also might be used; for instance, 2,4-D BEE applied to resistant plants such as *Cabomba* provides a phytotoxic stress, so that a follow-up treatment with copper sulfate at only 8 lb/acre provides excellent control.[434]

Integration of biological and chemical control methods offers considerable promise in minimizing environmental impact. It has been shown that the use of the *Agasicles* beetle coupled with low dosage 2,4-D or diquat can provide adequate control of alligator weed in pool tests.[381,382] A biocontrol measure might be used against one aquatic weed in combination with a herbicide for control of a second noxious plant at the same treatment site. This has been successfully accomplished where the problem plants are alligator weed and water hyacinth.[381,435]

Environmental Impact

Aquatic weed control methods have various effects on the ecology. Although environmental impact is almost always associated with the use of chemical agents, mechanical and cultural control techniques can have considerable biological effect, e.g., manipulation of water level as a weed control technique can considerably alter the biosphere involved.[436,437] Biological control methods also affect the environment, usually through destroying the dominant target plant and thereby enhancing the growth of subordinate species. Mechanical harvesting or chopping of aquatic weed mats tends to change the problem site, e.g., floating plant segments downstream establishes populations elsewhere, or creates more favorable insect habitats.

Herbicides, by their nature, will lead to undesirable effects on nontarget members of the biota. Since benefits may be great and the readily observed side effects few, there is a general tendency to ignore the considerable contamination that may occur. In one sense herbicides can be considered antipollutants in that the target weed controlled may have a deleterious effect on piscine and other favorable life forms.[399,438]

Many chemical agents used at the dosage rates necessary for adequate plant control destroy fish, and residues may be long lasting. Water and hydrosoil samples have shown Silvex residues for 7 months post treatment, with fish kill.[421] Such kill may be selective, with the more susceptible populations wiped out while tolerant species may be little affected.[439,440] Herbicides also kill various invertebrate species,[441] but are generally short lived in the aquatic environment, so the impact on lower-life populations often arises from alteration of the ecology in such a way as to render habitats less suitable for life.[442] Treatment method can play a significant role in the nature of the herbicidal effects on nontargets, especially as regards the points of application of the agent.[443]

Herbicides applied to irrigation waters may reach growing crops or end up in the tissue of animals that drink the water,[444-446] while those applied to natural watercourses may also affect growing crops.[447]

The literature is replete with reports of herbicide damage to nontarget members of the immediate and remote biota. However, there have also been many investigations demonstrating that a given herbicide properly applied has little or no effect on aquatic fauna.[384,407]

Often, nontarget aquatic fauna show a slight-to-moderate decline in the immediate herbicide post-treatment period, followed by rapid resurgence to pretreatment levels.[448]

Terrestrial Herbicides

Herbicidal materials are in widespread use for the control of weeds in lawns and turf, forests, along rights of way, and especially in agriculture. The economic loss to the American farmer from weed infestation may be as high as $10 billion annually.[449] Weeds are far more costly to control than other plant pests. In the early 1970s estimates were $1500 million for weeds compared to $400 million for insects and $200 million for plant diseases.[450] Without the use of herbicides, yields would dramatically decrease. It is estimated that bushels per acre of corn would drop from 104 to 64![451]

There can be little doubt that world-wide population growth coupled with an increasing energy shortage will bring about an unprecedented food crisis within the near future. It is neither socially nor politically feasible that the world's population rate will decrease, barring perhaps general warfare involving weapons of mass destruction, so the immediate solution lies in increasing the amount of readily available foodstuffs. This, in turn, can be done only by increasing acreage or increasing the yield of a given crop per acre. Both will, of course, be done. However, neither approach bodes well for environmental concerns. Pesticide usage will increase, and alternatives such as biological control and improved land management are either in the distant future or not practical on a world-wide basis. Since weeds represent the major detriment to most crop yields, as evidenced by expense and tonnage used (compared to insecticides or nematacides, for instance), it is reasonable to assume that terrestrial herbicide usage will increase vastly.

When population growth reduces the annual 6 to 10% food surplus to zero, restrictions on the use of weed control agents in the United States will likely be relaxed or abandoned. Even if this does not occur, protection of the U.S. environment in the long run would be of little avail as other nations rapidly increase pesticide use.

It is not within the scope of this text to describe the terrestrial weed problem and pertinent control methods to the extent provided in the discussion of aquatic weeds, for space and lack of knowledge on the author's part does not so allow. Basically, though, controlled release aquatic herbicides are available, whereas the tremendous area encompassing terrestrial agents remains only an extremely fertile field for future research. A basic intent of the author is to urge others to initiate such endeavor.

Many excellent texts and reports concerning all aspects of herbicide use in agriculture are readily available. DeOng, Bishop, and Bishop provide a general description of weed control methodology in various field crops, grains, vegetables, fruit orchards, turf, and noncrop lands.[452] A more detailed description of weed ecology and control is available from the National Academy of Sciences.[417] Excellent weed control handbooks have been published that provide information on herbicide selection for specific target weeds, application, and dosages.[453,454]

The use of herbicides represents one of several vital factors in high-yield agricultural production. Tillage in combination with herbicides, for instance, provides greater benefits than the use of either alone.[455] Insecticides can be applied simultaneously with select herbicides, although care is necessary to assure that the combination chosen is not antagonistic.[456] Herbicide/herbicide and herbicide/insecticide mixes can, of course, be synergetic.[456–458]

Terrestrial herbicides, since they are used in far greater quantity than aquatic weed control agents, present a sizable menace to a quality environment. Herbicides are meant to control weeds without injuring the protected crop. Unfortunately, agents applied in one locale may be dispersed through spray drift, wind, ground water, etc. and reach other crops. Turf grass, ornamentals, lawns and nontolerant crops may be affected.[445,459] Toxic effects of herbicides on man, domestic animals, and fauna in general *appear* to be minimal.[460] Politically inspired furor arising from misinformation and misinterpretation of scientific data such as in the case of 2,4,5-T serves no useful purpose.[461] However, terrestrial herbicide residues are found in trace amounts in water bodies and nontarget flora and fauna.[104,462]

Nonchemical methods for terrestrial weed control, other than proper land management and cultural practices, are not as advanced as those alternatives developed by the U.S. Army Corps of Engineers reported in the previous section. Biological control of weeds using insects and phytopathogens is under laboratory investigation.[429]

The well-known concept of allelopathy, the ability of certain plants to release chemicals antagonistic to competing plants, has not been investigated as a biological control weapon.[463]

CONTROLLED RELEASE AQUATIC HERBICIDES

The development and evaluation of controlled release aquatic herbicides is a natural outgrowth of the antifouling and molluscicide endeavors of the author and the work with plastic/pesticide systems under the direction of Dr. William Barnes, then at the U.S. Army Environmental Hygiene Agency. In each case the chemical agent is slowly released from a binding matrix into the pest-infested aquatic environment. Since elastomer/herbicide and plastic/herbicide systems are fundamentally different, they will be treated separately in this section.

Elastomeric Formulations

Development

The initial development work keyed to the creation of controlled release aquatic herbicides was predicated on two concepts: (1) that the 2,4-dichlorophenoxyacetic acid, its esters, and its amines could be incorporated in elastomers and be released slowly over a period of time, and (2) that formulations could be so made and applied to treat only the target phytozone. The physical properties of 2,4-D indicated solubility in chloroprene-type elastomers as based upon the state of the art in 1969. Theory, developed from the antifouling and molluscicide programs, predicted that a diffusion-dissolution type mechanism should be establishable.

By their nature, elastomers can be processed in a variety of forms, making evident much more versatility than that achievable with conventional impregnated clays, vermiculite, and so on. Thus, it was conceivable that floating, sinking, and suspending materials could be made. Aquatic weeds, in turn, occupy certain ecological niches. The water hyacinth, duckweed, and water lettuce float, so a floating herbicide formulation ought to be more effective than one that sinks. In contrast, elodea and milfoil can root, but often float freely within the water body, thus being susceptible to suspending formulations. Alligator weed, polygonum, and so on, as rooted plants, would conceivably be more affected by a sinking material. Also, reproductive parts of floating plants may be in the hydrosoil and therefore more susceptible to a sinking material. Obviously, if the mode of toxicant entry is through the roots, a pellet of sufficient density to penetrate bottom mud might be of greater value.

The phytozone concept looks to the delivery of the toxicant to the target in the most expeditious manner. This should decrease dosage in that the total water body is not treated – assuming that, at least in static systems, molecules of the control agent stay put; thus, environmental impact ought to be substantially decreased.

The 1969–1970 program sponsored by the U.S. Army Corp of Engineers at the University of Akron encompassed the development of slow release herbicides active in the phytozone of interest. Originally, the butoxyethanol ester and dimethylamine of 2,4-D were selected as toxicants. Five concepts relevant to phytozone release, sinkers, floaters, suspenders, time release floaters, and so-called bouquets, were advanced. The time release concept consisted of encasing controlled release herbicidal pellets in an acrylic-clay binder that easily broke foliage and sank to the bottom. As the binding substance gradually disintegrated, pellets (or suspenders) of the floating variety were released and rose to the water surface.

Tests indicated that 2,4-D BEE was soluble in natural rubber (35 to 40%), ethylene-propylene elastomers (15 to 21%), and styrene-butadiene polymers (10 to 12%);[464] 2,4-D DMA showed solubilities of less than 2% in the same materials, so further effort was discontinued.

Compounding studies followed the guidelines set forth in the molluscicide programs. Carbon black absorbed 2,4-D BEE, thus extending the diffusion path length and slowing migration to the rubber/water interface. Table 4–4 illustrates the variance in absorption by the various blacks. Particle size is the main determinant, but structure also plays an important role; thus, absorption cannot be based on the size alone (compare ASTM N231 with N550, for instance).

SAF black would appear to be the most effective in retarding internal movement of 2,4-D, while FT is the least effective. Of course, the amount of black used, i.e., the "loading," will influence migration.

Formulations prepared in this program were evaluated under laboratory conditions against the water hyacinth, *Elodea canadensis*, and Eurasian

TABLE 4–4[464]

2,4-D BEE Absorption by Carbon Black

Black type	ASTM code	Average particle size (nm)	2,4-D BEE absorption (g/10 g black)
SAF	N110	20	8.2
SCF	N294	22	6.4
CF	N293	27	6.2
ISAF	N231	29	6.0
FEF	N550	79	6.0
ISAF	N220	29	5.8
ISAF	N219	30	5.2
HAF	N326	40	4.8
HAF	N330	45	3.3
SRF	N774	160	3.3
SRF	N770	128	3.2
MT	N990	462	2.8
FT	N880	256	2.5

watermilfoil. Results indicated that 2,4-D BEE was released and effectively destroyed target plants. Carbon black and various oils could be used to adjust loss rate. Natural rubber proved to be the superior matrix material; chloroprenes and other matrices were proved to be inferior on the basis of agent solubility limits and effective loading. Simple cure systems were practical. Later recipe modifications were made to enhance processibility – for instance, partially cured, premasticated, stabilized natural rubber was used. Superior candidate recipes are shown in Table 4–5.

Suspenders were prepared by molding a 14ACE-B anchor on a 14ACE-A floating strand (Figure 4–4). Once immersed, a "top" suspender floats vertically with the tip of the 14-A portion just above the water surface, and the 14-B anchor downward. By slightly increasing the length of the

TABLE 4–5[464]

Controlled Release Elastomer-based Herbicide Materials

Ingredient[a]	Code (ACE series); recipe in parts								
	11-D	11-E	14-A	14-B	15-B	17-B	17-C	18-B	18-D
RSS #4	100	100	–	–	–	–	–	–	–
PA-80	–	–	75	75	75	75	75	75	75
SMR-5	–	–	25	25	25	25	25	25	25
BEE	20	30	30	30	63	75	75	63	100
ZnO	2	–	1	60	1	2	2	1	1
S	2.5	2.5	0.5	0.5	0.5	0.5	–	0.5	0.5
Altax®	1	1	–	–	–	–	–	–	–
PBNA	1	1	–	–	–	–	–	–	–
S300	–	2	–	–	–	–	–	–	–
MBTS	1	1	–	–	–	–	–	–	–
CBTS	–	–	2	2	2	2	2	2	2
Stearic acid	–	–	0.5	0.5	0.5	0.5	0.5	0.5	0.5
TMTDS	–	–	1	1	1	1	1	1	1
HAF	–	–	–	–	–	20	50	–	–
BJO	–	–	–	–	–	–	–	20	50
% active	15.8	21.8	22.2	18.0	37.5	37.5	32.4	33.5	39.1

[a]Ingredient code:
RSS #4 = natural rubber, rib smoked sheet, grade 4
PA-80 = 80% precured pale crepe natural rubber
SMR-5 = refined rib smoked sheet with castor oil added as a lubricant
BEE = butoxyethanol ester of 2,4-D
Altax = benzothiazole disulfide
PBNA = phenyl-β-napthylamine (antioxidant)
S300 = HAF, carbon black type
MBTS = mercaptobenzothiazole disulfide
CBTS = *N*-cyclohexyl-2-benzothiazyl sulfenamide
TMTDS = tetramethylthiuram disulfide
HAF = high abrasion furnace carbon black
BJO = phenolic microballons (BJO-0930, Union Carbide Corporation).[75]

14-B segment, a "bottom" suspender is created wherein the anchor end lies on or slightly in the bottom hydrosoil while the 14-B tail hangs vertically in the water.

Time release materials were prepared by encapsulation of a given pelletized formulation in a clay-acrylic binder, using carboxylated polyacrylics marketed by the B. F. Goodrich Chemical Company under the trade name Carboset. In the uncured state these materials pass slowly into solution when exposed to alkaline waters and disintegrate through exfoliating processes in acidic water. By varying the Carboset: clay composition the rate of solution or disintegration can be varied.

Laboratory bioassays were performed by enclosing the test plant in various size containers, conditioning under appropriate light and temperature conditions, and treating with the several formulations at several dosages. Table 4–6 is a compendium of water hyacinth test results in 5-gal containers, and Table 4–7 shows outdoor pool tests (752 gal). Averages of five or more tests are given, and mortality is adjudged visually by observation of browning and disintegration. Control mortality ranged from 5 to 10% over an 8-week test period.

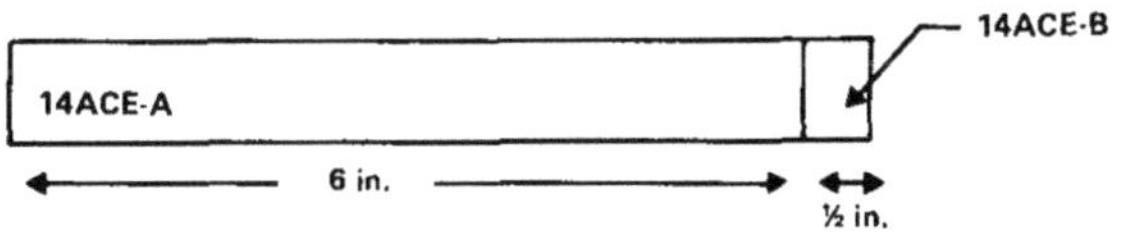

FIGURE 4–4. 14ACE-A/B design.

The phytozone concept received new credence when it was noted that the top suspenders, which entangle in the hyacinth root system, provide more effective control than either sinking or floating pellets. It is believed that under static water conditions floating materials release 2,4-D BEE in molecular form, creating a surface film so that relatively little downward movement occurs, and the target plant is inadequately exposed to the toxic agent in the susceptible root area.

Bioassay against *Elodea canadensis* showed that sinking formulations provided better control than that observed with suspending or floating materials. At least 4 weeks were required to reach 100% mortality in static laboratory tests; pool tests provided similar results.[464] Both suspending and sinking formulations evaluated against *Myriophyllum spicatum* provided nearly 100% mortality after 5 to 6 weeks' exposure. Probably 11ACE-E is the superior control formulation for Eurasian watermilfoil and alligator weed.

The 2,4-D BEE release rates were determined for a number of candidate formulations. Release curves, unlike those determined for organotin elastomers, indicated that the rate controlling factor is diffusion within the matrix structure and not dissolution at the rubber/water interface. Figure 4–5 depicts the 14ACE-B curve; note the

TABLE 4–6[464]

Water Hyacinth Laboratory Bioassay

Compound	Rubber dosage (ppm)	% BEE	Accumulative mortality by weeks (%)						
			Week 1	Week 2	Week 3	Week 4	Week 5	Week 6	Week 7
11ACE-E	50	21.8	15	20	38	55	90	100	–
11ACE-E	25	21.8	15	20	30	45	75	100	–
11ACE-E	5	21.8	5	8	12	18	30	45	70
14ACE-B	100	18.0	2	22	55	80	100	–	–
15ACE-B	250	37.5	10	18	50	70	100	–	–
15ACE-B	100	37.5	3	22	40	65	75	100	–
15ACE-B	50	37.5	2	5	10	12	18	25	55
15ACE-B	25	37.5	2	5	8	15	20	50	100
15ACE-B	5	37.5	0	2	5	10	15	30	65
15ACE-B	2.5	37.5	0	0	3	5	12	22	65
17ACE-B	50	37.3	5	20	30	40	45	70	100
18ACE-B	100	33.5	5	15	65	80	100	–	–
18ACE-D	250	39.1	5	15	65	100	–	–	–
18ACE-D	50	39.1	2	15	35	40	60	90	100

initial peak, indicating loss of surface and near-surface adsorbed toxicant molecules, and then the extended plateau where molecules lost are those that migrated from the matrix interior to the surface and then underwent dissolution. As expected, agent loss is proportional to the water-exposed surface area. Figure 4–6 shows 2,4-D BEE loss from two 11ACE-C pellets, one possessing twice the surface area of the other.

Vertical distribution of 2,4-D BEE released from sinking and floating materials as determined in a plexiglass static test chamber constructed so as to permit water withdrawal from various depths.[464] Results indicated that 2,4-D tends to layer out in still water. The floating pellet evaluation showed a heavy 2,4-D concentration in the top 12 in. (2.13 mg from a 30-g dosage at 48 hr) compared to only 0.024 mg at a 4.5-ft depth. In contrast, a sinking pellet provided higher concentrations at lower depths, although the concentration gradually increased in the upper water layers over a several-week test period.

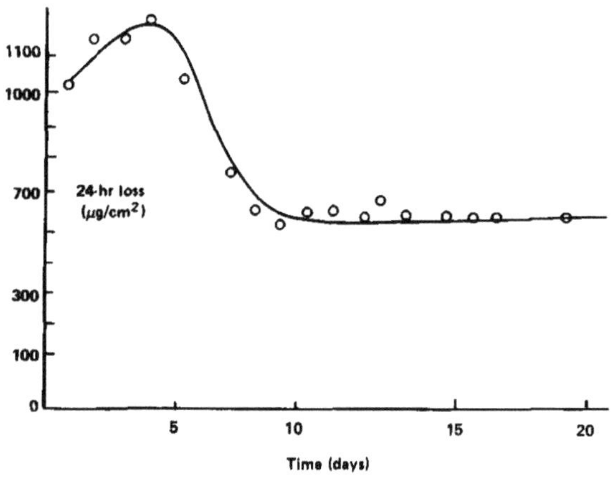

FIGURE 4–5. Release rate of 14ACE-B.[464]

Release longevity based on extrapolation of the loss curves is shown below:[464]

14ACE-B: 21 months
15ACE-B: 27 months
17ACE-B: 22 months
17ACE-C: 27 months
18ACE-B: 9 months.

Calculations assume a continuous loss at the equilibrium level until depletion. This does not mean that the given compound will be biologically effective throughout the computed loss period. Eventually the rate of diffusion must fall below that of dissolution and the loss curve will show a sudden drop to a rate below the phytotoxic level.

2,4-D BEE/elastomer formulations of the 11 ACE series were evaluated by the U.S. Department of Agriculture Laboratory, Ft. Lauderdale, Fla., and found effective against Eurasian watermilfoil at dosages as low as 1 ppm,[465] but the same materials were ineffectual, even at 5 ppm, against southern naiad, hydrilla, and the water hyacinth. Pool tests of various 2,4-D BEE/elastomers were conducted by the same laboratory during 1970–1971; results, shown in Table 4–8, were encouraging, and led to the decision to initiate field evaluations. It may be noted that 2.5 ppm of the controlled release material provided the same phytotoxicity as 20 ppm of the nonformulated material.

Several other organic aquatic herbicides have

TABLE 4–7[464]

Water Hyacinth Outdoor Pool Tests (752-gal Containers) for 30-day Period

Compound	Dosage (ppm)[a]	Type	E.R.[b]	Days to given % mortality 10%	25%	50%	80%	99%
14ACE-A/B	20[b]	Suspender	2	6	12	16	22	28
15ACE-B	20	Sinker	18	30+	–	–	–	–
11ACE-D	50	Floater	5	8	30+	–	–	–
Control 1	0	–	30+	–	–	–	–	–
Control 2	0	–	30+	–	–	–	–	–

[a]Dosage in parts per million of *formulation*, i.e., 20 ppm of 15ACE-B has a total 2,4-D content of 7.5 ppm.
[b]E.R. = epinastic response.

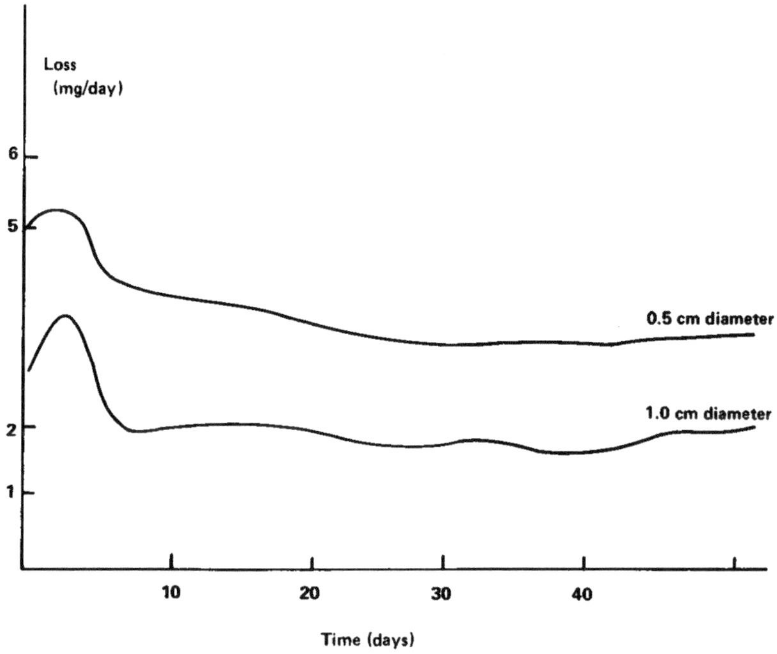

FIGURE 4–6. 11ACE-C release curves based on pellet size.[464]

TABLE 4–8[466]

Pool Evaluations (100 gal) of 2,4-D BEE Controlled Release Elastomers Against Aquatic Weeds

			Percent control (weeks post treatment)							
Formulation	Plant	Rate	1	2	3	4	5	6	8	10
14ACE-B	Milfoil	5.0 ppm	2	42	83	97	100	–	–	–
		2.5 ppm	2	38	68	96	100	–	–	–
11ACE-E	Milfoil	5.0 ppm	2	35	83	96	100	–	–	–
		2.5 ppm	2	17	77	97	100	–	–	–
2,4-D BEE	Milfoil	20 ppm	4	53	88	99	100	–	–	–
		10 ppm	3	40	70	97	100	–	–	–
Control	Milfoil	0	0	3	22	43	81	–	–	–
14ACE-A	Hyacinth	10 lb/acre	1	7	–	40	–	47	60	68
18ACE-B	Hyacinth	10 lb/acre	3	10	–	65	–	97	100	–
2,4-D BEE	Hyacinth	4 lb/acre	25	50	–	98	100	–	–	–
Control	Hyacinth	0	0	0	0	0	0	0	0	0

been incorporated in elastomeric formulations. These include 2,4-D acid (2,4-dichlorophenoxyacetic acid), diquat [6,7-dihydrodi pyrido (1,2-a:2,1-C) pyrazidiium dibromide], and silvex [2-(2,4,5-trichlorophenoxy) propionic acid]. Pertinent recipes are shown in Table 4–9.[467]

Formulations were immersed in distilled water and periodic aliquots taken and analyzed for herbicide residue. Water was changed after each sampling. Toxicant loss was established and release longevity extrapolated. The approximate release lifetimes are as follows:[467]

27ACE-C: 75 months
27ACE-D: 32 months
28ACE-C: 48+ months
28ACE-D: 48+ months
29ACE-C: 14 months

TABLE 4–9

Slow Release Aquatic Herbicide Formulations

Ingredient	Code (parts per hundred rubber)							
	27ACE-C	27ACE-D	28ACE-C	28ACE-D	29ACE-C	29ACE-D	30ACE-C	30ACE-D
Natural rubber	100	100	100	100	100	100	–	–
ZnO	1	1	1	1	1	1	–	–
CBTS	2	2	2	2	2	2	–	–
Stearic acid	0.5	0.5	0.5	0.5	0.5	0.5	0.5	0.5
TMTDS	1	1	1	1	1	1	1	1
HAF	15	15	15	15	15	15	–	–
Diquat	30	50	–	–	–	–	30	50
Silvex®	–	–	30	50	–	–	–	–
Sulfur	–	–	0.5	0.5	0.5	0.5	1.25	1.25
2,4-D acid	–	–	–	–	30	50	–	–
EPDM[a]	–	–	–	–	–	–	100	100
MBT[b]	–	–	–	–	–	–	0.75	0.75
Sulfads®[c]	–	–	–	–	–	–	0.75	0.75
SRF[d]	–	–	–	–	–	–	10	10

[a]Epcar® 5465 ethylene-propylene polymer (B. F. Goodrich Chemical Co.)
[b]Mercaptobenzothiazol.
[c]Dipentamethylene thiuram hexasulfide (R. T. Vanderbilt).
[d]Coarse carbon black; particle size averages 160 nm.

29ACE-D: 9 months
30ACE-C: 11 months
30ACE-D: 6 months.

Preliminary laboratory bioassays demonstrated feasibility at dosages between 3 and 30 ppm (pellet) against Eurasian watermilfoil, *Vallisneria, Elodea,* and *Cabomba.*[467] The diffusion-dissolution method appears to predominate as the release mechanism.

It is well known that various copper-containing materials are useful herbicides. Bartley has shown that daily applications of low concentrations of copper sulfate control various aquatic weeds in irrigation canals.[468,469] From the standpoint of crop injury, this method was felt to be desirable over the typical use of acrolein and other organic herbicides.

Copper-sulfate-bearing formulations compounded for snail control were evaluated against select water weeds and algae.[470] In replicated laboratory bioassays several aquatic weeds were treated with varying dosages of E-51 and copper ion applied at an ultralow daily dosage rate. Results with *Vallisneria americana* are illustrated in Table 4–10, Eurasian watermilfoil in Table 4–11, water hyacinth in Table 4–12, *Cabomba* in Table 4–13, and duckweed in Table 4–14. Green algae of mixed single cell species were used in bioassay with E-51; with results shown in Table 4–15.

Chronicity Phenomenon

Bioassay of controlled release 2,4-D BEE herbicides disclosed that the treatment dosage made little difference regarding the LT_{50} or LT_{90}.[464,471] Pool tests tended to confirm this observation.[466] Efforts to establish minimal dosage requirements prior to field evaluation showed that, if plant exposure is continuous, relatively low herbicide concentrations will destroy the target weed as readily as higher dosages. The Ct (concentration × time) concept apparently did not hold for ultralow continuous concentration of 2,4-D BEE. The so-called "chronicity"[a] phenomenon, discovered in 1970 was examined in detail over the 1971–1973 period.

A series of laboratory bioassays were performed wherein various herbicides were added daily to test containers at an ultralow to low treatment rate. In one series the herbicide concentration was allowed to accumulate, while in a parallel series a somewhat constant concentration was maintained by periodic water change prior to addition of the control agent. Table 4–16 illustrates results of bioassays with Eurasian watermilfoil, and Table

[a]The term, in all due modesty since it was coined by the author, is particularly apt. The phenomenon relates both to exposure *time* and chronic intoxication.

TABLE 4–10[470]

Laboratory Bioassay: E-51 vs. *Vallisneria americana*

Post-treatment day	Control (0 ppm)	0.03 ppm/day Cu^{++}	Mortality at a given time (percent) Copper-content pellet 1 ppm	10 ppm	50 ppm	100 ppm
5	2	16	0	8	16	17
10	4	28	13	27	34	41
20	9	52	44	67	55	70
30	10	77	62	89	61	85
40	10	84	68	94	77	96
50	9	91	69	100	95	99
60	8	100	74	–	100	100

TABLE 4–11[470]

Laboratory Bioassay: E-51 vs. *Myriophyllum spicatum*

Post-treatment day	Control (0 ppm)	0.03 ppm/day Cu^{++}	Mortality at a given time (percent) Copper-content pellet 1 ppm	10 ppm	50 ppm	100 ppm
5	3	10	7	6	9	7
10	6	32	18	14	17	19
20	9	50	45	34	41	45
30	4	60	33	33	42	52
40	6	60	33	40	39	55
50	7	70	40	51	53	72
60	7	80	74	74	74	79

TABLE 4–12[470]

Laboratory Bioassay: E-51 vs. *Eichornia crassipes*

Post-treatment day	Control	Mortality at a given time (percent) Copper-content pellet 50 ppm	100 ppm
5	5	10	3
10	5	15	10
20	18	37	40
30	34	60	75
40	40	83	80
60	42	85	89

TABLE 4–13[470]

Laboratory Bioassay: E-51 vs. *Cabomba caroliniana*

Mortality at a given time (percent)

Post-treatment Day	Control (0 ppm)	0.03 ppm/day Cu^{++}	Copper-content pellet 10 ppm	50 ppm	100 ppm
5	3	10	23	37	26
10	10	30	33	60	40
20	30	40	66	85	78
35	40	65	93	100	100
40	30	90	100	–	–
45	25	100	–	–	–
50	25	–	–	–	–

TABLE 4–14[470]

Laboratory Bioassay: E-51 vs. *Lemna* (Duckweed)

Mortality at a given time (percent)

Post-treatment day	Control	Copper-content pellet 50 ppm	100 ppm
5	0	15	20
10	0	48	65
15	15	78	73
20	15	93	85
25	15	100	95
35	20	–	100

4–17 shows *Vallisneria* tests results. Table 4–18 summarizes the chronicity study performed at the Creative Biology Laboratory under sponsorship by the U.S. Army Corps of Engineers (Contracts DACW 73-72-C-0031 and DACW 73-73-C-0042).

The extremely low effective herbicide concentrations observed indicated that the use of controlled release herbicides is feasible. From a practical standpoint it is impossible to achieve and hold a conventional high concentration using slow release pellets, i.e., reliance on acute intoxication would preclude a controlled release approach to the aquatic weed problem. Treatment would literally require hundreds of pounds per acre-foot. However, the possibility of inducing a terminal chronic intoxication via long-term weed exposure to ultralow toxicant concentrations indicated that control using 14ACE-B and other herbicide/elastomer systems should be possible. Table 4–19 presents extrapolated LdD_{90}s (lethal daily dose for 90% mortality in a given population) for a number of herbicide/plant systems, illustrating the very low concentrations hypothesized as providing adequate weed control. It is not likely that continuous dosages in excess of 0.05 to 0.1 ppm would be economically practical. Thus, Table 4–19 permits the selection of potential controlled release herbicide candidates if the desired agent, release rate, etc. are known.

If the Ct relationship were valid, then the ratio between concentration and LT_{100} would be 1:0.1:0.01:0.001 ppm = 1:10:100:1000, which it is not. Janes has demonstrated the ratios that are observed in laboratory bioassays.[473] For instance, using diquat against Eurasian watermilfoil, the 1:0.1:0.01:0.001 ppm equivalency (LT) is 1:1:1.1:1.8. Fenac-*Cabomba* bioassays provide 1:1.9:2.5:2.8, etc.

The chronicity phenomenon is apparent in much of the aquatic herbicide literature, but data are not interpreted in this light.[406,468]

The mechanism underlying the chronicity effect is unknown. It is not an "expected" phenomenon. Indeed, it has long been known that many herbicides are growth regulators, and when applied in very small dosages to various plants enhance growth.[474] Experimental work keyed to the elucidation of the mechanism involved is in progress at the author's university laboratory, and presumably in other research establishments.

The basic hypothesis is that ultralow herbicide concentrations fail to trigger the exposed plant's

TABLE 4–15[470]

Laboratory Bioassay: E-51 vs. Green Algae

E-51 dosage (ppm)	Algae mortality at a given time (%)								
	Day 22	Day 24	Day 29	Day 30	Day 34	Day 39	Day 44	Day 51	Day 54
200	37	45	60	57	90	100	–	–	–
100	33	35	67	80	91	90	100	–	–
50	15	28	32	50	50	53	57	73	87
25	10	33	35	40	47	47	53	87	100
Control	0	4	12	10	10	13	10	10	10

TABLE 4–16[472]

Eurasian Watermilfoil Bioassay

	Accumulative		Constant	
Herbicide	Dosage (ppm/day)	Time to 100% mortality (days)	Dosage (ppm/day)	Time to 100% mortality (days)
Diquat	1.0	14	1.0	11
	0.1	14	0.1	19
	0.01	19	0.01	16
	0.001	38	0.001	32
2,4-D BEE	1.0	15	1.0	14
	0.1	17	0.1	22
	0.01	18	0.01	27
	0.001	22	0.001	–
Fenac ®[a]	1.0	21		
	0.1	43		
	0.01	48		
	0.001			
Silvex®	1.0	14	1.0	15
	0.1	43	0.1	
	0.01	21	0.01	
	0.001			
2,4-D acid	1.0	20	1.0	20
	0.1		0.1	
	0.01		0.01	
	0.001		0.001	
2,4-D BE[b]	1.0	20	1.0	7
	0.1	22	0.1	10
	0.01	24	0.01	10
	0.001		0.001	
2,4-D OA[c]	1.0		1.0	7
	0.1		0.1	12
	0.01		0.01	17
	0.001		0.001	

[a]Sodium salt of 2,3,6-trichlorophenylacetic acid.
[b]Butyl ester of 2,4-dichlorophenoxyacetic acid.
[c]Oleylamine of 2,4-D.

TABLE 4–17[473]

Vallisneria Bioassay

Herbicide	Accumulative: Dosage (ppm/day)	Accumulative: Time to 100% mortality (days)	Constant: Dosage (ppm/day)	Constant: Time to 100% mortality (days)
Diquat	1.0	6	1.0	4
	0.1	12	0.1	20
	0.01	31	0.01	64
	0.001	60	0.001	–
2,4-D BEE	1.0	18	1.0	42
	0.1	45	0.1	45
	0.01		0.01	51
	0.001	56	0.001	51
Fenac®	1.0	34	1.0	19
	0.1	35	0.1	23
	0.01	56	0.01	42
Silvex®	1.0	34	1.0	41
	0.1	60	0.1	44
	0.001	65+	0.01	58
2,4-D acid	1.0	12	1.0	22
	0.1	59	0.1	90+

TABLE 4–18[473]

Chronicity Phenomenon

Agent	Aquatic plant[a]: Watermilfoil	*Vallisneria*	*Cabomba*	*Elodea*	Water hyacinth	Southern naiad
2,4-D BEE	+	+	+	+	+	+
2,4-D BE	+	+	N	+	–	N
2,4-D DMA	N	N	N	N	–	N
2,4-D OA	+	N	N	N	+	N
Diquat	+	+	+	+	+	+
Silvex®	+	+	–	+	–	N
Fenac®	+	+	+	–	N	+
2,4-D acid	+	–	+	+	N	N
Acrolein	N	N	N	N	–	N
Hydrothol®[b]	N	N	N	N	–	N
Dichlobenil[c]	N	N	N	+	N	+
Endothall[d]	–	–	+	–	N	+
Fenuron[e]	N	–	+	+	N	N

[a]+ = chronicity phenomenon present; – = chronicity phenomenon not observed; N = not evaluated.
[b]di (*N,N*-dimethylaklylamine) salt of endothall.
[c]2,6-Dichlorobenzonitrile.
[d]7-Oxabicyclo (2.2.1) heptane-2,3-dicarboxylic acid.
[e]3-Phenyl-l,l-dimethylurea.

TABLE 4–19[473]

Estimated Accumulative Dosage Required for 30- and 60-day 90% Weed Kill

Herbicide	Exposure period (days)	LdD_{90} [a] (ppm/day) *Vallisneria*	*Cabomba*	Milfoil	*Elodea*	Southern naiad
2,4-D BEE	30	0.006	0.2	0.0004	0.0006	–
	60	0.002	0.008	0.0003	0.0003	–
Diquat	30	0.006	0.08	0.0008	0.03	0.01
	60	0.0004	0.015	0.0005	0.0006	0.001
Silvex®	30	0.9	0.13	0.007	0.05	–
	60	0.09	0.08	0.002	0.008	–
Fenac®	30	0.07	0.0008	0.4	0.5	0.09
	60	0.02	0.0004	0.008	0.08	0.04
2,4-D acid	30	0.09	–	0.6	0.0009	–
	60	–	–	0.4	0.0005	–
Endothall	30	0.09	0.2	–	–	0.11
	60	–	–	–	–	0.009
Fenuron	30	–	0.2	–	0.08	–
	60	–	0.09	–	0.007	–

[a] Lethal daily dose for 90% mortality in a given population.

defense mechanisms against intoxication.[475] Chemoreception in plants is well known, and even the presence of a nervous system has been suggested.[476] Herbicide molecules must enter the subject plant and be translocated to the site of killing action unchanged, or, if chemically altered, still toxic, in amounts sufficiently large enough to destroy the plant. In order to accomplish this the given material must overcome various defenses and detoxifying factors raised by the subject. Although the mechanisms involved can be complex, considerable knowledge concerning terrestrial plant/herbicide interactions is available. Unfortunately, very little work has been done with aquatic plants, and the validity of drawing analogies is dubious.[477]

In the poisoning of a terrestrial plant the herbicide must enter the treated portion (leaf, root, etc.) and move to the site of action. In entering the leaf, the herbicide penetrates the cuticle via an aqueous or lipoid route and enters the epidermal cells. It then migrates across the mesophyll and enters the phloem via intercellular protoplasmic connections called plasmodèsmata. Once in the phloem the herbicide moves by mass flow to areas of active growth, e.g., meristems. In entering the root, the herbicide penetrates the epidermal cell and moves to the xylem, usually via cell walls. Once in the xylem it moves upward in the plant via the transpiration stream and accumulates in areas of water utilization, mainly leaves. Various species of terrestrial plants have differential properties that can influence the ability of a herbicide to kill one species and not another. These mainly include differential rates of absorption, translocation, and degradation of the herbicide molecule into inactive forms. In general these defenses are not induced by the herbicide, however some defenses may be induced, e.g., closing of the stomata. It has been suggested that biological stress may induce a resistance to that stress in the subject plant by raising chemical or metabolic barriers.[478]

The defenses against phytotoxic activity raised by aquatic plants are little known. Present studies in the author's laboratory have indicated, so far, that the epidermal action of 2,4-D BEE and copper ion against *Elodea canadensis* and *Vallisneria americana* shows little difference in regards to one-time exposure to a "conventional" dose, or continuous exposure to an ultralow dosage. In both instances congealing of the protoplasm with loss of differentiation of cell components occurs. The rate of action is greater at the higher dosage, as could be expected.[479]

Stochastic considerations indicate that a concentration threshold exists, below which herbicidal effects on the plant cell are not deleterious.[480] Below this limiting value toxicant reaction occurs and cells are injured; however, the normal rate of replacement of cell constituents overcomes the herbicide-induced destruction. Determination of the threshold concentration, if such exists, would allow the computation of a null dosage, or long-term concentration that is not biologically significant. Just above this point the plant would be injured to some extent or its growth or other vital functions (reproduction, etc.) retarded.

FIGURE 4–9. Eurasian watermilfoil test area prior to controlled release herbicide application.[481] (Photograph courtesy of U.S. Army Corps of Engineers, New Orleans District.)

FIGURE 4–10. Eurasian watermilfoil test area shown in Figure 4–9 4 months post treatment. Circular clear areas are test sites. Note luxurious growth in surrounding untreated areas.[481] (Photograph courtesy of U.S. Army Corps of Engineers, New Orleans District.)

Field Evaluation

Field test applications of controlled release aquatic herbicides have been made in the New Orleans area. In accordance with pool test determinations, 14ACE-A/B suspenders were applied to heavy water hyacinth mats in 1/2-acre plots at a rate of approximately 16 lb/acre and 32 lb/acre active 2,4-D BEE (Figure 4–7).* Plants were slowly destroyed, and, by the 7-month inspection, the sites were clear (Figure 4–8).* Control areas remained covered by dense mats. After 18 months the sites remained free of the subject weed.

A lake containing a heavy infestation of Eurasian watermilfoil was treated with a sinking 2,4-D BEE formulation,[481] applied to nine 60-ft-diameter circular test plots. Within 4 months seven of the nine sites were clean of vegetation[481] (Figures 4–9 and 4–10).

Plastic Formulations

Development

In 1968 scientists at the U.S. Army Environmental Health Agency (Edgewood Arsenal, Md.) successfully incorporated 2,4-D in plastic matrices.[482] Polyvinyl chloride (PVC) plastisol formulations incorporating 2,4-D BEE, 2,4-D butyl ester, simazine, endothall acid, and dichlobenil were prepared and evaluated against Eurasian watermilfoil.[483] Plastosols were prepared by mixing 100 g of Geon® 135 (B. F. Goodrich Chemical Co., Cleveland, Ohio) in 100 g of dioctylphthalate (DOP) plasticizer, adding the herbicide and curing in pellet molds at 125 to 130°C for 30 min. Recipes are shown in Table 4–20.

N-butyl ester of 2,4-D, endothall, dichlobenil, and simazine loaded pellets were bioassayed with the target plant. In five successive 2-week test periods 100% mortality was achieved, except for the simazine formulation, which showed no phytotoxicity.

Release rates of C^{14}-labeled 2,4-D BEE varied with agent loading; after 20 weeks, Young and Nelson reported the following losses:[483]

*For Figures 4–1 and 4–2 see Page 90.

10% active: 25.92% of total 2,4-D
20% active: 31.29% of total 2,4-D
30% active: 57.20% of total 2,4-D.

Analysis of the data according to the procedures of Higachi[91] and Quacenbos[484] indicated that 2,4-D migration rate within the polymer matrix was the release-rate controlling factor.

Since 2,4-D BEE shows no appreciable solubility in PVC, the plasticizing element apparently acts as a carrier as it slowly diffuses towards the surface. DOP loss after immersion was not monitored in the described study.

Pool tests were conducted on a number of PVC/herbicide formulations,[465] and bioassay results of PVC/2,4-D BEE against several weed targets at varying dosages are shown in Table 4–21.

Laboratory tests at Fort Lauderdale indicated that PVC/endothall was effective against *Hydrilla;* PVC/Fenac provided good control of southern naiad[465] and showed long-term release characteristics.[485]

Nelson concluded that PVC/2,4-D BEE materials, although showing long-term release and control of aquatic weeds, were not superior to simple 2,4-D on attaclay granules, and further work ought to be discontinued.[485] The 14ACE natural rubber/2,4-D BEE formulations demonstrated better control and a longer effective lifetime than either PVC or clay granule carriers. PVC/ Fenac release life exceeded 46 weeks at effective control levels.[485,486]

Fenac acid and its sodium salt have also been incorporated in polyethylene matrices.[487] Thermal fusion and herbicide loading significantly affected the release rate. Fenac salt migration rate within the pellet was found to be release-rate controlling.[90,487]

Fenac salt and copper sulfate have been formulated within biodegradable ethylcellulose matrices,[488] with the former releasing very rapidly upon immersion and the latter showing poor release due to the formation of a surface precipitate.

Controlled release methods other than those dependent upon leaching or diffusion-dissolution phenomena have been investigated. 2,4-D has apparently been polymerized and upon environmental exposure exhibits a slow depolymerization and release of the phytotoxic monomer,[489] but the author and his colleagues have not been able to verify this work.[490] Also, 2,4-D chloride has been reacted with unmodified, cyanoethylated, and cross-linked corn starches to produce 2,4-D esters of high acyl content.[491] Compounds are insoluble, but undergo hydrolysis that releases 2,4-D and 2,4-D esters at rates indicative of potential value as controlled release herbicides.

TABLE 4–20[483]

Plastic-based Herbicide Formulations

Herbicide	% herbicide	% Geon® 13	% DOP	% inert
2,4-D BE	12.5	43.75	43.75	–
2,4-D BE	25.0	37.5	37.5	–
2,4-D BE	50.0	25.0	25.0	–
2,4-D BE	66.0	34.0	–	–
Simazine	24.0	35.0	35.0	6.0
Endothall acid	24.0	35.0	35.0	6.0
Dichlobenil	25.0	25.0	25.0	25.0
2,4-D BEE	30.0	35.0	35.0	–
Fenac® acid	30.0	35.0	35.0	–

TABLE 4–21[465]

Evaluation in Outdoor Pools of 2,4-D BEE/PVC Formulation, 30% Acid Equivalent

Treatment rate (ppm)[a]	Target plant	% control at given time					
		2 weeks	4 weeks	6 weeks	8 weeks	10 weeks	22 weeks
0.25	Watermilfoil	13	23	20	38	25	5
0.5	Watermilfoil	20	38	53	68	53	38
1.0	Watermilfoil	38	90	97	97	68	35
2.0	Watermilfoil	45	85	52	52	52	50
5.0	Watermilfoil	74	98	99	100	99	100

[a]Conventional treatment against watermilfoil with 2,4-D BEE would be about 1.5 ppm.

In another meritorious advance in controlled release technology, Harris, Feld, and Post prepared various vinyl polymers containing hydrolyzable herbicides (2,4-D and Silvex) as pendant side chains.[492] High molecular weight materials were realized and some polymers were synthesized having in excess of 80% herbicide.[493]

Theoretical Considerations

Controlled release based upon diffusion dissolution has been mathematically defined in Chapter 2. However, in considering herbicides formulated in plastic materials the presence of true solution is not likely. Consequently, diffusion if it occurs, is not driven by solution pressure, and other release mechanisms must necessarily be present.

One approach is based upon the behavior of a given polymer in water. Allan et al., have considered release from formulations composed of a pesticide in water-soluble polymer.[494] Although a definite release end point is reached, and release may well be constant from such a material, water-soluble polymers can be expensive, usually are sensitive to water pH and mineral content, and would probably not provide *long-term* controlled release.

Beasley and Collins have approached controlled release wherein the binding element is a water degradable polymer.[495] Considering a spherical pellet geometry, if the rate of release is proportional to the surface area and falls to zero at the end of a time interval, then

$$v = -\frac{3V}{t}\left(1 - \frac{T}{t}\right)^2$$

where v = pellet volume at end of release period, V = original volume, t = release time, and T = time.

Assuming that the herbicide released has an environmental half-life, $T_{1/2}$, both release and decay of the agent can be illustrated in one equation:[496]

$$C = \frac{3V}{T}\int_0^t \left(1 - \frac{t'}{T}\right)^2 e^{-\lambda(t-t')}\,dt'$$

where $t \leqslant T$, C = concentration, and $\lambda = \ln\frac{2}{T_{1/2}}$ (first order rate constant), solving

$$C = \frac{3V}{(\lambda T)_3}\,[\lambda^2(T-t)^2 + 2\lambda(T-t) + 2 - e^{-\lambda t}(\lambda^2 T^2) + 2\lambda T + 2]$$

when $t \leqslant T$ or

$$C = CTe^{-\lambda(t-T)}$$

when $t > T$.

In order to achieve a constant loss rate the pellet volume would have to remain unchanged through the period of release. Then

$$v = -\frac{V}{T}$$

and

$$C = \frac{V}{T}\int_0^t e^{-\lambda(t-t')}\,dt' = \frac{V}{\lambda T}\,(1 - e^{-\lambda t})$$

when $t \leqslant T$ or

$$e = C(T)e^{-\lambda(t-T)}$$

when $t = T$.

In his theoretical treatment Collins considers the relative efficiency of conventional release, controlled release from a degrading pellet, where the volume decreases continuously, and a pellet of constant volume.[496] It is assumed that a given herbicide concentration in water is needed to provide weed control. Above this concentration, C*, control is present but the chemical is wasted, while below this value the desired effect is not present. Efficiency is the ratio of the minimum requirement to that used. For the ideal case where C* is applied initially and the losses, $dC^* = C^* \ln 2\, d\bar{t}$, are made up continuously, the total amount of herbicide needed for protection over a period $\bar{t}$ is

$$C_0 = C^*(1 + \bar{t}\ln 2),$$

which is 100% efficient use of pesticide.

A conventional herbicide with multiple applications spaced at $T_{1/2}$ intervals requires

$$C_0 - C^+(1 + \bar{t})$$

Also, a conventional one-time application requires

$$C_0 = C^* e^{\bar{t} \ln 2}$$

This is a wasteful approach in that the initial concentration must be very high in order to maintain a control concentration for a length of time.

In an expanded theoretical treatment Collins examined the consequences of the several controlled release methodologies for rate of release, active concentration ensuing in the presence of first order degradation, and efficiency.[497] He showed that the limiting efficiencies are

Microcapsules ≅ 38%
Constant surface area (slabs, etc.) ≅ 30%
Decreasing surface area (sphere, etc.) ≅ 22%.

In controlled release systems lacking solubility of the agent in the polymeric matrix, loss is dependent, in general, upon a leaching mechanism. Unlike diffusion-dissolution where the agent molecule diffuses to the water/polymer interface, the exterior water must penetrate the matrix and contact a fairly immobile agent molecule; dissolution occurs and the toxicant passes via watery paths out of the matrix following a solution gradient. Excluding the use of a third-phase carrier material, agent migration must depend upon the porosity of the binding element. A quantitative consideration of the number of pores or total pathways through the matrix is not, in itself, adequate. The tortuosity of the pore structure plays a key role in water transport.

Harris et al. have shown that exposing a pressed pellet to heat decreases porosity and thus decreases the loss of Fenac from a polyethylene matrix, as shown in Figure 4–11.[498] It was further shown that increasing the agent concentration in the binding pellet increases loss rate and that the effect of water turbulence is likely minimal regarding loss rate. Also, additives such as sucrose which will substantially increase polyethylene porosity significantly increased loss rate. Harris et al. concluded that the Higuchi leaching equation represented thus:

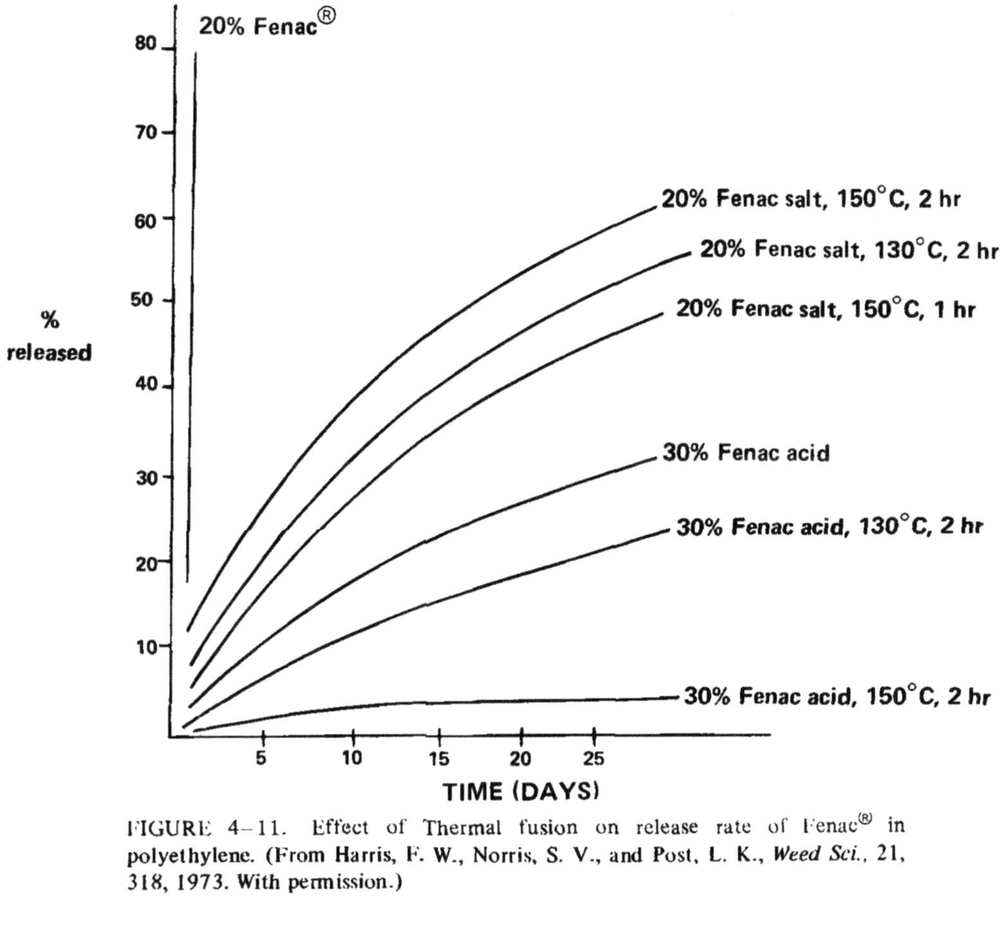

FIGURE 4–11. Effect of Thermal fusion on release rate of Fenac® in polyethylene. (From Harris, F. W., Norris, S. V., and Post, L. K., *Weed Sci.*, 21, 318, 1973. With permission.)

$$Q = \sqrt{D \frac{\epsilon}{\tau} (2A - \epsilon C_s)C_s t}$$

where Q = quantity released, D = diffusion coefficient of agent in the release medium (water), ϵ = porosity of matrix, τ = tortuosity of matrix, A = amount of agent in the matrix per unit volume, and C_s = agent solubility in release medium substantially defines the experimental data for the Fenac/polyethylene system.[498]

Pellet porosity can be determined by measuring the density of each component, calculating the weight percentage of each, and measuring the pellet volume. Assuming that prior to immersion the pores contain trapped air, then the volume of that air can be computed as follows:[488]

$$V_{air} = V_{pellet} - \frac{\text{component weight}}{\text{component density}}$$

$$E_{air} = \frac{V_{air}}{V_{pellet}}$$

However, porosity increases as the agent is slowly leached out, so the agent loss volume must also be considered.

Tortuosity is a much more complex parameter to determine. The Desai et al. equation

$$\tau = \frac{4\epsilon^2 C_0{}^2 D}{\pi(\text{slope})^2}$$

where ϵ = porosity, C_0 = initial agent solubility in matrix, and D = diffusion coefficient of agent in release medium can be used where the slope of the Q vs. $t^{1/2}$ plot is used.[499]

Harris views the migration of a herbicide from a polymeric matrix as being governed by the principles underlying transfer of matter,[90] i.e., the migrating herbicide meets two resistances: (1) resistance to movement within the matrix, and (2) resistance in escaping from the matrix into the matrix/water interface. Rate of loss of the agent depends upon the rate at which either or both of these resistances are overcome. In systems studied to date, including antifouling and molluscicidal rubber, one resistance is loss rate controlling.

If diffusion or migration within the matrix is rate controlling, then the quantity of herbicide lost to the external medium is proportional to the square of time. If surface resistance (dissolution) is rate controlling, the system follows first order kinetics, and loss is proportional to time in accordance with the following equation:[484,500]

$$Q = A(1 - e^{kt})$$

where Q is the amount of herbicide released after time t per unit exposed area, A is the amount of herbicide present in the matrix per unit volume, and k is the rate-loss constant.

Harris has categorized the rate-controlling step into four general types and presented the pertinent equations:[501]

1. Consider diffusion through a nonporous polymer membrane and the rate of release for a spherical geometry:

a. For steady state conditions,[501]

$$\frac{dM_t}{dt} = 4\pi D(K_1C_1 - K_2C_2)\left(\frac{r_0 r_1}{r_0 - r_1}\right)$$

where $\frac{dM_t}{dt}$ = net flow of permeant at time t (g/sec), D = diffusion coefficient of permeant in the membrane (cm²/sec), C_1 = concentration of permeant in reservoir fluid (g/cm³), C_2 = concentration of permeant in the surrounding medium (g/cm³), K_1 = partition coefficient between membrane and the reservoir fluid, K_2 = partition coefficient between membrane and surrounding medium, r_0 = outside radius of sphere, and r_1 = inside radius of sphere.

Diffusion coefficients can be determined using the time-lag equation developed by Daynes[502] and modified by Barrer.[503]

b. For nonsteady state conditions,[504]

$$\frac{dM_{1t}}{dt} = -\frac{M_\infty ADK}{hV_1} e^{\frac{-ADKt}{hV_1}}$$

where M_{1t} = mass of permeant in the reservoir at time t, M_∞ = total mass of the permeant, A = membrane area, and V_1 = reservoir volume.

2. Next, considering diffusion through a nonporous, solid polymer matrix, we have:

a. dissolved permeant,[501,505]

$$\frac{dM_t}{dt} = 3M_\infty\left(\frac{D}{r_0{}^2\pi t}\right)^{1/2} - \frac{3D}{r_0{}^2}$$

when $\frac{M_t}{M_\infty} < 0.4$ and

$$\frac{dM_t}{dt} = \frac{6DM_\infty}{r_0^2} e^{\left(\frac{\pi^2 Dt}{r_0^2}\right)}$$

when $\frac{M_t}{M_\infty} > 0.6$.

b. If the permeant[91,501,506] is dispersed and release is from a slab geometry,

$$\frac{dM_t}{dt} = \frac{A}{2}\left[\frac{DC_s(2C_0 - C_s)}{t}\right]^{1/2}$$

where C_0 = initial concentration of permeant release from a spherical goemetry:

$$\frac{dM_t}{dt} = \frac{3C_s DM_\infty}{r_0^2 C_0}\left[\frac{\left(1 - \frac{M_t}{M_\infty}\right)^{1/3}}{1 - \left(1 - \frac{M_t}{M_\infty}\right)^{1/3}}\right]$$

3. In the mechanism of diffusion through a porous, solid polymer matrix[501] the agent is leached from the matrix by a permeating fluid.

$$\frac{dM_t}{dt} = \left[\frac{D_t \epsilon (2C_0 - \epsilon C_f) C_f}{4 \tau t}\right]^{1/2}$$

where D_f = diffusion coefficient of the agent in the permeating fluid, C_f = solubility of agent in permeating fluid, ϵ = porosity of matrix, and τ = tortuosity factor.

4. In the loss of agent through dissolution of the matrix,[501,507]

$$\frac{dM_t}{dt} = \frac{3M_\infty\left(1 - \frac{t}{t_\infty}\right)^2}{t_\infty}$$

where t_∞ = time for which release is complete.

Environmental Impact

Application of herbicides to watercourses has some degree of effect on nontarget biota, but studies conducted over the past 5 years indicate that the contamination problem is minor.

2,4-D residues and effects in water bodies have been investigated by a number of scientists. The biotic factors influencing 2,4-D residues are rate of treatment, temperature, dilution by depth, and water flow rate. Reduction of 2,4-D residues is mainly through biodegradation[508] and plant metabolism.[509,510] Photodegradation and hydrosoil absorption are present but likely of little consequence. Gangstad and Averitt have shown in a number of pond, lake, and bayou tests that at a standard treatment rate, 4 lb 2,4-D acid equivalent per surface acre, the average dissipation rates are[511]

1. 58 ppb decrease in residue for each 2-ft increase in water depth
2. 115 ppb decrease in residue for each 10°F temperature increase above 60°F
3. 53 ppb decrease in residue for each 7-day interval of time after treatment.

Novosad and Nailon have measured 4 lb 2,4-D acid per acre residues as 189 ppb 1 day after spray treatment and 4 ppb after 29 days.[512]

Steward has evaluated controlled release formulations under laboratory conditions against various piscine species in short-term studies, results of which are depicted in Table 4–22.[486] Although confirming field tests are necessary, it would appear that both 2,4-D and Fenac are significant hazards to fish where the detoxification factors, incident to natural bodies of water, and not present.

Copper ion at the ultralow concentrations required for aquatic plant control does not appear to constitute any significant hazard to nontarget biota. The author has examined copper as a one-time dose in laboratory microenvironments consisting of *Lebistes reticulatus* as the nontarget fish, and algae, *Vallisneria*, and *Cabomba* as the target weeds.[470] Results are shown in Table 4-23.

Both copper and copper-triethanolamine complex (Cutrine®) at 1 ppm are considered to have low hazard potential in irrigation water.[445] Copper is toxic to fish and shellfish at levels of 100 ppm and above,[513-515] so controlled release copper at levels necessary for aquatic weed control ought not to exceed 0.1 ppm per day. Short-term exposure at such low rates, hardly above natural copper background in many areas, will not affect fish. Continuous exposure and copper accumulation in fish tissue could conceivably result in

TABLE 4–22[486]

Toxicity of Controlled Release Aquatic Herbicide Formulations to Various Fish Species

Fish	Formulation	Agent	Treatment rate (ppm)	Number of fish used	Mortality (%) 24 hr	48 hr	96 hr	168 hr
Bass[a]	(30% PVC)	Fenac®	5	78	0	0	0	–
	(30% PVC)	Fenac	10	78	1	1	1	–
	(Control)	0	0	32	0	1	1	–
Bluegill[b]	11 ACE-E	2,4-D BEE	5	20	5	35	–	–
	11 ACE-E	2,4-D BEE	10	20	8	23	–	–
	LV-4/PVC	Fenac	5	20	5	23	–	–
	LV-4/PVC	Fenac	10	20	3	23	–	–
	Conventional	2,4-D BEE	10	40	5	25	–	–
	Conventional	2,4-D BEE	20	39	3	13	–	–
	14 ACE-B	2,4-D BEE	5	40	3	18	–	–
	14 ACE-B	2,4-D BEE	10	39	3	13	–	–
	Conventional	Fenac	5	40	3	18	–	–
	Conventional	Fenac	10	40	8	33	–	–
	Encapsulated	2,4-D alkanolamine	5	40	5	40	–	–
	Encapsulated	2,4-D alkanolamine	10	40	0	23	–	–
	Control	0	0	95	12	65	–	–
Catfish[c]	11ACE-E	2,4-D BEE	5	20	0	10	10	10
	11ACE-E	2,4-D BEE	10	20	10	18	18	18
	LV-4/PVC	Fenac	5	40	8	8	8	8
	LV-4/PVC	Fenac	10	40	0	0	0	0
	2,4-D/PVC	2,4-D BEE	10	40	5	10	13	13
	2,4-D/PVC	2,4-D BEE	20	40	8	20	23	23
	14ACE-B	2,4-D BEE	5	40	0	0	0	0
	14ACE-B	2,4-D BEE	10	40	0	13	13	13
	Conventional	Fenac	5	40	5	13	15	25
	Conventional	Fenac	10	40	18	18	23	23
	Encapsulated	2,4-D alkanolamine	5	40	13	13	13	13
	Encapsulated	2,4-D alkanolamine	10	40	0	0	0	0
	Check	0	0	96	6	16	18	22

[a]*Micropterus punctulatus.*
[b]*Lepomis machrochirus.*
[c]*Ictalurus punctatus.*

chronic poisoning. Lin and his colleagues have discovered an ameliorating anomaly in rainbow trout, one long suspected to be present in molluscs exposed to copper: There is no apparent relationship between exposure duration and copper residue in fish tissue,[516] nor is there a relationship between the exposure concentration and tissue residue.

Copper is a synergist with diquat,[517-519] endothall,[520] and 2,4-D over a narrow concentration range.[521] Thus, one can foresee the combined use of controlled release copper with other herbicides, which could result in less overall contamination of treated waterways.

It is apparent that the use of controlled release aquatic herbicides will result in decreasing the total amount of chemical introduced, assuming reliance on the chronicity factor.[522] However, effects on nontarget biota subjected to long-term exposure are unknown, and considerable study will be needed before reasonable statements regarding environmental impact can be made. The basic question can be stated as follows: Which is the greater hazard to the biota – one-time exposure to a massive dose with detoxification in a relatively short period of time, or continuous exposure to an ultralow toxicant concentration? Just as we hypothesize that aquatic weeds succumb at very low concentrations due to inactivation of natural defense mechanisms, might not the

TABLE 4–23[470]

Microenvironmental Evaluation of Copper

	Post-treatment day	Mortality dosage (percent) 200 ppm	100 ppm	50 ppm	25 ppm	Control (0 ppm)
Lebistes reticulatus	1	3	0	0	0	0
	2	7	14	10	7	0
	3	100	100	100	21	3
	6	–	–	–	100	7
	8	–	–	–	–	10
	10	–	–	–	–	10
	13	–	–	–	–	10
	27	–	–	–	–	10
	46	–	–	–	–	10
Algae	1	0	0	0	0	0
	6	80	70	53	25	0
	8	100	94	70	60	0
	10	–	100	77	100	0
	13	–	–	100	–	0
	27	–	–	–	–	0
Cabomba	1	0	0	0	3	0
	2	10	14	14	10	3
	3	30	30	27	10	3
	6	53	70	37	30	10
	8	73	100	84	45	13
	10	80	–	90	70	13
	13	100	–	100	100	17
Vallisneria	1	17	17	10	0	10
	2	33	30	24	7	10
	3	37	30	33	30	10
	6	43	40	53	50	17
	10	63	63	53	63	13
	15	73	67	67	60	13
	27	77	77	80	60	13
	30	84	84	88	60	13
	33	90	88	90	63	13
	43	100	100	100	67	13

same situation hold for fish, daphnia, phytoplankton, and so on?

Finally, in addressing the use of aquatic herbicides one must face the fact that man is balancing chemical contamination against natural contamination. Most of the problem weeds are not native to the area of infestation, but are foreign imports that find themselves in a favorable environment lacking natural enemies. Alligator weed is but one prime example. If this noxious alien is not controlled chemically the harm done does not affect man alone. Decaying alligator weed mats reduce the oxygen content of water and increase hydrogen sulfide output, both factors destroying fish.[523] The control literature is replete with analogous examples. Thus, we can pose another question: Is the degree of contamination arising from the use of chemical control agents (conventional or controlled release application) any greater than that resulting from noncontrol of the same weed infestation?

CONTROLLED RELEASE TERRESTRIAL HERBICIDES

Introduction

The extension of controlled release methodology into the area of terrestrial weed control is the next logical step in the continual development of this technology. The global imperative to increase agricultural yield, when balanced against the environmental threat of increasing herbicide use, makes it mandatory that methods of reducing chemical usage and persistence be found and exploited. Food output would decrease 25% or more if herbicides were removed from the commercial market, an intolerable situation to humanity, which is now very close to the precipice of mass starvation. Yet increased use of agricultural chemicals may create imbalances in the earth's biosphere that could constitute a grave menace to human life.

It is believed that the delivery of terrestrial herbicides via controlled release systems will serve both needs. Collins succinctly observed, "Controlled release offers a middle ground between the economic needs of farmers for persistence and the justifiable concern of the people about the ultimate fate of the active materials. The concept of two persistencies, long at the site of original application and short following release into the environment, can in principle be solved by the careful application of controlled release."[497]

Controlled release terrestrial herbicides will predictably offer the following advantages:

1. Substantial decrease in the quantity of chemical added to the environment
2. Improved dispersal technique
3. Increased duration of control, thus increasing the cost effectiveness
4. Safer handling characteristics
5. Substitution of nonpersistent agents for persistent ones.

Potential disadvantages lie basically in the necessity of changing agricultural methods, especially in regard to land management and cultural practices.

Very little effort has been expended to date on terrestrial weed control through the use of long-term controlled release herbicides. Thus, this section is appended to the work on aquatic weed control not to present data (we have almost none), but to provide information and pertinent references to those who will be working in this area within the next few years.

Conventional Terrestrial Weed Control

Over 80 herbicides are in widespread agricultural usage in the United States. Inorganic materials such as arsenicals, cyanates and cyanamides, sulfamates, borax, and various chlorates remain in use, but organic substances, often with auxin properties, have supplanted most inorganic usages. Both phytotoxic and nonphytotoxic materials are employed. In the latter case environmental factors cause the agent to undergo chemical and/or biological reactions converting it into a toxic substance. For example, 2,4-DB, 4-(2,4-dichlorophenoxy) butyric acid, is nontoxic per se, but once absorbed by the target plant β oxidase reaction changes it chemically to 2,4-D acid having considerable toxicity. With Sesone® a reaction involving OH^- in soil causes it to convert to 2,4-dichlorophenoxyethanol, which is attacked by various soil microbes, causing further conversion to 2,4-D.

Organic herbicidal materials are conveniently classified by structure. Table 4–24 illustrates the several basic groups and their modes of action.

The agricultural problem, of course, is to destroy the competing weeds without adverse effects on the crops. Selectivity of herbicidal action is through timing and/or placement of the agent, or by characteristic differences in metabolism between the plant to be protected and the weed to be destroyed. In some instances the crop plant can absorb phytotoxic materials and transform them into nontoxic substances. Simazine® and Atrazine,® both potent phytotoxins, are readily transformed by corn into innocuous chemicals. Such transformations can be made within the plant through metabolic or nonenzymatic reactions, or they can occur outside the plant tissue through photochemistry, reaction with soil components, or attack by soil microorganisms. Metabolic reactions can be classified as follows:[525]

Oxidation (mainly oxidative dealkylation)
Hydroxylation
Synthesis of conjugates (glycosides predominate)
Hydrolysis
Reduction

Herbicides can be classified on the basis of application, the conditions of which may be (1) preplanting, (2) preemergent, (3) post emergent, and (4) preharvest. Application may be made directly to foliage or to the soil; foliar application can be further broken down into *contact phytocides* and *selective* (translocated) *phytocides.* One agent may be used in several ways depending on the crop involved and the weed to be controlled.

Many excellent publications are readily available describing the nature of the weeds,[417]

TABLE 4–24[477,524]

Terrestrial Herbicide Classifications: Organic Agents

Classification	Examples	Mode of action
Chlorophenoxy and chlorobenzoic acids	2,4-D; 2,4-DB	Auxin-like; upsets hormone balance
Chloroaliphatic acids	Trichloroacetic acid	Denatures proteins and enzymes
Amides	Propanil	Inhibits Hill reaction
Substituted ureas	Monuron (3-*P*-chlorophenyl)-1,1-dimethylurea); fenuron (3-phenyl-1,1-dimethylurea)	Inhibits photosynthesis
Carbamates	IPC (isopropyl *N*-phenylcarbamate); CIPC (chlorinated IPC)	Interrupts mitotic cycle
Symmetrical triazines	Simazine®	Inhibits photosynthesis
	Atrazine®	Inhibits CO_2 fixation
Chloroacetamides	Randox® (α-chloro-*N,N*-diallylacetamide)	?
Thiol and dithiocarbamates	Vegedex® (2-chloroallyl diethyldithiocarbamate)	?
Aminotriazole	Amitrole	Interferes with chlorophyll synthesis; affects carbohydrate metabolism

herbicide selection based upon the crop to be protected and the pest weed,[418,451,526,527] techniques of applying herbicides, and methods of evaluation.[530] It is not the intention of this text to devote space to the prosaic, but rather to point out specific considerations useful to the concept and development of controlled release terrestrial herbicides.

Biological control of noxious terrestrial plants is in its infancy. Plant pathogens, some fairly selective, are known, and methods of inoculation via insect vectors have been postulated.[451,528] Some plants are allelopathic, competing with other plants by secreting phytotoxic exudates that are antagonistic to nearby vegetation,[529] and the use of this factor in weed control is conceivable in very specific circumstances. Weed control through the use of predatory insects is under evaluation.

Conceptual Considerations

Conceptually, controlled release terrestrial herbicides have two distinct modes of application: foliar and soil applied. A number of basic questions arise regarding each category. In each case, what chemical, biological, and/or environmental advantage is to be gained? Foliar applications must obviously be sprayed, but how do we create fairly long-term adhesion of the (probable) film matrix to plant tissue? Will agent transfer occur from matrix to the plant, i.e., what penetration pathway can be used? If the agent is applied to soil, can a controlled low concentration release agent move through the intervening soil barrier and reach entry sites in the root system? How does one create short-term controlled release when the preferred herbicide, used as a pre-emergent soil sterilant, is gone prior to the germination of the food crop?

Probably the crucial question centers upon the chronicity phenomenon. Can chronic intoxication of terrestrial target weeds be induced by continuous exposure to ultralow herbicide concentrations?

Foliar Penetration Considerations

The spraying of target plants with herbicide solutions or emulsions is one of several conventional weed control techniques. The spray droplet must adhere to the foliage long enough for some amount of the agent to penetrate the cuticle and move across the mesophyll. This process requires a definite duration of time dependent upon several factors, e.g., 1 to 2 hr are required for 2,4-D to penetrate sunflower leaves.[531] Generally, the smaller the droplet size the more rapid the penetration and the greater the toxic activity.[531,]

[532] This indicates that the smaller the droplet, the less herbicide concentration necessary for control – thus, greater economy and decreased environmental impact ought to result. Unfortunately, small droplets, even of relatively nonvolatile materials, vaporize fairly rapidly. Volatilization is obviously antagonistic to penetration, so an optimum droplet size probably exists for each herbicide/solvent concentration, considering external factors such as temperature and air flow.

Cuticular penetration can follow either of two routes, aqueous or lipoid, depending upon the relative humidity within the stomata and the nature of the phytotoxic molecule. 2,4-D-type materials enter leaves most rapidly through a lipoid route, so use of an oil carrier enhances penetration.[533] In contrast, dalapon (2,2-dichloropropionic acid) enters leaves through an aqueous route; the sodium salt is the most effective form.[477]

If the herbicide fails to penetrate the leaf no killing action occurs. Improper formulation can result in droplets that remain on the outer leaf until volatilization is complete without entry into the stomata, or the toxicant can penetrate the cuticle and remain there in solution without further effective movement toward the kill sites.

The relative importance of the lipoid and aqueous routes through the cuticle depend upon several factors: (1) nature of the toxic molecule, i.e., whether lipoid soluble or water soluble, (2) water condition within the plant, i.e., whether at saturation or under a hydrodynamic stress, and (3) the formulation itself in terms of surface tension and compatability with the cuticle pore geometry. Foliar uptake rate of a given formulation can vary widely with pH,[534] and leaf structure may also play a crucial role. Large spray droplets of MCPA[a] shatter upon impact with the stellate hairs of *Eremocarpus setigerus* Benth., with some material reaching the cuticle, whereas smaller spray droplets from low-volume applications do not shatter but lodge on the hairs, thus having little effective phytotoxic activity.[535] The droplet distribution pattern on the leaf can be distinct and relatively independent of herbicide concentration.

Formulation substantially assists in overcoming external environmental factors that tend to detoxify. Additives can be used to increase resistance to rain removal of the herbicide droplet, although the material of choice is probably dependent upon the leaf characteristics of the target species.[536] Low surface tension and high wetting power are desirable properties that can be enhanced through the use of additives. Adjuvants are used to enhance toxicity, increase spreading properties, promote leaf retention and cuticle penetration, and assist in emulsification.

Adjuvants that reduce interfacial tension (between spray liquid and leaf, agent and carrier, formulation ingredients, etc.) are termed *surfactants.* Such additives possess an unsymmetrical molecule, one part being highly polar and thus hydrophilic, the other part being nonpolar and attracted to oils, fats, and waxes. Each part of the molecule orients itself to the attractive phase, thus bonding repellent surfaces.

Other additives include thickening agents to modify viscosity, sticking agents to increase leaf adhesion (most of them consisting of a water-insoluble polymer that dries to a water resistant film), penetration agents that solubilize the waxy part of the cuticle, and activators (synergists).

Incorporation of the agent in a film-forming controlled release polymer will dramatically affect all the above factors, influencing the conventional foliar spray technology. Undoubtedly, the use of a controlled release matrix element would greatly reduce agent loss through volatility, and vastly increase water resistance; longer contact time should allow more effective herbicide delivery, thus reducing the total quantity of chemical needed. However, one can postulate various problems that may arise: Herbicidal films may bridge over stomata, thus reducing actual plant-surface-to-herbicide contact; the use of oil or water carriers to enhance penetrability of the cuticle is probably not feasible; the surface tension and wettability factors, so crucial to penetration in conventional systems,[537] are unknown when considering the diffusion of phytotoxic molecules from a controlled release polymer into the polymer/leaf interface.

Movement through Soil

Chandrasekaran et al. have mathematically analyzed the kinetics of transport of biologically active materials from controlled release systems to the soil, assuming that diffusion is the dominant mechanism.[538] Assuming that the releasing pellet

[a] [(4-Chloro-*o*-tolyl) oxy] acetic acid.

is spherical, then diffusion within the sphere is given by

$$\frac{\partial C_1}{\partial t} = D_1\left(\frac{\partial^2 C_1}{\partial r^2} + \frac{2}{r}\,\frac{\partial C_1}{\partial r}\right)$$

where C_1 is the agent concentration, D_1 the diffusion coefficient, r the radius, and t the time.

The following boundary conditions must be satisfied:[538]

$$C_1 = \alpha C_2, \text{ at } t \geqslant 0, r = a \text{ (radius of the sphere)} \quad (1)$$

$$\frac{dC_1}{dr} = 0 \text{ at all } t, r = 0 \quad (2)$$

$$J_1 = J_2 \text{ at } t > 0, r = a \quad (3)$$

where α is the partition coefficient, C_2 is the agent concentration in the soil, a is the pellet radius, and J is the flux.

Likewise, diffusion of the agent in the soil can be represented by [538]

$$\frac{\partial C_2}{\partial t} = D_2\left(\frac{\partial^2 C_2}{\partial r^2} + \frac{2}{r}\,\frac{\partial C_2}{\partial r}\right) - KC_2$$

where C_2 is the agent concentration in the soil, D_2 is the diffusion coefficient in the soil, and K is the degradation rate constant.

It is assumed that degradation is a first order process and that no nonlinear binding of the agent to soil particles exists. The following boundary conditions must be satisfied.[538]

$$C_2 = 0 \text{ at } t = 0, a < r \leqslant \infty \quad (1)$$

$$C_2 = 0 \text{ at all } t, r \to \infty \quad (2)$$

$$\frac{dC_2}{dr} = 0 \text{ at all } t, r \to \infty \quad (3)$$

For large values of D_2/D_1, agent transport is controlled by diffusion within the matrix. As the partition coefficient decreases, the rate of agent movement from matrix to soil increases.

Collins and Doglia, employing classical diffusion theory, considered pesticide movement in soil.[539] Assuming that the detoxification rate follows first order kinetics, it follows that the concentration of released agent "approximates that of an active material which decays more slowly." Collins and Doglia conclude (and properly so if one may use analogies from molluscicide work) that "controlled release mechanism allows short persistent chemicals to masquerade as persistent chemicals."[539]

It is well accepted that the conventional application of herbicides is wasteful, in that the initial concentrations used must be very high so that later activity is sufficient for proper control despite continuous detoxification processes.[507]

Soil diffusion and degradation rates are known for many herbicides and probably measurable for all.[540-543] Diffusion varies with soil moisture content, bulk density, and temperature,[541,542,544,545] but the effect of each variable differs with the nature of the herbicide.[541] Degradation likewise is greatly affected by temperature[541] and, to a lesser extent, by soil moisture content.[541] Diffusion rate is not a simple function of moisture content;[541] with trifluralin, for instance, diffusion increases with water content to a maximum value at 15% moisture, and then decreases with increasing water content as the air-filled fraction of soil void space is reduced below about 40%.[542]

Detoxification or degradation rate of a given herbicide in soil varies with temperature, moisture, and often microorganism content.[477,543,546]

Predictably, controlled release herbicides functioning through diffusion-dissolution, leaching, solubilization, or exfoliation will release in soil, although it is expected that moisture must be present to serve as a transport medium from pellet to target root system. Controlled release 2,4-D BEE, formulation 14ACE-B, has been examined in soil surface tests with subirrigation.[547] Bioassay against mustard (*Brassica* spp.) showed a lethal radius around the individual granule with significant effects over the 6-week test period.

Synergetic Considerations

By definition, a synergetic effect results when two herbicides applied in combination provide a greater plant response than the sum of the two used alone. Another commonly observed effect, "enhancement," is defined as an increase in plant response when a nontoxic adjuvant is used in combination with a herbicide. Both effects are advantageously known to weed control science, and methods of determining synergetic effects are available.[548-551] Various mechanisms are postulated with considerably experimental evidence, e.g., nitrite accumulation arising from the uncoupling of those factors controlling nitrate reduction.[552] Enhancement phenomena have also been elucidated for various herbicide-surfactant combinations.[553]

Synergetic, antagonistic, additive, and enhancing activity can likely arise from the use of controlled release herbicides. Obviously, two or more phytotoxins can be applied as controlled release granules simultaneously, but whether it is possible to combine several herbicides in *one* matrix is open to question, although a less complex formulation scheme would be to provide several different materials for simultaneous application. A more subtle situation may well arise from the fact that a controlled release system contains other possibly active materials such as the binding element, curatives, processing aids, and so on, which become available as the matrix itself degrades. Substances such as zinc oxide, mercaptothiazoles, free sulfur, etc. could be synergetic or antagonistic to agent activity. The dividing line between dosages that destroy the target and the higher concentrations that affect the nontarget crop can be dangerously narrow, so that synergetic activity leads to decreased crop yield. Plant response to environmental stress can be complex and not easily determined without lengthy field evaluation.[554]

Environmental Impact

Terrestrial herbicides are not major contaminants of waterways in that their effects are masked by the much more dramatic activity of insecticides and phosphate fertilizers.[104] As a rule, herbicide persistency is measured only in weeks, although a few used in large dosages as soil sterilants may show activity for several months. Biomagnification does occur and may enhance impact, but this is of minor concern compared to the vastly greater insecticide problem. Conventional spray operations can lead to undesirable damage to nontarget species, due usually to drift, and long-term soil residues, especially under cool, dry conditions, have been found.

Those involved in the development of controlled release technology obviously prefer to adopt an optimistic view, i.e., smaller amounts of less persistent materials. As Allan et al. concisely summarized, "Controlled release combinations, in comparison with free pesticide, provide either an extended period of protection at equivalent levels of application or the same period of protection at greatly reduced levels of application. Both of these alternatives tend to reduce the environmental hazards associated with the continued use of conventional pesticides."[555] The first statement can be amply supported by copious data from antifouling and molluscicide experimental and field work. Although one would like to accept it at face value, the second statement is open to serious question. Data, favorable or unfavorable, are lacking. Nontarget plants and animals exposed to nonpersistent toxicants, even at very low levels, may be affected through bioaccumulation, long-term exposure to toxic breakdown products, and mutagenic or teratogenic effects arising from the continuous ingestion of minute amounts of the control agent over a greatly extended duration; even the complete eradication of a target or target species may radically and detrimentally alter the bioenvironment, in the same sense that if one eliminates the coyote a jack rabbit population explosion results and sheep production falls due to rabbit competition for forage.

The biological fate of low persistency pesticides in soil is not well understood. In studies based upon this property the investigator seeks acute rather than chronic intoxication effects. Chemical analytical techniques for residue analysis in soils exist but often cannot reach the ultralow concentration levels expected with controlled release, indicating that refinement in technique is needed. Likewise, the biological effects of the "inert" matrix and compounding elements are not known, nor are methods readily available for their chemical or biological assessment.

Lastly, it should be noted that the auxin-like character of 2,4-D and other materials when applied at continuous ultralow levels may lead to prolific weed growth, and the concentration boundary between enhanced growth and destruction may be quite narrow.

Chapter 5

ALGICIDES, BACTERICIDES, AND FUNGICIDES

ALGICIDES

Nature of the Problem

Algae can be classified into four basic groups: pigmented flagellates, diatoms, green, and blue-green. Algae are normal constituents of both marine and fresh surface waters, where they exist either as single-celled or multicellular organisms. Most have no power of locomotion, so their movement is dependent upon water flow. In small numbers algae are beneficial, but in quantity they create or are implicated in numerous problems. Marine algae are prominent members of the fouling community; the control techniques used were discussed in Chapter 2.

Algae infestations change the pH, alkalinity, color, turbidity, and dissolved gas content of fresh waters. Large numbers of filamentous algae may bond together in "mats" that destroy fish by acting as a barrier to oxygenation, impede navigation, clog filters and reservoir screens, and reduce flow through channels, pipes, irrigation canals, drainage ditches, and so on. Infestations of single-celled forms, termed "blooms," result in fish kill and often impart undesirable odor and taste to potable water.

In large numbers, single-celled algae, usually of the blue-green variety, are termed "slime." Heat transfer in cooling systems is reduced by slime deposits, and a serious industrial problem arises when slime coats develop on paper mill rolls, resulting in a blemished product.

A few types of algae are toxic. *Gonyaulax* (marine) contains a toxin ten times as potent as strychnine, and humans who have consumed contaminated clams have become seriously ill. *Gymnodinium* (marine), "red tide," is toxic to fish, shellfish, and echinoderms. *Anabaena,* a fresh-water algae, causes dermatitis in man. However, the most serious toxic problem arises from contamination of potable water with species of the blue-green genus *Anacystis.* Horses, dogs, cattle, sheep, and rabbits have been fatally poisoned after drinking from ponds having dense *Anacystis* blooms.

Algae, by changing pH and liberating oxygen, greatly accelerate corrosion in metal tanks and basins open to sunlight.

Fresh-water algae constitute the foremost hull-fouling problem in fresh-water lakes and ponds; the ship owner is forced to periodically remove growth or suffer drastic curtailment of speed and increased fuel consumption.

Probably the most widespread algae problem arises from the impartation of odor and unpleasant taste to drinking water supplies. Odor can arise from dying or dead algae, or from large quantities of living algae. Palmer has compiled a list of responsible groups, examples of which are shown in Table 5–1.[556] Descriptions of the various algae, their physiology, ecology, etc. can be found in many excellent texts.[557,558]

Conventional Control

Algae control can be by mechanical means (through filtration and disposal), or by the use of algicides. Unlike most chemical control regimens, rapid kill of algae is *not* desirable in that dying leads to oxygen depletion and generation of hydrogen sulfide, often resulting in fish kill. It is usually better to slowly destroy algae, which necessitates concise monitoring of the control agent. Chlorine and copper sulfate are predominantly used for algae control in natural waters.

Nonmetallic algicides such as various ethylenic sulfilimines, naphthoquinone derivatives, and stilbenes have been reported.[559] Insoluble copper carbonate can be used to control *Chara* at 15 to 25 lb/acre without fish damage.[560] Conventional herbicides have been evaluated against semicellular algae, with mixed results.[561] Cuprous-oxide- and TBTO-containing antifouling paints have been

TABLE 5–1[556]

Odor- and Taste-producing Algae

Genus	Type	Odor	Taste
Ceratium	Flagellate	Septic	Bitter
Cladophoria	Green	Septic	–
Peridinium	Flagellate	Fishy	–
Stephanodiscus	Diatom	None	Oily
Anacystis	Blue-green	Septic	Sweet
Chara	Green	Garlicky	–
Nitella	Green	Septic	Bitter

TABLE 5–2[470]

Green Algae Mortality from Incracide® E-51 Granules

Cu^{++} dosage (ppm)	Post-treatment time (days)								
	22	24	29	30	34	39	44	51	54
	Algae mortality (percent)								
200	37	45	60	57	90	100	–	–	–
100	33	35	67	80	91	90	100	–	–
50	15	28	32	50	50	53	57	73	82
25	10	33	35	40	47	47	53	87	100
Control	0	4	12	10	10	13	10	10	13

applied to concrete-lined irrigation canals[562] and algae control for up to 6 years has been achieved.

The constant search for an algicide superior to copper sulfate and chlorine for use in natural water bodies has not been particularly fruitful. Copper sulfate has long remained the most economical and safe material for algae control.[563] Recently, other copper-containing chemicals have shown promise. Cutrine, containing a copper triethanolamine complex, has been found useful against *Chara* (at 0.4 ppm), a particularly difficult problem, and other algae.[564]

Various materials such as bis(trichloromethyl) sulfanes, peroxide and acetyl compounds, thiocyanate and formaldehyde mixtures, and nitroso aryloxy quaternary ammonium compounds have been recently patented as slime control agents.[559] Under certain conditions, acrolein is more cost effective than chlorine as a cooling water slimicide.[565] Slime control in paper and pulp mills is achieved through the use of organotin chemicals,[566] polychlorophenates, quaternary ammonium compounds, methylene bisthiocyanate, and others.[567]

Controlled Release Algicides

The role of copper sulfate in algae control is well known.[563,568-570] Cupric ions inhibit respiration and photosynthesis in algae, leading to mortality.[571-573]

Copper sulfate has been encapsulated in a microporous polymer (Envirocap®) so that upon immersion[574] release is gradual (said to last several months). Recommended application rates are unusually high, 40 to 100 lb/acre.

Controlled release copper sulfate, Incracide® E-51, was evaluated against blue-green algae,[470] and the observed mortality data are given in Table 5–2.

Gradual destruction of algae using E-51 is thus observed under controlled laboratory conditions, but whether this will translate into biological efficacy under field conditions remains to be seen.

BACTERICIDAL POLYMERS

Relatively little advance has been made in the development of long-term controlled release bactericides. Bactericides have been historically incorporated in polymeric films with the view towards protecting the *material* from microbiological attack. In recent times the development of germicidal fabrics has been directed toward hospital use, as a means of preventing infection. In one of the earlier studies, streptomycin and other germicides were incorporated through an ion-exchange technique into polyvinyl alcohol and cellulose fibers.[575] The need for such materials in the form of white goods, wall coverings, floor polishes, etc. to prevent the transfer of pathogenic bacteria in hospitals has been well indicated.[576]

Controlled release bactericides are under intensive study for medical use in and on the human body, e.g., bactericidal contact lens solution based upon the incorporation of thimerosal in polymers of vinyl pyrrolidone.[577] A number of drug release systems have been or are currently being developed.[578] However, a description of the very active area of controlled release drug delivery systems does not fall within the purview of this text.

Various organotins, including TBTO, are known to be bactericidal.[579] The successful use of organotins in controlled release antifoulants and molluscicides prompted the evaluation of the various formulations against bacteria.[70] In a U.S. Air Force study, several controlled release

TABLE 5–3[580]

Biocidal Formulations Evaluated Against Bacteria

Code[a]	Base elastomer	Toxicant	Toxicant loading
350A	Neoprene WRT	TPTC[b]	10.0 pph[c]
351B	Neoprene WRT	TBTO®	12.0 pph
378E	Paracril®	TBTO	5.0 pph
419A	Neoprene WRT	TBTAd[d]	8.0 pph
419B	Neoprene WRT	TBTR[e]	8.0 pph
443A[f]	Neoprene WRT	TBTO	8.0 pph
802C	Neoprene WRT	TBTA	8.0 pph
802D	Neoprene WRT	TBTF	8.0 pph
819B	EPDM	TBTO	6.5 pph
819D	Neoprene WRT	TBTS	6.5 pph
324C	Neoprene WRT	TBTO	17.5 pph
1112H	Neoprene WRT	TBTF	20.0 pph
1112HX	Neoprene WRT	TBTF	15.0 pph
1102A	Paracril	TBTO	11.0 pph
B5	Natural rubber	TBTO	8.0 pph
397C	Neoprene	TBTO	6.5 pph
1120B	Neoprene	TBTAd	10.0 pph

[a]B.F. Goodrich Co. designation.
[b]Triphenyltin chloride.
[c]Parts of toxicant per hundred parts of base elastomer.
[d]Tributyltin adipate.
[e]Tributyltin resinate.
[f]Commercial designation, Nofoul®

organotin loaded elastomers were evaluated against six species of bacteria, with the general objective of control in potable water regenerated from human urine.[580] The biocidal rubber formulations shown in Table 5–3 were examined. Rubber specimens were inoculated with each test organism and incubated for 48 hr. Table 5–4 illustrates the results. Bactericidal activity in urine and in a dynamic water flow system were positive, with candidate biocidal elastomers 351B and 324C superior.

In unpublished work at the University of Akron, various organotin-containing polymers were found active against the six bacteria listed in the above study, as well as against *Pseudomenas aeruginosa, Aerobacter aerogenes, Bacillus megaterium, Serratia marcescens,* and *Streptococcus lactus.* The generalized bactericidal rubber formulation is as follows:

Ingredient	Formulation (parts)
Chloroprene (neoprene WRT or GT)	100
Carbon black	15–30
Zinc oxide	5
Magnesium oxide	4
Phenyl-β-naphthylamine	0.5–2
Mercaptobenzolthiazole	1
Fatty acid[a]	0–6
Organotin[b]	0.5–12

[a]Palmitic acid is probably superior, followed by lauric acid. Analysis indicates that TBTO and possibly other organotins combine with fatty acids within the matrix, thus providing substances such as tributyltin dilaurate. The environmental implications could be serious. Registration based upon the organotin added may be meaningless when a different material is released into the external environment.
[b]TBTF is probably superior, followed closely by TBTO, and then TBTA, with the TBT-adipate, resinate, and carbonate being inferior.

TABLE 5–4[580]

Bactericidal Properties of Organotin Elastomers

Elastomer code	Bacteria[a] *Alcaligenes faecalis*	*Bacillus subtilis*	*Escherichia coli*	*Proteus vulgaris*	*Staphylococcus albus*	*Streptococcus pyogenes*
Control I[b]	4+	4+	4+	4+	4+	4+
Control II[c]	4+	4+	4+	4+	4+	4+
350A	3+	0	4+	4+	4+	1+
351B	0	0	0	0	0	0
378E	0	0	0	0	0	0
419A	4+	0	0	4+	3+	2+
419B	3+	0	2+	4+	1.5+	0
443A	0	0	0	0	0	0
802C	0	0	3+	0	0	0
802D	0	0	1.5+	0	0	0
819B	4+	0	1.5+	4+	0	2+
819D	0	0	0	0	0	0
B5	0	0	0	0	0	0
397C	0	0	3+	0	0	0
324C	0	0	1+	2+	0	0

Note: Materials continued to provide bactericidal activity after aging.

[a] Code:
4+ = excellent profuse growth similar to control.
3+ = growth obvious, little or no extension of initial inoculum radius.
2+ = growth present but spotty and generally poor.
1+ = very little growth.
0 = no growth, total lysis.

[b] Agar only.

[c] Code 107A, nontoxic neoprene formulation similar to 443A except for toxicant.

Contact lysis is evident, but would organotin-releasing elastomers destroy bacteria over an intervening air space through vapor release? Table 5–5 depicts results when the rubber specimen is separated from the test flora by a 1.5-cm air gap. Edge growth, in terms of inhibitory effect, indicates that 802D was the best material, while 443A and 419A were inferior. Compound 802D has demonstrated 9 months of vapor release, and compounds 351B and 443A have demonstrated 24 months (at which time evaluation was terminated). Based on efficacious release in antifouling evaluation, all three ought to exceed 48 months of bactericidal vapor activity.

FUNGICIDAL MATERIALS

Various fungi constitute agricultural pests, lawn and garden threats, and, in some instances, serious menaces to health. In the agricultural area, sulfur was used as the first foliar fungicide, followed by sulfur-lime mixes and then copper salts. Copper, such as in Bordeaux mixture, $Cu_4(OH)_6SO_4 \cdot XCa(OH)_2$, seems to inhibit various sulfhydryl-containing enzymes essential to fungi viability. A number of organic substances, chorophenols, nitrophenols, quinones, dithiocarbamates, organomercuries, and organotins, are in use today. The active mechanism varies. Phenols uncouple oxidative phosphorylation, quinones react with the sulfhydryl (–SH) groups in amylase and carboxylase, and mercury compounds are probably biocidal due to the Hg^{2+} moiety reacting with –SH.[522]

The organotin toxic mechanism likely lies in the R_3SnX-type molecule, preventing the incorporation of inorganic phosphate into adenosine triphosphate.

A number of antibiotics used in human and animal therapy, griseofulvin, streptomycin, cyclohexamide, blastocidin, and others, can also be utilized in treating fungi-infected higher plants.

TABLE 5–5

Bioactive Growth Characteristics of Bacteria Exposed to Organotin Vapors from Controlled Release Rubber

Organism	Growth on controls[a]			Growth on bioactive compounds[a,b]							
	Temp. (°C)	Blank	Nontoxic rubber	324C	315B	278E	419A	443A	802D	819D	B5
Alcaligenes faecalis	25	4+	4+	–	–	–	–	–	–	–	–
Bacillus subtilis	37	4+	4+	–	–	–	–	–	–	–	–
	25	4+	4+	–	–	–	–	–	–	–	–
Escherichia coli	37	4+	4+	–	–	–	–	–	–	–	–
	25	4+	4+	–	–	–	–	–	–	–	–
Proteus vulgaris	37	4+	4+	4+	4+	4+	4+	4+	4+	–	–
	25	2+	2+	–	4+	4+	4+	4+	4+	–	–
Staphylococcus albus	37	4+	4+	–	–	–	–	–	–	–	–
	25	2+	2+	–	–	–	–	–	–	–	–
Streptococcus pyogenes	37	2+	2+	–	–	–	–	–	–	–	–
	25	1+	1+	–	–	–	–	–	–	–	–

[a]Code:
– = no growth, complete inhibition.
1+ = very little growth.
2+ = some growth.
3+ = considerable growth.
4+ = excellent growth.

[b]Compound list:

Code (proprietary B.F. Goodrich Co. formulations)	Base rubber	Toxicant	Toxicant concentration (parts per hundred)
315B	Neoprene	TBTO® (dilaurate?)	12
278E	Acrylonitrile	TBTO	5
419A	Neoprene	TBTAd	8
443A	Neoprene	TBTO (dilaurate?)	8
802D	Neoprene	TBTF	8
819D	Neoprene	TBTS	6.5
B5	Natural rubber	TBTO	8
324C	Neoprene	TBTO	17.5

TABLE 5–6[a]

Growth Characteristics of Fungi on Bioactive Polymers

Organism	Growth on controls		Growth on bioactive compounds[b]				
	Blank	Nontoxic rubber	351B	378E	419A	443A	802D
Mucor hiemalis	4+	4+	0	0	0	0	0
Phycomyces blakesluanus	4+	4+	0	1+	0	0	0
Cladosporium	4+	4+	0	0	0	0	0
Trichoderma	4+	4+	2+	1+	0	1+	0
Aspergillus	4+	4+	0	0	0	1+	0
Arthrobotrys musiformis	4+	4+	0	0	0	0	0

[a]From unpublished work of D. Goldstrohm and N. F. Cardarelli, University of Akron. Since this is the first communication of results to the scientific community, a somewhat detailed description is presented.

[b]Code:

0 = no growth.
1+ = very little growth.
2+ = some growth.
3+ = considerable growth.
4+ = complete overgrowth.

Fungicides can be used as a contact phytocide or as a systemic, such as the use of Benlate® benomyl in Dutch elm disease (*Ceratocystis ulmi*).[581,582] Systemic materials are gradually replacing contact fungicides in many areas, including protection of turf.[583]

Fungicidal chemicals are used as additives in paints, coatings, plastic goods, and other consumer products.[584]

The incorporation of fungicides in polymeric substances keyed to long-term controlled release was reported by Ugi and Fetzer, who used polyisonitriles as base materials.[585]

Organotin-containing antifouling elastomers were evaluated against various fungi by Dr. Donald Goldstrohm at the University of Akron. It was known from the Major and Cardarelli work that several such formulations were toxic to *Hermodendrum cladosporium*, both as a contact fungicide and through vapor action.[580] Early tests against *Aspergillus niger* and *Rhizopus nigricans* showed excellent contact kill. This work was then extended to cover a wide range of fungi, and to evaluate efficacy in terms of contact (with the fungicidal rubber sheet) or by vapor kill.

In one experiment, a sterile 2.5-cm^2 test specimen, a B.F. Goodrich Co. antifouling formulation, was suspended above a petri dish partially filled with medium (glucose-yeast-agar-water) which was inoculated with a given fungus. Incubation was for 5 days at 23°C, with alternating 12 hr/12 hr periods of light/dark. Controls with no rubber and nonbiocidal rubber were used, with the results shown in Table 5–6.

In another experiment, rubber samples from a TBTO/chloroprene graded series were tested against seven fungi (Table 5–7). The medium, method of inoculation, and conditions of incubation were the same as that described for the experiments recorded above. In samples 351G through 351C, increasing concentrations of TBTO resulted in increased inhibition, as would be expected. However, in samples 443A through 324C, this trend did not continue, i.e., increased concentration of biocide did not always result in decreased growth. Perhaps at increased concentrations of TBTO, products are formed which bind the biocide or are stimulatory to the test organism. It is possible that TBTO is not the inhibitory material but gives rise to toxic derivatives during the processing of the rubber. Pure TBTO and TBTF droplets 0.75 in. below the inoculum gave inhibition of the colony diameter, as shown in Table 5–8.

It is of note in the foregoing and following experiments that the slow release rubber disc is separated by an air gap from the inoculated media.

In more elaborate experimentation, elastomeric

TABLE 5–7

The Effect of 18-mm Discs of the TBTO®/Chloroprene Graded Series on Colony Diameter (in Millimeters) of Seven Different Fungi After 5 Days' Incubation

Fungus	Rubber samples									
	105A (0%)[a]	351G (0.43%)	351F (0.86%)	351E (1.7%)	351D (3.4%)	351C (5.1%)	443A (6.8%)	351B (10.2%)	324A (15.0%)	324C (17+%)
	Colony diameter (millimeters)									
Cladosporium sp.	26.5	23	19	0	0	0	0	0	0	0
Mucor sp.	88	88	88	45	6	0	0	23.5	0	0
Arthrobotrys sp.	76	72	41	0	23	0	0	0	0	0
Phycomyces sp.	88	88	88	34	21	3	19.5	6.5	14	11.5
Trichoderma sp.	81.5	88	82	60	16	0	15	0	11	9
Alternaria sp.	58.5	48	40	9	5	0	5	0	0	0
Aspergillus sp.[b]	(4)	(4)	(4)	(4)	(2)	(0)	(1)	(1)	(1)	(2)

[a]Percentages in parentheses indicate biocide content.
[b]*Aspergillus* sp. was not rated on colony diameter but on 0 to 4 basis; 0 = no growth, 4 = growth equivalent to control.

From Goldstrohm, D. and Cardarelli, N. F., unpublished data.

TABLE 5–8

The Effect of TBTO® and TBTF on Colony Diameter of Seven Different Fungi After 9 Days' Incubation at 25°C

	Colony diameter (millimeters)		
	Control	TBTO	TBTF
Phycomyces sp.	88	0	15
Cladosporium sp.	88	0	0
Mucor sp.	88	0	35
Arthrobotrys sp.	88	0	0
Trichoderma sp.	88	7	18
Alternaria sp.	80	–	10

From Goldstrohm, D. and Cardarelli, N. F., unpublished data.

formulations containing varied TBTO concentrations were bioassayed with *Trichoderma* (Table 5–9).

In later definitive experiments, biocidal rubber was placed from 6 to 24 in. from various fungi genera in open glass cylinders. Vapor intoxication was apparent by complete lack of growth in some instances and severe inhibition in others.

Further tests were performed with additional fungi from the American Type Culture Collection (ATCC):

Number	Name
ATCC # 9348	*Aureobasidium pullulans*
ATCC # 9569	*Stemphylium*
ATCC # 9643	*Aspergillus flavus*
ATCC #10052	*Fusarium moniliforme*
ATCC #16022	*Cladosporium herbarum*
ATCC #16423	*Alternaria tenuissima*

Results were similar to those of the previous tests, indicating that TBTO release in vapor form from biocidal rubber compounds will destroy many fungal species. Also, preliminary attempts were made to assess the biocidal-rubber-to-air volume necessary to destroy fungi. An 18-mm-diameter by 2-mm-thick disc gave excellent inhibition of all test fungi in a 15,010-cm^3 volume.

The zone of inhibition noted in the above experiments (see Figure 5–1)* was examined in many instances to determine whether the fungi were actually destroyed or merely dormant. As a general observation, biocidal rubber proved to be fungicidal rather than fungistatic in its mode of action. There is a small fungistatic area at the boundary between inhibition and the growing colony.

Increasing the relative humidity to 100% and

*For Figure 5–1 see Page 90.

TABLE 5–9

The Effect of Treated Rubber Samples Containing Increasing Concentration of Biocide on Growth of *Trichoderma* sp.

Sample	Biocide concentration (percent)	Zone of inhibition (millimeters)
610G	0	0
351G	0.43	0
351F	0.86	20
351E	1.7	40
351D	3.4	50
351C	5.1	68
443A	6.8	50
351B	10.2	42
324A	15.0	50
324C	17+	60

From Goldstrohm, D. and Cardarelli, N. F., unpublished data.

the incubation temperature to 30°C for a 2-week period did not appreciably alter the results previously observed.

Extension of all tests to include TBTF (tributyltin fluoride)-containing rubber in various base matrices produced results similar to those experienced with other organotins. However, one can generally conclude that TBTF vapor release occurs at a lower rate than that observed with TBTO biocidal rubber. Hence, TBTF compounds, for effective long-term control, require a greater toxicant concentration, i.e., 2% TBTF in chloroprene rubber is about equivalent to 0.8% TBTO in the same matrix.

In one experiment a screening test was performed against *Chlorella pyreniodosa* (algae). The characteristic zones of inhibition appeared with TBTO at 0.86% or higher loadings.

Chapter 6

CONTROLLED RELEASE INSECTICIDES

INTRODUCTION

Insect and arachnid pests constitute a major detriment to the health and economic well being of much of the world's human populace. Widespread endemic diseases of immense public health impact are vectored by insects. Bacterial and viral pathogens as well as parasites are transmitted. The mosquito is the Number One menace, spreading malaria, yellow fever, filariasis, Dengue fever, hemorrhagic fever, and several forms of encephalitis. Up to one third of humanity suffers from one or more of these ailments. Other major insect vectors are the tick (which spreads meningo-encephalitis, relapsing fever, tularemia, tick typhus, louping ill, Russian spring-summer encephalitis, Rocky Mountain spotted fever, and others), black flies of the genus *Similium* (which transmit onchocerciasis), the sandfly vectors leishmaniasis and bartonellosis, fleas (which spread bubonic plague from rodents to man, plus one form of typhus fever), the Tsetse fly (which is responsible for "sleeping sickness" or trypanosomiasis), Reduviidae such as *Triatoma* (which are the vectors of Chagas disease), the louse (which spreads typhus fever and trench fever), and the mite (which transmits scrub typhus). There are others, but these constitute the major dread diseases of humanity spread by arthropods.

Man suffers considerable annoyance and inconvenience to his person arising from insects, such as attacks by blood-sucking flies, fleas, and mosquitoes, lice infestations (Pediculosis), wasp stings, and the like. Allergic reactions terminating in death have been brought about by insect bites and stings. Several spider species possess toxins in sufficient quantity to be fatal to man if bitten.

A large group of insects attack livestock, fowl, and other domestic animals, sometimes with fatal results, as with mosquito-vectored equine encephalitis, or more commonly by decreasing the economic worth of the animal by reducing meat, milk, or fur output.

Infestation of man's dwellings by ants, cockroaches, house flies, and other vermin are aesthetically unappealing, even though the health hazard involved is minor. Wooden structures are damaged or destroyed by termites and to a lesser extent by carpenter ants.

Agricultural enterprises are faced with a host of insect enemies that damage crops, orchards, and pasturage not only by devouring fruit and foliage, but by transmitting various bacterial, fungal, and viral pathogens. Stored crops are likewise menaced by a number of insect pests that not only consume, but also spoil.

Lawns and ornamental plants serve as foodstuffs to thrips, mites, aphids, Japanese beetles, and many others. Forests are denuded by the gypsy moth, orchards by the codling moth. The list is seemingly endless, the only surprising element being how man managed to survive and share the same environment with a countless host of insects of inimical life styles!

The development of controlled release insecticides, although in its infancy, has taken a number of distinct routes and can be best categorized by examining each distinct group of insects, rather than following the scheme of "problem – conventional control – controlled release applications" as followed in the preceding chapters. A *brief* description of the problem alone would be an immense undertaking covering many volumes, so is best left to the entomologist and other specialists. In this chapter the insects of public health importance will be considered under separate headings: mosquito, tick, black fly, etc.; those of agricultural importance will be discussed through control methodology, emphasizing the newer approaches. Since insecticides play a leading role in environmental contamination, more space will be devoted to this subject than in the case of antifoulants, molluscicides and herbicides, which are of far less importance in this aspect.

MOSQUITO CONTROL

Medical Considerations

The taxonomy, physiology, and ecology of the hundreds of mosquito species of public health importance have been amply described in various reference works,[586] as have the methods of laboratory rearing and evaluation of pesticides.[587] Standard texts and reference works on medical entomology and the epidemiology of mosquito-borne diseases are readily available.[588,589] Consequently, these facets of the mosquito problem are not subjects of this text.

World-wide medical and economic aspects of mosquito-vectored diseases render these insects the prime arthropod enemy of mankind. Prior to the vast control and eradication campaigns that were aimed at malaria and yellow fever following World War II, large areas of the world were nonproductive or underproductive due to the large numbers of victims. In India alone, somewhere between 75 and 100 million persons suffered the scourge of malaria, with one million or more dying annually as a result.[588] In the United States over 100,000 cases of malaria were reported yearly during the 1930s.[590]

Malaria control programs initiated between 1949 to 1953 in South America, Africa, and Asia enjoyed outstanding success. Between 1952 and 1969, malaria victims in India dropped from about 20% of the total population to less than 0.1%.[591] Vast strides were made against yellow fever in South America during this period, and public health officials believed that total eradication was possible.[592] Dengue fever outbreaks such as the 1922 Texas epidemic, which claimed over one-half million cases, appeared to be a thing of the past.[590,593] In general, mosquito control was based upon the use of DDT (dichlorodiphenyl trichloroethane).

However, other problems have arisen, and the optimism of the 1950s has become rather tarnished. The demographic implications are vast. In India the 27.4/1000 annual mortality rate of the 1940s dropped to 16 to 18/1000 by 1967, with mosquito control assuming an important aspect.[591] Huge population increases have tremendously burdened agriculture, and world-wide food shortages have resulted (although it may not be meaningful to the victim as to whether his fate is sealed by starvation or by malaria!).

DDT is an extremely persistent insecticide, and its unrestricted use has led to serious environmental contamination. Absorption of DDT by fish can lead to mortality;[594,595] reservoir contamination has been observed;[596] mallard ducks accumulate DDT and its metabolites;[597] DDT toxicity has been noted in blue jays, bobwhites, house sparrows, and cardinals;[598] pheasant exposure to DDT shows *decreasing* tolerance with time;[599] grouse egg loss results from DDT spraying;[600] some piscivorous birds accumulate DDT from fish;[601] and robin mortality can be as high as 85% from 10 lb/acre DDT treatments.[602] Mammals are also adversely effected: long-term cumulative effects in carnivores have been noted.[603,604] DDT has been found in most areas of the world, including the contamination of subarctic soils, plants, and animals.[605] Biomagnification is often evident.[606]

The large-scale widespread usage of DDT has led to increased mosquito resistance,[607,608] so effective control requires ever-increasing dosages. Also, sublethal exposure of mosquitoes to DDT induces an increase in reproductive potential.[609,610] Since less persistent and more effective mosquito insecticides are available, and obvious environmental risk is present, DDT has been banned for most usages.

Resistance of malaria vectors to DDT and other insecticides has led to a resurgence of the disease in various areas of the world.[611,612] Intensive use of DDT in Greece eliminated malaria-transmitting species for a while, but now one of the principal vectors, *Anopheles sacharovi*, has returned to preeradication densities, and is resistant to DDT and dieldrin.[612] Malaria has returned to the United States via infected civilian and military personnel.[613]

Yellow fever endemicity is increasing, and epidemics are being noted in areas previously thought to be free of mosquito vectors.[614]

Recent events have dramatically underscored the fragility of man's control of mosquito vectors. The summer 1974 outbreak of St. Louis encephalitis in Memphis, Tenn., resulted in 15 deaths and as many as 265 cases.[615] Venezuelan equine encephalitis epidemics claimed 30,000 to 60,000 equine deaths and caused an estimated 20,000 human cases in Ecuador in 1969; the disease then moved northward, through Central America, reaching Texas by spring 1971.[616] Several thousand horses succumbed, and a number of suspect human cases were noted.

Filariasis is a disabling disease arising from nematode parasites in the human body. Its common name, elephantiasis, aptly describes the extreme swelling of arms, legs, breasts, and genitalia. The disease is endemic in parts of the western Pacific, Southeast Asia, Africa, and the eastern Mediterranean. Transmission is through the bite of mosquitoes of the *Anopheles, Culex, Aedes,* and *Mansonia* genera.[617] Unlike malaria, where most of the vector species are domestic and can be destroyed through the chemical treatment of dwellings, filariasis vectors are generally not found in the proximity of human habitation, so

TABLE 6–1

Larvicides Used on Mosquito-breeding Water

Name	LD_{50} (rat) (mg/kg)	Formulation	Use rate (lb/acre) (a.i.[c])
Abate®[b]	2,000	e.c.[c]	0.05
		Granules	0.10
Pyrethrum	1,500	Oil solution	0.01
Flit®MLO	10,000	Oil	(1–5 gal/acre)
Bromophos[d]	4,000	e.c.	0.05 –0.1
Gardona®[e]	4,000	w.p.[f]	0.10 –0.25
Dursban®[g]	163	e.c.	0.025–0.05

[a]a.i. = active ingredient.
[b]*O,O,O′,O′*-tetramethyl *O,O′*-thiodi-*p*-phenylene phosphorothioate.
[c]e.c. = emulsifiable concentrate.
[d]*O*-(4-bromo-2,5-dichlorophenyl) *O,O*-dimethyl phosphorothioate.
[e]2-chloro-1-(2,4,5-trichlorophenyl) vinyldimethyl phosphate.
[f]w.p. = wettable powder.
[g]*O,O*-diethyl-*O*-3,5,6-trichloro-2-pyridyl phosphorothioate.

wide-area spraying campaigns are essential to control. The exact number of human victims is unknown, but is significant among the several hundred millions living in endemic areas.[618]

Conventional Control Methods

Mosquitoes can be controlled with some success by proper water management.[619] The expedient of deepening the shoreline of a water body will reduce mosquito populations.[620] Obviously, the draining of static water bodies will reduce the number of sites for ovipositing. Aquatic vegetation influences behavior of the female mosquito and larvae, e.g., lack of aquatic plants means increased wave action and a decrease in both egg laying and larvae survival.[621] Improved water management arising from modernization of agriculture can reduce mosquito populations. The change from a rice agriculture to pineapple and sugar on Okinawa led to a reduction in anopheline and culicine populations.[622] In some instances cultural control methods may be feasible.[623] Occasionally, mosquito populations will undergo significant reduction in numbers for reasons so far unknown.[624]

The major method of controlling mosquito populations lies in the use of chemical agents applied either as larvicides to the infested water-courses or as adulticides. Adulticides can be applied via air or ground vehicles by spraying or fogging. In recent years ultralow volume (ULV) techniques have been developed, permitting the use of an ounce or less of insecticide per acre.[616,625] Cold or hot fogging apparatuses or mist blowers can be used with ground equipment. Larvicides are typically sprayed over the infested water, although application by air is feasible.[626,627]

Relatively few control agents are in widespread use. Table 6–1 presents information regarding larvicides, while Table 6–2 covers most of the in-use adulticides.

Biological Control

Recent years have witnessed an increased emphasis on the control of pest insects by non-toxic methods. Techniques can be classified into four generalized categories: predation, sterilization, growth regulants, and behavior-modifying substances, i.e., pheromones. The main thrust in mosquito control has been keyed to the use of predatory life forms.

The use of fish as predators on mosquito larvae has been justly considered the major biological control method. The *Gambusia affinis* species complex ("mosquito fish"), *Lebistes reticulatus* (the common guppy), and the Southeast Asian *Panchax* have been extensively evaluated and used in control programs in the United States, Japan, and other nations.[628,629] Use of fish in suitable habitats has led to substantial reductions in larvae populations.[629,630] However, there are pitfalls

TABLE 6–2

Adulticides Used in Mosquito Control

Name	LD_{50} (rat) (mg/kg)	Formulation	Use rate (lbs/acre) (a.i.[a])
Malathion[b]	1,375	e.c.[c]	0.5
		Tech. ULV[d]	0.45
Fenthion (Baytex®)[e]	500	e.c.	0.1
		Tech. LV	0.1
Chlorpyrifos (Dursban®)[f]	163	e.c.	0.5
Baygon®[g]		e.c.	0.07
Naled[h]	430	Tech. ULV	0.1
		Dust	0.2
Dichlorvos[i]	56–80	e.c.	0.05
Pyrethrins	1,500	Dust	0.01
		e.c.	0.01

[a]a.i. = active ingredient.
[b]*O,O*-dimethyl dithiophosphate of diethyl mercaptosuccinate.
[c]e.c. = emulsifiable concentrate.
[d]Tech ULV = ultralow volume technique.
[e]*O,O*-dimethyl *O*-(4-(methylthio)-*m*-tolyl) phosphorothioate.
[f]*O,O*-diethyl-*O*-3,5,6-trichloro-2-pyridyl phosphorothioate.
[g]*O*-isopropoxyphenyl methylcarbamate.
[h]1,2-dibromo-2,2-dichloroethyl dimethyl phosphate.
[i]2,2-dichlorovinyl dimethyl phosphate.

involved. *Gambusia* will displace indigenous fish populations,[631] and cannot tolerate low temperatures.[632] Likewise, the guppy cannot survive winter conditions; though it is fairly resistant to pollution,[629] it is unfortunately not as effective as *Gambusia* in general, nor can it survive long in the presence of piscivorous fish.[631]

Unlike the aquatic herbicide program, where the integrated use of biological and chemical control techniques has been proven effective, mosquito-predatory fish are not particulary tolerant to many commonly used insecticides. Abate® is tolerated at fairly high water concentrations by *Gambusia affinis,*[633] while chlorpyrifos,[633] DDT,[634] parathion,[634] and other larvicides have considerable effect upon fish populations.

Agamornermis culicis, a nematode parasite of various mosquito species, can have a dramatic effect on host populations. Petersen et al. report parasitism at levels as high as 91% in *Aedes sollicitans.*[636] Parasitism by mermithid nematodes is geographically widespread and affects a number of mosquito species.[637,638] Nematode parasite release in limited field evaluations has demonstrated the feasibility of using this technique in mosquito control.[639]

Insect pathogens such as *Bacillus thuringiensis*[640] and the fungi *Entomorphthora coronata*[641] and *Beauveria bassiana*[642] may constitute another control method if application techniques can be developed.

Radiation sterilization of male mosquitoes represents another, target-selective, approach to mosquito control.[643] Field studies have indicated feasibility.[644]

A recent approach to mosquito control lies in the use of juvenile hormone mimics (JHM) to prevent maturation. In 1966 Spielman and Williams demonstrated under laboratory conditions that a JHM had a lethal effect on larvae of *Aedes aegypti.*[645] Initial limited outdoor trials provided favorable results.[646,647] Isler and Vogel conducted small-scale field tests of five JHMs with mixed results. Stauffer® R-20458, 6,7-epoxy-1-(*p*-ethylphenoxy)-3,7-dimethyl-2-octene, and Zoecon® ZR-515, isopropyl-11-methoxy 3,7,11-trimethyl-2-cis/trans-4-trans-dodecadienoate, provided meritorious results that strongly indicated value as nontoxic mosquito control agents.[648] ZR-515 now constitutes the active agent of Altosid SR-10®, a commercial nontoxic (to man and mammals) formulation that, when applied to larvae-infested water, prevents adult emergence from the later pupae stage of the life cycle.[649] Various studies have confirmed the value of this material in mosquito control.[650,651]

Controlled Release Mosquito Larvicides

The development of controlled release mosquito larvicidal formulations has taken three parallel paths based upon the release matrix: elastomeric incorporation functioning through a diffusion-dissolution mechanism, plastic incorporation resulting in a leach-type release process, and encapsulation, wherein an internal agent reservoir is retained by means of a plastic or elastomeric membrane.

Elastomer-based Formulations

In the fall of 1964, the author, in seeking a rapid bioassay method for determining the toxicity and rate of release of organotins from antifouling elastomers, exposed various aquatic insect larva. Rapid culicine mosquito mortality was observed. Chloroprene-based materials releasing TBTO, TBTAd, TBTA, and TBTR were effective larvicides against *Culex pipiens*, *C. quinquefasciatus*, and *Aedes taeniorhyncus*, provided that several days' continuous exposure was allowed.[330,652]

The U.S. Army Environmental Hygiene Agency examined various organotin-releasing elastomers as both larvicides and adulticides. In the latter case, high vapor toxicity to mosquitoes, body lice, houseflies, and cockroaches was observed.[653] Sample 443A caused complete mortality to *C. quinquefasciatus* after a 10-min exposure period;[653] 72-hr *C. quinquefasciatus* second instar larvae exposed to TBTC/neoprene, TBTF/neoprene, TBTO/neoprene, and TBTA/neoprene gave initial 90 to 100% mortality, but in repeated exposures failed to provide the long-term efficacy desired.[654] Boike and Rathburn evaluated 23 organotin/elastomer compounds against various mosquito larvae and ova in a repeated flooding-type experiment.[655] Results in clear tap water indicated poor ovacidal properties but high larvicidal activity, while in water containing organic matter results were generally poor against larvae.

These studies and others indicated that the hydrophobicity of organotins and their rapid absorption by organic constituents of natural waters, coupled with the logarithmic decay curve, rendered them unsuitable as larvicides.[655,656]

Indirect evidence from controlled release organotin bioassays against mosquito larvae gives rise to the hypothesis that release from an elastomer is molecular in form.[657] Laboratory studies by various agencies have shown that the LD_{90} for "pure" TBTO dispersed in water is at least ten times higher than when this agent is released from an elastomer. Simple preparation of a TBTO concentration in water through serial dilution *probably* results in dispersed aggregates rather than molecules, a direct and expected result of the extreme hydrophobic nature. Release from an elastomer, based on dissolution kinetics, ought to be molecular. The statistics of insect contact and penetration will vary considerably between the two types of "solution."

Problems associated with the use of organotins in a natural setting led to the incorporation of conventional larvicides in various rubber formulations. Table 6–3 lists insecticides which were discovered to have some degree of solubility in various elastomers and several pertinent properties. Most materials of this group have never been subject to rigorous testing. Chlordane/EPDM,[a] Dibrom®/chloroprene, Abate/acrylonitrile, and Abate/chloroprene were ineffectual.[654] However, three Abate/EPDM materials provided 90 to 100% mosquito larvae mortality through 26 consecutive 72-hr exposure periods.[654] The larvae LC_{95} for Abate is around 0.6 ppb,[658] a level achieved by exposure of an Abate/EPDM pellet (0.045 g) to water at 25°C for 48 hr.[659]

Abate/EPDM *floating* formulations were subsequently patented as controlled release larvicides.[660,661] Reduction in the amount of carbon black used in organotin and organophosphorus elastomeric compounds provided a specific gravity of less than one and thus a floating material. Since mosquito larvae inhabit the infraneustral zone in stagnant or near-stagnant water, it is likely that the Abate molecules spread as a film across the water surface and the target contacts the toxicant when it surfaces to breathe. Sinking Abate/EPDM is ineffective for this use, although it may be apropos for bottom-dwelling insect larvae. The preferred EPDM elastomers are of the nonconjugated diolefin interpolymers described in U.S. Patents 2,933,480 and 3,000,866.

Similar phosphorothioates, Baytex®, chlor-

[a]Ethylenepropylene-diene terpolymers.

TABLE 6–3

Insecticides Soluble in Elastomers

Toxicant (common name)	RT [a]state	bp/mp (°C)	Oral LD_{50} (mg/kg) (rat)
Chlordane[b]	Solid	350(sublimes)	570
DEET[c]	Liquid	297	1950
DDVP[d]	Liquid	192	56
Dieldrin[e]	Solid	123	47
Endrin[f]	Solid	217	5
Heptachlor[g]	Solid	95	90
Lindane[h]	Solid	114	125
Malathion[i]	Liquid	196	1375
Dibrom®[j]	Liquid	203	430
Abate®[k]	Liquid	213	2000
Dursban®[l]	Solid	43	163
Baytex®[m]	Liquid	196	500

[a]Room temperature.
[b]Octachloro-4,7-methanotetrahydroindane.
[c]*N,N*-diethyl-*m*-toluamide.
[d]2,2-dichlorovinyl dimethyl phosphate.
[e]Hexachloroepoxy-octahydro-endo,exo-dimethanonaphthalene.
[f]Hexachloroepoxy-octahydro-endo,endo-dimethanonaphthalene.
[g]Heptachlorotetrahydro-4,7-methanoindene.
[h]Benzene hexachloride (gamma isomer).
[i]*O,O*-dimethyl dithiophosphate of diethylmercaptosuccinate.
[j]1,2-dibromo-2,2-dichloroethyl dimethylphosphate.
[k]*O,O,O′,O′* -tetramethyl *O,O*-thiodi-*p*-phenylene phosphorothioate.
[l]*O,O*-diethyl-*O*-3,5,6-trichloro-2 pyridyl phosphorothioate.
[m]*O,O*-dimethyl *O*-[4-(methylthio)-*m*-tolyl] phosphorothioate.

pyrifos (Dursban®), and malathion are incorporable and soluble in EPDM, and can be expected to release upon immersion. The basic formulation found effective in U.S. Army Environmental Hygiene Agency tests is as follows:[661]

Ingredient	Formulation (parts by weight)
EPDM	100
Zinc oxide	2.0
Stearic acid	1.0
Benzothiazyl disulfide	0.5
Sulfur	1.5
Tetramethylthiuram disulfide	1.5
ISAF carbon black	5.0
Paraffin oil	22.0
Larvicide	3.0–22.0

To increase loss rate, the amount of carbon black can be reduced or a less structured black such as EPC or FEF used. In order to decrease loss rate a more structured black such as SAF or a greater amount of ISAF can be used.

No difficulties, save that of vapors, are encountered in the mixing and curing of Abate, Malathion, and other phosphorothioates. Mill mixing proceeds best at 120°F and temperature ought not to exceed 135°F. Internal mixers generate considerably higher temperatures and small batch sizes only should be used. It is convenient to pre-mix all ingredients, dump, and then repass, adding the phosphorothioate. Cure conditions can vary, but the author noted that a 30- to 40-min cycle at 280 to 300°F provided an adequate formulation. In mixing and curing, adequate ventilation is a must to eliminate human health hazards arising from vapor exposure. Autoclave-type vulcanizations are probably superior, in that a postcure cold water flush can effectively purge vapors prior to opening.

Abate is an extremely effective larvicide in concentrations as low as 0.005 ppm. It has a low mammalian toxicity, but shows considerable avian toxicity.[598] Although Abate is considered nonpersistant,[104,663] it may be transferred through the food chain and concentrated at high trophic levels.[664]

Abate used at dosages commensurate with the destruction of mosquito and midge larvae is safe for shrimp amphipods and various microorganisms.[664]

The *Aedes aegypti* control problem may be especially amenable to the use of controlled release Abate formulations. *Aedes* tend to breed in artifacts of man such as discarded tin cans, bottles, other containers, tires, water barrels, ditches, and so on.[665] Abate has been shown to be an effective agent for treatment under these conditions,[666] with both granular and emulsifiable concentrates providing up to 34 weeks of control.[667] Also, Abate is far more effective than organotins under polluted water conditions.[668]

One potential use for the Abate/EPDM formulations would lie in permanently attaching, with adhesives or mechanical means, a strip to the inside of a water barrel or other *A. aegypti* breeding container. Clearance for use at 1 ppm or lower in potable waters makes this usage attractive.

A number of chemicals have been discovered that mimic juvenile hormones in insects.[669] By proper application to the pest-infested environment, it is possible to interfere with the biological cycle of the insect and thus effectively destroy it at some stage prior to its becoming a health or agriculture hazard. Since these materials are extremely target specific, they generally affect no other member of the general biota, and are considered to be "third generation" insecticides.

Altosid®, commonly termed "methoprene," has been incorporated in a controlled release polyurethane foam matrix.[670] The unformulated material was effective for only 1 to 2 days, whereas the controlled release product exhibited up to 53 days' efficacy in catch basin tests. Altosid formulation SR-10 has been registered by the Environmental Protection Agency for specific uses.[671] Methoprene is essentially nontoxic to mammals and birds, and apparently nonteratogenic and nonmutagenic to all life forms. It is somewhat toxic to fish at levels far higher than those necessary for effective mosquito control.[649] *Culex pipiens* has developed resistance to methoprene, and cross resistance with hydroprene has been noted.[672]

Plastic-based Formulations

The leading program in the development of plastic-based controlled release mosquito larvicides has been conducted at the U.S. Army Environmental Hygiene Agency (USAEHA) at Edgewood Arsenal, Md. Initial work was with plaster briquettes impregnated with Abate or malathion.[673] Outdoor small pool tests showed good control with minimal ecological disturbance of nontarget organisms.[674] This effort was followed by the development and evaluation of Abate, Dursban, naled, and malathion larvicides impregnated in plastic pellets of polyamide, polyvinyl chloride, and polyurethane.[675]

Briquettes failed to provide long-term control for various reasons, and dispersal also represented a problem.[676] malathion, for instance, washed out fairly rapidly, so only a few days' release was evident. Of those insecticides evaluated in this manner, Dursban appeared to be the superior choice.[676-678]

Polyurethane pellets failed within several weeks.[676] Dursban/PVC proved to be superior in terms of longevity over Abate/PVC, Dursban/polyethylene, etc., and consequently was selected for further study.[676,678,679]

Dursban (common name, chlorpyrifos) has long been recognized as an effective mosquito larvicide, often finding use in those areas where resistance to chlorinated hydrocarbons and other organophosphorus materials has developed.[680,681] It is active in septic water where larvae feed on organic absorbants.[682] Dursban has demonstrated greater longevity as a residual spray than various other insecticides;[683] due to its high target toxicity, it provides a wide margin of safety to even the more susceptible fish species,[684] and probably has less environmental impact than Abate.[663]

Dursban/polyethylene 9.9% and 11.5% concentrations, and Dursban/polyvinyl chloride 10.0% were evaluated in artificial outdoor pools.[685] Three formulations provided effective control for the entire 24-week test period. One material, PVC-10, was prepared at USAEHA by mixing Geon® 135 (B. F. Goodrich Chemical Co., Cleveland, Oh.) PVC resin at 1:1 (w:w) ratio with di(2-ethylhexyl) phthalate plasticizer; Dursban was added (10.0%) and cured at 125°C for 30 min in a pellet mold. Further testing with Dursban/polyethylene against all stages of the life cycle demonstrated at least 21 weeks of control in artificial pools.[686] This material was also studied regarding its effect on nontarget populations of gerrids, chaoborids, and chironomids.[687] In all cases, populations were reduced but not eliminated at treatment levels providing effective mosquito control.

One-time applications of 5 ppm active of two formulations of Dursban/polyethylene and one of Dursban/PVC to artificial pools provided effective mosquito control for 18 months or longer.[688] The large amount of each formulation required, 131 to 153 kg/ha, would probably render large-scale use impractical. Reduction of dosage in laboratory tests showed that continuous exposure of larvae to 0.23 ppb, as compared to the conventional LC_{90} of 0.9 ppb, gave 99.7% control.[689]

Dursban/chlorinated polyethylene formulations were prepared by the Dow Chemical Co. and evaluated at USAEHA. It was shown that increasing the surface area of a test pellet results in a geometric increase in release rate, while increasing the percentage of active agent results in a linear increase in loss rate.[690] Materials were ineffective at 0.1 ppm active, and only marginally effective at 0.5 ppm. By proper tailoring of pellet size and Dursban concentration, it should be possible to achieve long-term control at practical treatment rates. Use of the better material, a 9.7% active ingredient (a.i.) formulation in a pellet of S/V (surface/volume) ratio of 4.4 would require about 31.1 kg/ha.[691]

Encapsulated Materials

Barrett Green prepared the first gelatin microcapsules in 1942 by using a colloid-rich gelatin coacervate to envelope a dispersed droplet of an organic phase.[692] A number of preparatory methods are known: dehydration,[692] spray drying,[693-695] fluidized bed,[696-698] mechanical,[699-701] electrostatic,[702-704] and *in situ* polymerization.[705,706] Microencapsulation theory, techniques and applications, has been amply described in the current literature.[707]

Capsule walls can be made solvent resistant[709] or solvent degradable so that exposure to water, body fluids, etc., results in the release of the agent.[710] The polymeric membrane, or encapsulant, may release its contents through rupture,[711] diffusion outward[712] or inward,[713] or solution.[714] Variation of the membrane material or geometry wall will alter the permeability and hence the loss rate of a microcapsule. Fanger has generalized this as follows:[692]

Polymeric wall material: property	To decrease permeability:
Density	Increase
Crystallinity	Increase
Orientation	Increase
Cross-linking	Increase
Plasticizer level	Decrease
Fillers	Increase (depending upon type of filler)
Solvent used in film preparation	Use strong (vs. poor)

Capsular parameters	For lower permeability
Size	Increase
Wall thickness	Increase
Configuration	As spherical as possible
Conformity	As regular as possible
Cross-linking (post treatment)	Utilize
Multiple coatings	Utilize

Encapsulation techniques have been used in drug delivery systems,[715] cosmetic preparations,[716] dyes,[717] flavors,[717] etc. Polymers used as the coating material or membrane wall include carboxymethylcellulose, ethyl cellulose, ethylene vinyl acetate, gelatin, nitrocellulose, polyvinyl alcohol, various waxes,[717] poly (lactic acid),[718] polyamides,[719] and other substances.

Possibly the first commercially available encapsulated insecticides was Tossits® (Wyco International Co., West Palm Beach, Fla.), a gelatinous *macro* capsule containing a DDT/oil mixture that ruptured in time after immersion, releasing its contents. Over the last few years a number of encapsulated pesticides have appeared. Miller et al. examined proprietary encapsulated malathion and Abate,[720] and found that, even at 220 ppm, all formulations failed to achieve mosquito control beyond the 7th week in field tests. Since various emulsions of these larvicides will provide up to 3 weeks of effective mosquito control, the extension of control duration with these experimental materials was probably insufficient on a cost-effective basis.

The first commercial encapsulated insecticide which can be utilized against mosquito larva is the Pencap-M® formulation marketed by the Pennwalt Corp., Monrovia, Cal. The active ingredient is methyl parathion, and the prescribed usage is against insect pests of alfalfa, sweet corn, and cotton.[719,721] Pencap-M displays a number of advantages, perhaps the most important being the decreased mammalian toxicity observed. The

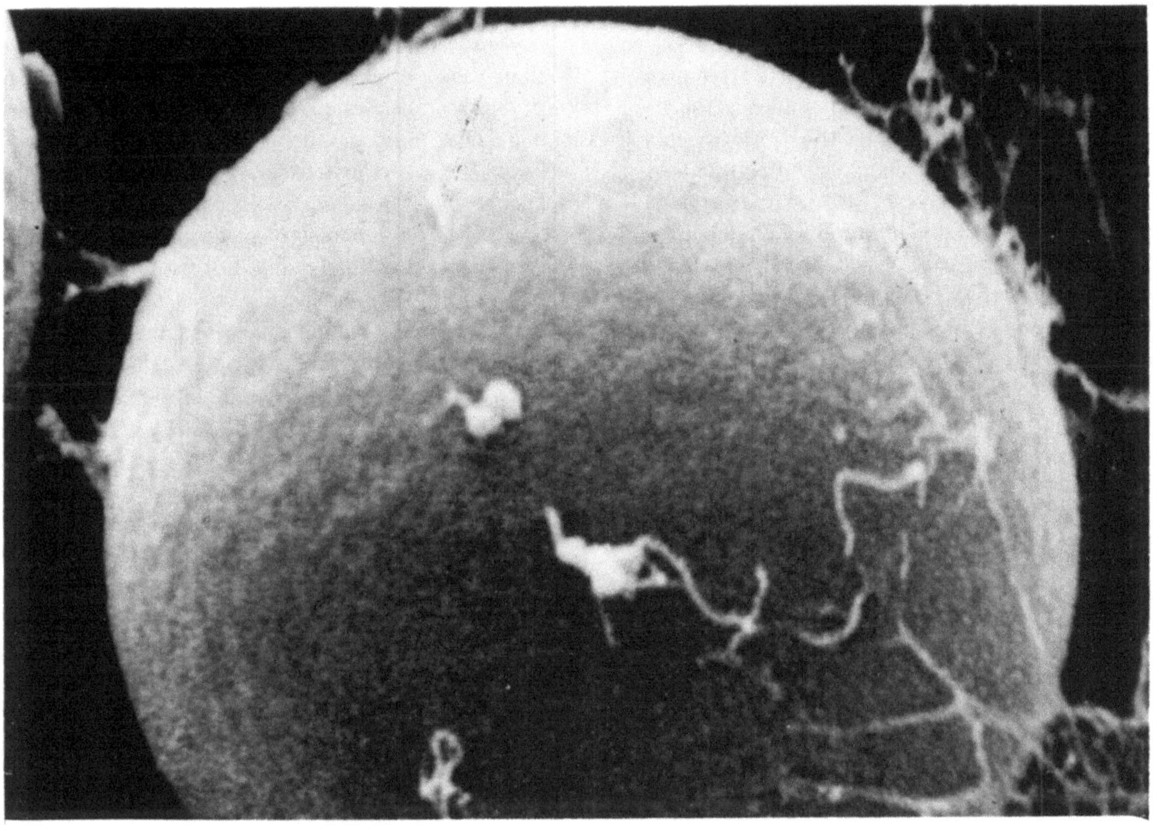

FIGURE 6–1. Pencap-M® microcapsule magnified 1,000 times. (Photograph courtesy of the Pennwalt Corp., Monrovia,

LD_{50} (mice) for nonencapsulated methyl parathion is 10 mg/kg, while the encapsulated material has a LD_{50} (mice) of 90 mg/kg.[722] Likewise, the dermal LD_{50} (rabbit) increases from 100 mg/kg for the emulsifiable concentrate to 2250 mg/kg for Pencap-M. All indications are that the encapsulated formulation can be handled with far less toxic hazard to agricultural workers, etc. The actual extension of residual activity and, thus, the effective period of control, is not particularly exciting. Emulsifiable concentrate (e.c.) formulations provide 1 to 3 days' residual, while Pencap-M at the same application rate appears to give 5 to 8 days' effective control. Other experimental organophosphorus encapsulated formulations are reported as providing up to 12 weeks' control.[719]

The capsule wall of Pencap-M is a cross-linked polyamide, and is created by the reaction of sebacoyl chloride and polymethylenepolyphenylisocyanate with an ethylenediamine/diethylenetriamine mixture at the interface of a microdroplet of the agent in water (Figure 6–1).[719]

Controlled Release Mosquito Adulticides

The control of adult mosquitoes is an important adjunct to the use of larvicides as a means of preventing disease transmission and removing annoyance to man and his domestic animals.[590] Materials in wide use by pest control operators have been described elsewhere.[723] Various application methods are used,[724] including ground spraying and fogging and aerial spraying. In recent years the use of ULV techniques have become increasingly popular. The cost effectiveness of a controlled release adulticide would likely be comparable to the efficacy of ULV for a given task. Droplet size is a critical parameter for ULV, so the design and use techniques of application equipment and devices become vital.[724,725] ULV methodology may involve ground spraying[726] as well as applications by aircraft.

Mosquito control in and around dwellings is an important aspect of public health, especially in yellow fever endemic areas; it is also desirable to prevent annoyance. Pyrethrins, usually synergized, lindane, and malathion are generally used as space sprays, and feature quick kill and short residual life.[727,728] Other materials such as DDT having a relatively long environmental life have been used as residual sprays in dwellings, but residual sprays are often inadequate, depending upon the target species and other factors.[729,730]

The first known long-lasting controlled release adulticide arose from the observation that dichlorvos (dimethyl-2,2-dichloro-vinyl phosphate) can be simply incorporated in various polymeric materials and will release slowly as a vapor.[731,732] Dichlorvos in wax, polyvinyl chloride resins, and permeable plastic dispensers showed long-lasting control in storm sewer catch basins.[733] The commercial product manufactured by the Shell Chemical Co. is the No-Pest® strip, a formulation composed of around 20% dichlorvos in a polyvinyl chloride base (probably B.F. Goodrich Co. Geon 101 E.P.) with appropriate plasticizers and stabilizers.[731] Plasticizers may be those common to the art such as dioctyl phthalate, dibutyl phthalate, tri(butoxyethyl) phosphate, and others. The subject patent claims biological efficacy for dichlorvos as a volatile insecticide released from a number of plastic materials. Materials similar to dichlorvos such as thionodichlorvos may also be used as the active agent.[734]

Dichlorvos is an effective adulticide against mosquitoes, houseflies, and other annoying flying insects. It has been used in killing jars,[735] space sprays,[736,737] emulsifiable concentrates for larviciding, and baits. Dichlorvos is considered to be nonpersistent,[738] hydrolyzing rapidly through natural environmental factors, and dichlorvos is also hydrolyzed readily in mammals through hepatic metabolic pathways without persistent tissue residues.[739-741] Whether a human health hazard from long-term exposure to dichlorvos vapors emitted from various formulations exists remains unknown, although such effects, if they do exist, are subtle.[742]

Controlled release dichlorvos has been successfully used in several other commercial products: flea collars and flea tags for cats and dogs, and as a broad-spectrum anthelmintic for humans and domestic animals.[743,744]

The No-Pest product and its predecessors have been demonstrated as being effective in catch basins,[745] as residual fumigants in huts,[746] and in enclosed areas such as barns, homes, and various business establishments. Longevity varies from 2 to 4 months, depending upon temperature, ventilation, and other factors. The loss mechanism probably relies upon the plasticizer as a carrier material, moving the incorporated dichlorvos liquid to the plastic surface where it volatilizes. Dichlorvos is not soluble in polyvinyl chloride or any other known plastic material, so a diffusion-volatilization mechanism is not possible. Examination of the No-Pest strip discloses that vapor emission ceases, in the practical sense, after about one third of the quantity of dichlorvos incorporated has been released. The residue agent migrates very slowly, if at all. However, physically cutting the strip to disclose a new surface provides a temporary reemission of dichlorvos. One explanation for the observed behavior assumes that a three-phase system exists: a material that serves as a physical carrier or a solution path for the dichlorvos molecules, and can migrate through the plastic. It is well known that plasticizing chemicals incorporated in vinyl and other plastics slowly move to the surface, a phenomenon termed "blooming." At the surface, loss is through volatilization, dissolution, mechanical washing, or other processes. The movement and resulting loss continues until a thermodynamic equilibrium is reached. Consequently, the plasticizer is never totally lost, so any agent dependent upon its migration will also never be totally lost – one reasonable explanation for the approximately 30 to 35% of the incorporated dichlorvos loss from No-Pest strips.

Repellents

Mosquitoes and other noxious insects and arachnids can be repelled from the human body or human artifacts using various chemical agents. Repellency as a protective device has been approached by three generalized means. Considerable elaborate work has been performed in an attempt to develop an orally effective insect repellent, but the desired goal of an insect repellent prophylactic drug has not been achieved.[747] Space repellents, which are designed to prevent insect entry into a protected zone, and individual repellents that are applied directly to the human dermis or animal fur constitute the other areas of investigation.

The use of space repellents is quite old. The burning of sulfur and various oils has been an effective repellent process for hundreds of years.

Kerosene-type solvents absorbed on vermiculite and broadcasted as granules will provide up to 1 week of area repellency.[748,749] It has been noted that various insecticides applied as a spray can repel or attract mosquitoes, e.g., BHC (benzene hexachloride) repels, Dibrom is neutral, and malathion emulsion attracts.[750] Additives such as the deodorant Deodall® (Sindar Corp., New York, N.Y.) will render an attractive adulticide such as malathion repellent.[751] Space repellents applied in the conventional manner as sprays, dusts, etc. give only very short-term protection, a few days at most.[751-754]

Area-wide space repellency is probably an easier achievement than protecting an individual or a small area such as a room in a dwelling. Bar-Zeev and Sternberg have pointed out very clearly that the distance of effective repellency from a focus of repellency (e.g., single granule) is only a few centimeters.[755] Preparations used on human skin essentially prevent biting and not insect approach, landing, and exploration, so an area repellent must have characteristics markedly different from personal repellents. That is, to spray a room with DEET (diethyl-*m*-toluamide) or to incorporate this agent in a plastic dispenser similar to that of the No-Pest strip would provide little if any protection to humans or animals in the vicinity.

Repellent materials applied to wide-mesh netting will provide repellency for varying periods of time. Treatment of the fibers of 1/4-in. mesh netting prevented mosquito passage for up to 92 days, depending upon the agent used.[756] In a subject study dipentyl tartrate was superior, while topical repellents such as DEET (21 days) were markedly inferior in use. Treatment of cloth with repellents can provide up to 4 weeks' protection[757]

Although hundreds of candidate repellents have been topically applied to humans, only four have seen extensive practical usage: DEET, ethyl hexanediol (Rutgers 6-12), dimethyl phthalate, and Indalone® (butyl-3,4-dihydro-2,2-dimethyl-4-oxo-1,2 H-pyran-6-carboxylate). It is generally observed that DEET, applied as a 5 to 75% lotion to human skin, is superior to other repellents in terms of protective longevity,[758] safety, and wash and abrasion resistance. DEET is effective not only against anophelene,[758,759] culicine,[759,760] and *Aedes*,[758,761] mosquitoes, but also against deer flies (*Chrysops atlantica* and *Chrysops discalis*),[758,762,763] black flies,[764,765] stable flies (*Stomoxys calcitrans*),[758,763,766] and sand flies (*Culicoides canithorax*),[758] and is used as a clothing impregnant against rat fleas (*Xenopsylla cheopis*),[758] chiggers (*trombicula splendens*),[758,767] a few tick species,[768,769] and biting midges.[763] It is not particularly effective against biting gnats[770] and other flying insects and most arachnids.

DEET and other repellents applied to skin and clothing are lost by volatilization, the vapors either establishing a repellent zone through which the target organism will not enter, or possibly by masking the normal attractants (in the case of human skin) so that the noxious insect does not perceive the presence of a suitable host. However, loss from human skin is not only outward but also inward via absorption, with probably the major loss occurring in this fashion.

Indalone provides excellent protection against mosquitoes, black flies, house flies (*Musca domestica*), chiggers, gnats, stable flies, and sand flies.[771] It is undoubtedly superior to DEET in providing tick and mite protection. Although Indalone is generally more persistent than DEET, it is not nearly as cosmetically acceptable, depositing a very oily or greasy feel to the skin and acting as a catcher of dirt and dust particles.

The general disadvantage with all repellents lies in their lack of persistency after treatment. On the human dermis any practical evaluation of DEET under normal environmental conditions provides 2 to 4 hr protection. With perspiration this drops to 1 hr or so, and under tropical conditions with profuse perspiration or in rain or ground water perhaps a half hour at best. Ethyl hexanedial is decidedly inferior to DEET in this respect, while Indalone is somewhat superior.[763] In terms of abrasion resistance, all three materials leave much to be desired. The rubbing or wiping action of clothing quickly removes ethyl hexanedial; in this respect DEET is intermediate and Indalone superior. Other repellents are available, but even the best of experimental compounds will hardly provide over 7 hr protection under idealized conditions.[772]

DEET conventionally applied as a topical lotion suffers from a number of disadvantages: it smells bad, tastes very unpleasant, and dissolves paints and plastics, including eyeglass frames and watch crystals. Physiological gross effects on humans are rare, although cases of blistering have been reported.[773] (The fact that DEET adsorption through the skin may not be physiologically desirable seems to have escaped the attention of investigators.) The major shortcoming is that a relatively large amount of material must be applied for any reasonable repellent longevity. The minimum effective dose is around 50 to 80 $\mu g/cm^2$, probably 10 to 100 times greater than desirable.

In 1968 the author observed that DEET, Indalone and, to a lesser extent, ethyl hexanedial are soluble in the carboxylated polyacrylates discussed in Chapter 3. Since the Carboset materials are known to form films upon topical application, it was conceivable that the volatility and dermal absorption of a given repellent might be decreased if formulated with Carboset, thus increasing longevity. Methods of preparing lotions and ointments generally rely upon the use of non-film-forming materials,[774] although additives are used to enhance adhesion of the agent to the skin.[775] Likewise, insect repellent creams that are non-film formers have been produced.[776] Medicinal delivery from carboxypolymethylene polymers, the B. F. Goodrich Co. Carbopol® group, in cream and gel-type preparations showed sustained release.[777]

DEET was incorporated in Carboset-type formulations by Conrady and Stockman at levels far in excess of the solubility limit.[778] Application to humans of a DEET-Carboset-alcohol lotion is said to have provided mosquito repellency for up to 24 hr. In elaborate studies using both meta-DEET and para-DEET in formulated Carboset-based film-forming lotions, Kurtz et al. found that the presence of the polymer enhanced wash and abrasion resistance and longevity.[773] Unfortunately, the materials requested and used contained the repellents at levels far in excess of that critical to proper film formation, so that the resultant candidates were cosmetically unacceptable.

When the agent solubility limit in Carboset is exceeded, i.e., above about 8% DEET or 12% Indalone by weight, reversion occurs and the polyacrylate solubilizes in the agent. Then, as the *agent* volatilizes, a rather poorly adhering topical film is formed. By *not* exceeding the solubility limit an excellently adhering film is formed. The following generalized formulation used by the author will provide 12 to 18 hr mosquito protection under rigorous conditions of heat, perspiration, and rainfall and considerably in excess of that against black flies:

Film-forming polymer	100 parts
Carboset 525	
Carboset 526	
Solvent	450–600 parts
Ethanol (95%)	
Ethanol/water 60/30	
Adhesion promoters	
Carboset 515	3–6 parts
Ethyl acetate	0.5–1 part
Stabilizer	0.1 part
Ammonium hydroxide	
Curative (optional)	0.01–0.06 parts
Calcium ion (from $CaCl_2$)	
Agent	0–12 parts
DEET	4–8 parts
Indalone	9–12 parts
DEET/Indalone mix	6:6 parts

Silicone oil *decreases* wash resistance, and all emollients evaluated decrease longevity in the above, which, while not cosmetically elegant, is certainly acceptable.

Whereas the author's experience has been essentially with the Carboset materials, there is ample evidence that long-lasting mosquito repellents may be based upon the incorporation of DEET etc. at or below the solubility limit in polysiloxanes,[779] natural or synthetic elastomers including natural rubber,[780] acrylic and vinyl latex,[781] and polyvinyl pyrrolidone.[782]

The Problem of Resistance

The increased use of both chlorinated hydrocarbons and organophosphorus insecticides has led to the development of increasing tolerance by target mosquito species. Resistance has been noted with DDT,[626,783,786] toxaphene,[626,787] ethyl parathion,[626,784] methyl parathion,[626] Baytex,[626,787] malathion,[784,785,789] dieldrin,[786] BHC,[786] and fenthion.[788] Mosquito abatement districts and investigators have attempted to overcome the development of tolerance in mosquitoes by various means, including using a given material only for larviciding and a different one for adulticiding,[790,791] using synergists with a given insecticide,[791] decreasing the use of chemical agents through improved water management,[792] using nontoxic oils, expecially Flit® MLO,[626,784,793,794] developing new compounds,[795–797] and improving methods of using older materials.[798]

The use of controlled release materials, probably as larvicides, will not alleviate the growing resistance problem but, to the contrary, may well enhance it. Sublethal dosages of insecticides tend to lead to the development of more resistant populations.[610,799] Since controlled release larvicides will conceptually be used at ultralow concentrations, and long-term release will be achieved for cost effectiveness, considerable sublethal dosing can be expected, resulting in many generations of the same species in the same locale exposed to the same larvicide. The possibility of enhancing tolerance through the use of controlled release methods has yet to be investigated.

Mosquitoes, generally considered to be annoyances to humans, also vector various domestic-animal diseases such as equine encephalitis and heartworm disease in dogs,[800] and can kill cattle by concentrated attack.[801] The use of long-lasting repellents may be considerably beneficial to the health and well being of domestic animals, leading to increased productivity by beef and dairy cattle, fowl, and so on.

BLACK FLY CONTROL

Diptera Simuliidae, "blood-sucking black flies," are vectors of human and domestic animal diseases. During biting the insect can transmit avian leucocytozoons to domestic poultry and toxins to horses, cattle, and sheep. Venezuelan equine encephalitis and possibly Eastern encephalitis are vectored to man.[802,803] However, the transmission of the parasite *Onchocerca volvulus* to man represents the most important relationship between humanity and the Simuliidae.

Onchocerciasis is a chronic, usually nonfatal filarial disease. During biting, microfilariae are injected into the dermis, where they develop into adult worms at or near the locus of injection. Reproduction occurs and myriads of microfilaria are released, usually living in the epidermis, where they cause itching and other minor effects. There is some migration through tissues, with a number of the parasites reaching the eye and causing extremely serious effects. The major syndromes are various ocular disturbances and blindness. Infection is accumulative, the parasite burden increasing with each bite of the infected *simulium* flies.

Onchocerciasis is widespread through central Africa, with major foci in subsahara West Africa and less extensive areas in East Africa. Endemic areas exist in Mexico and Central and South America. The disease affects at least 20 million people in West Africa alone, and unknown, though substantial, numbers elsewhere.[804,805] Blindness rates of 20% or more are recorded for various severely infested areas of West Africa.[805-807]

Humans tend to avoid settling in areas where black fly populations are laıge, and when they accept the risk the high incidence of eye problems leads to serious socioeconomic effects. In wide areas of the African savanna, overall human productivity is significantly reduced and much good agricultural land abandoned. Hamon and Kartman of the World Health Organization have stated: "It is thus no exaggeration to say that the economic development of very extensive areas, particularly in the savannas of tropical Africa, cannot be reasonably undertaken without first controlling onchocerciasis."[804]

No method of conferring immunity exists, and drugs presently in use effectively destroy microfilariae but not all the adult parasites. Severe allergic reactions are not uncommon with diethylcarbamazine, and the other principal medicinal, suramine, must be given as a weekly injection for effectiveness. Of course, health services are not readily available in many of the endemic areas.

Black fly ova are deposited on partially submerged vegetation or stones on or near running water. Concrete dam spillways are a popular modern site.[808] Eggs hatch in 3 to 7 days and the emerging larvae migrate downstream seeking a favorable site for attachment. Adhesion to rocks or other solid features of the stream bottom is secured through exudation of a salivary "silk." Larvae then pass through six or seven instars over a 7- to 10-day period and pupate, with the adult emerging within several days. Only the adult female seeks blood, a requisite for proper egg development. During the course of biting and ingesting, human blood microfilariae from an infected individual are taken in, and those that survive the chemical rigors of the fly gut mature and multiply rapidly. The black fly, now infected, will transmit the parasite in future blood-sucking activity. The adult fly lifespan averages 3 weeks, although some individuals in a given population may survive up to 3 months (Figure 6–2).

Simuliidae live in fairly warm, heavily oxygenated water during the larval and pupal stages. Larvae are nonselective feeders. Water flow may be slow, moderate, or fast – anywhere between ½ cm^3/sec to 24 cm^3/sec, depending essentially upon the oxygen demands of a given species.

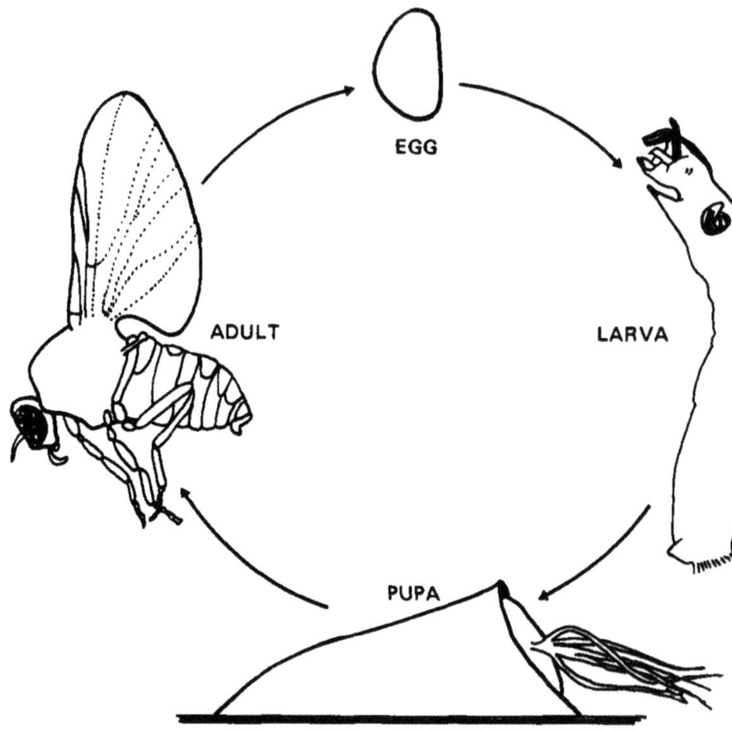

FIGURE 6–2. Black fly life cycle.

Several species adhere to various marine animals such as fresh-water crabs[804] and mayfly and dragonfly nymphs.[809]

Both adulticides and larvicides are used in the chemical control of black flies. In general, no practical methods of control through water management exist. Residual adult control for up to 10 days can be achieved through spraying operations, usually by aircraft. Space control using fogging equipment has been successful for short-duration protection. Although DDT is the adulticide of choice in most areas, effective control by this route has been realized in relatively few instances,[810] but attempts continue.[811]

Larviciding requires the application of the control agent to the infested watercourse. Treatment through repeated aerial spraying is probably the only feasible control method in regions possessing a multiplicity of flowing water bodies.[812] DDT has long been the preferred larvicide, but a number of other materials have shown promise. Abate,[813,814] methoxychlor,[814,815] and Dursban[816] have been successfully used in laboratory and field evaluations. While DDT remains the agent of choice for large-scale eradication campaigns,[810,815,817] the undesirable effects on fish and other members of the biota may well dictate the use of ecologically less hazardous materials,[104,522,595] e.g., methylthio analogues of DDT which are biodegradable and, unlike DDT, are not magnified in the food chain.[818] Juvenile hormone mimics such as Altosid (isopropyl 11-methoxy-3,7,11-trimethyl-2,4-dodecadienoate) have been successfully used against black flies under laboratory conditions and may represent a potent new control weapon.[819]

The basic difficulty associated with black fly control arises from the nature of the larval habitat. Water flow is rapid at most sites, so larvicides are quickly washed downstream from the application site. Although downstream carriage provides considerable control range, high larvae mortality being achieved many miles downstream,[820] incoming waters rapidly dilute the applied toxicant to sublethal concentrations. Repeated applications are essential in that stream intoxication lasts only a few hours at best and adult flies continue to lay eggs. The treatment scheme varies from area to area, ranging from 26 applications at 1-day intervals to 5, 6, 7, 10, or 12 applications at 5- to 14-day intervals.[821] The amount of agent necessary to maintain the conventional dosage of

0.1 ppm or better for 30 min at a given site can be excessively large. Dry season treatment of sections of the Victoria Nile would require 272 lb of DDT per application, while the Congo at Leopoldville (1 million cusecs discharge rate) would call for 8 tons!

The utilization of controlled release formulations against aquatic pest life is generally considered as being limited to static or slow-moving water.

A continuous 0.01 ppm appears to be an economical and effective controlled release concentration in aquatic weed and snail control. Black fly larvae exposed to this dose rate would require around 4 days for effective mortality. Whole-stream intoxication does not appear reasonable. Past experience with DDT incorporated in plaster blocks, wood shavings, and vermiculite has proven to be inferior to liquid doses.[822]

Since black fly larvae attach to a firm substrate, it is reasonable to assume that (1) only that portion of the substrate prone to attachment need be considered for treatment, and (2) maintaining a suitable toxicant concentration at the substrate/water interface is all that is necessary, not a lethal concentration throughout the depth and breadth of the watercourse. Snail control has been achieved by painting concrete irrigation systems during drawdown with organotin paints (work to be published by Dr. Clive Shiff, Blair Research Laboratories, Salisbury, Rhodesia); could the same technique be used in black fly control?

It has been shown that DDT, Abate, Dursban, and methoxychlor can be incorporated in film-forming polymeric substances. When formulated as paints and applied to rock surfaces, they will, after air drying, show at least several weeks' longevity in a laboratory dynamic water flow system.[823] Such a technique, if workable, would only be of value if the infested watercourse were dry at some time during the year. Treatment of a flowing river bed presents an extremely complex problem. One approach might be to incorporate the toxic agent in a dense hydrophobic film-forming polymer and pump it onto the substrate, allowing current flow to spread-coat the formulation in a downstream pattern!

OTHER INSECTS OF PUBLIC HEALTH IMPORTANCE

The application of controlled release technology as a means of eradicating or controlling most arthropods of public health importance has lagged far behind the antimosquito effort. Little or no work has been done in the development and evaluation of the methodology discussed in this text against a number of insect and arachnid pests that would appear to be vulnerable.

Several species of ticks are implicated as vectors of various acute febrile diseases. *Dermacentor andersonii* and *D. variabilis* transmit Colorado tick fever from a small mammal reservoir to man. Other tick species carry various *Rickettsia* of human importance: boutonneuse fever, Mediterranean tick fever, South African tick fever, Kenya typhus, Australian tick typhus, Central Asian typhus, and others. The human mortality rate approaches 30% in untreated cases. Area-wide eradication of ticks, especially in forested areas, does not appear to be feasible.[589] Ticks may rank second only to mosquitoes as a public health menace.[824] In general, tick populations do not respond well to control with chlorinated hydrocarbons, and where DDT etc. have been successfully used application rates are necessarily high. Tick control in dwellings and on domestic animals can be afforded by the use of Baygon, Diazinon,® Sevin,® and/or dichlorvos.

Unlike ticks, mite vectors transmit only a few ailments of note, rickettsialpox, in the United States and Equatorial and South Africa, and scrub typhus being the most important. Mites do, however, penetrate human skin, giving rise to scabies, an infectious nonfatal disease. The most effective means of guarding against mites is through personal protection, especially the wearing of insecticide-impregnated clothing; lindane and dibutyl phthalate are used for this purpose. Area treatment with DDT, dieldrin, aldrin, and lindane have been tried with a modest degree of success. One interesting new material, ZR-856 (hexadecyl cyclopropane-carboxylate), a phytophagous miticide produced by the Zoecon Corporation, may be of value against mites of public health importance.[825] The molecule ought to be soluble in various carboxylated elastomers, or at least compatible, so that a controlled release formulation is feasible. ZR-856 is essentially an ovacide, so a residual area treatment approach, rather than clothing impregnation, would likely be useful. The very specific nature of this control agent makes it environmentally attractive.

Chagas disease is the common name for a

Trypanosoma cruzi infection vectored by various blood-sucking reduviid bugs. *Triatoma infestans, T. brasiliensis,* and *Strongylus megistus* are the principal transmitters, although other vectors are known. The disease is endemic in various sections of South and Central America, 20% of the Venezuelan population being infected.[589] The disease may be acute or chronic, the latter form often being fatal.

T. cruzi is not transmitted by injection during biting, but rather in the excrement of the insect. The bitten victim literally scratches the infected material into the dermis. No effective drug prophylaxis is known, nor is there any permanent immunity. No drug therapy is of recognized value. Controlled release techniques as a means of protecting man from *Triatoma* attack are under investigation.[826]

To the author's knowledge, no application of controlled release methods has been attempted in the eradication or control of sandflies (*Phlebotomus* spp.), vectors of the dread tropical disease leishmaniasis. In some areas, domesticated dogs and various rodents, including the popular gerbil, serve as reservoirs. The use of DDT, benzene hexachloride, and malathion as adulticides was not particularly successful.[827,828] Larvae develop in soil, so larviciding activity becomes extremely difficult and ecologically hazardous.

All classes of insect vectors of major diseases have been mentioned, except the flea, *Xenopsylla cheopis,* that transmits plague from rodent to man. However, the control approach is generally keyed to the elimination of the rat, not the insect, so discussion is deferred until Chapter 7. Malathion, DDT, and dieldrin have been used to destroy the flea, especially in dwellings.

Problem fleas in the United States are those that annoy domestic animals, *Ctenocephalides canis* and *C. felis*, and often attack man. DDT and chlordane have proven ineffective, possibly due to the development of resistant strains.[829] Various organophosphorus compounds are in use as dusts, sprays, or dips. The very effective "flea collars," produced by various concerns, are usually formulated with dichlorvos as the active agent and a plastic, commonly plasticized polyvinyl chloride, as the binding material. These controlled release materials act similarly to the No-Pest strip.

Stinging and biting insects that are not known vectors of disease may constitute a localized, though serious, menace to humans prone to allergic reactions. The Brazilian bee, *Apis mellifora adansonii,* an aggressive strain of the honey bee prone to attacking humans, has received considerable publicity in its movement northward towards the United States.[830] While awaiting its arrival by 1985 or so, it might be meritorious to develop better methods of controlling the already present fire ant and other Hymenoptera (wasps, hornets, yellow jackets, etc.). Imported fire ants, *Solenopsis* spp., are spreading to more areas of the southern United States. Control with DDT and other chlorinated hydrocarbons has been an ecological disaster in some areas.[831] Mirex, dodecachlorooctahydro-1,3,4-metheno-2*H*-cyclobuta(cd)pentalene, is presently used in baits for fire ant and other ant control. Also, a 1% chlordane in canned cat food has been successfully used as a bait against *Vespula pennsylvanica* (Saussare), the yellow jacket.[833] Adults workers carry the bait back to the nest, poisoning the entire colony.

Few other examples of the use of controlled release baits or toxicants in public health work are known. Lyons reported that the guineaworm can be controlled by the use of Abate (1%) on sand granules applied to an infested pond, with about 6 weeks' efficacy being noted.[833] This might be a use for the Abate/EPDM or Abate/natural rubber long-lasting controlled release systems.

HOUSEHOLD AND STRUCTURAL INSECTS

A number of insects, especially the termite, destroy wooden structures and items of furniture, carpeting, etc. within the home. In some cases, insect infestation is of little consequence regarding disease-bearing or structural attack, but the insects are unsightly and thus seemingly unwanted; this group includes cockroaches, silverfish, spiders, centipedes, and so on. A fair-size service industry has developed whose main concern is to rid houses and business establishments of these unwanted pests.

The destruction of wooden structures through termite attack is well known, and the economic consequence may well exceed a billion dollars per year in the United States alone. Termites also attack various plastics and, importantly, buried cables. Once the sheathing is eaten away the cable shorts out, causing power outages and considerable expense in locating and repairing the damage.

Toxicants can be added to plastics to discourage termite burrowing. In the case of polyvinyl chloride, the use of tricresyl phosphate, which poisons termites, is preferable to the use of dioctyl phthalate, which is an attractant, as the plasticizer will greatly reduce termite damage.[834] Lindane incorporated in polyvinyl chloride prevented almost all termite attack and was superior to aldrin and dieldrin in this regard.[835] Other plastic and elastomeric materials tested against termites have indicated an obvious need for the incorporation of a toxicant or repellent.[836]

Termites will ingest all woods commonly used in construction, although some preference can be demonstrated.[837] The arrival of the Formosan subterranean termite, *Coptotermes formosanus* Shiraki, in the last 10 or 12 years, and the spreading of infestation beyond the Houston-New Orleans foci has accelerated attempts to develop a suitable antitermite weapon. The Formosan termite is able to penetrate asphalt, plastics, lead, mortar, and plaster in order to contact and ingest wood,[838] and is considerably larger and more prolific than native species.

Conventional termite control methods rely upon the application of insecticides to the soil surrounding the structure to be protected. Chlordane, aldrin, and other persistent materials have been used to provide long-term protection. Sevin and Dursban have shown efficacy in laboratory tests.[839] Poisoned soil apparently will not stop the Formosan termite,[840] and pest control operators have begun excluding them from their work guarantees.

One recent advance in termite control has been the development of mirex-impregnated wooden blocks as bait,[841] a technique that depends upon the foraging activity of termites and the fact that the toxicant is brought home to the nest to poison animals not directly contacting the bait. This method appears workable in practical situations.[842]

Organotin-dieldrin treatment of wood has been shown to be an effective preservative measure preventing termite attack.[843] The author has observed that antifouling elastomers containing TBTO and TBTA will destroy termites through contact intoxication if transit time is prolonged.

The German cockroach, *Blatella germanica*, is the most important household insect in most locales in the United States throughout the year. Other cockroaches, Oriental, American, Brown, etc., are of local or seasonal importance. Several species have been implicated in toxoplasmosis transmission, a usually mild or symptomless disease that can cause congenital defects in an unborn child if the mother is infected.[844]

Conventional control methods have relied primarily upon residual treatment of infested areas, particularly harborages. Materials used include Baygon, chlordane, Diazinon, dichlorvos, dieldrin, fenthion, malathion, and kepone (decachloro-octahydro-1,3,4-methano-2H-cyclobuta(cd)pentalen-2-one) baits. Chlordane resistance has been noted since 1957.[829] Resistance has been detected with Baygon,[845] pyrethrins,[846] and Diazinon.[847] Chlorpyrifos and other organophosphorus insecticides provide adequate immediate control but lack residual potency.[848] Dusting or spraying, including the use of ULV aerosols,[849] constitutes conventional treatment regimens. Recently, repellents and bait traps have been evaluated.[850,851] Encapsulated Diazinon at 0.5 and 1% concentrations provided excellent kill after 60 days' aging.[850] In residual tests propoxur (Baygon) outlasted chlordane, chlorpyrifos, and other in-use materials, providing 20 weeks of 100% control of the Cuban cockroach, *Panchlora nivea* (L.).[852] Residue loss is mostly due to volatility, and methods of relating the amount of residual chemical remaining after a given time to the control efficacy have been shown to follow first-order kinetics.[853]

In 1967, Simmons reported that the organotin vapors released by biocidal elastomers, especially compound 443A (Nofoul), was highly toxic to the German cockroach.[653]

The first long-term controlled release method of regulating the cockroach is a development of the Herculite Division of Health-Chem Corp., New York, N.Y. The product, Hercon® roach tape, is an adhesive-backed, multilayered plastic strip utilizing a conventional roach control agent such as Diazinon, Baygon, Dursban, malathion, or chlordane.[854] Dursban, Diazinon, and chlordane showed greatest effectiveness in long-term tests; even after 150 days, complete cockroach kill was achieved with only a few days' exposure. In some tests the dispenser strength remained undiminished after 15 months.

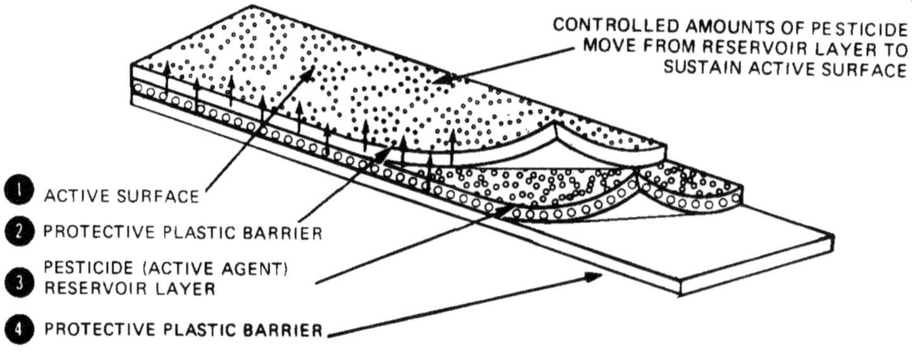

FIGURE 6–3. Hercon® insecticidal strip. (Photograph coursesy of the Health-Chem Corp., New York, N.Y.)

Cumulative toxicant loss from Hercon dispensers was less than 5% per month. Loss is not related to the volatility of the respective insecticides, indicating that diffusion through the polymeric matrix is rate controlling.[854]

The Hercon insecticidal strip is composed of four basic elements (Figure 6–3). A plastic barrier coated with a pressure-sensitive adhesive is used on the side facing the substrate. Above this layer, sandwiched between two protective barrier materials, is the pesticide reservoir. The toxicant moves slowly but continuously from the reservoir into and through the upper protective layer, creating an active toxic surface. Surface loss is through volatilization, radiative degradation, hydrolysis, or insect uptake, or mechanically by humans, wash water, etc.

The Hercon roach tape contains 10% insecticide (Baygon, Diazinon, or Dursban). The multilayered sheet can be reduced in size to flakes, confetti, or even a granular powder, the latter being in the sprayable 30-μm or so size.[855]

It is of note that the kill mechanism arises from contact of the cockroach with the tape, and is not due to a fumigant action. Contact exposure time is not an important factor, the kill rate (100%) being the same for 2 sec' contact as for 240 sec.

The roach tape is substantially less hazardous to nontarget organisms relative to the use of the toxicant in a conventional sense. The rat oral LD_{50} for Diazinon is 325 mg/kg. Roach tape was ground to a 100-μm powder and fed by intubation to rodents; in a 14 day evaluatory test no deaths occurred at 5000 mg/kg.[855]

Field trials of roach tape containing either of the three toxicants of choice were significantly more effective than the conventional spray application. The mean cockroach population reduction for residual spraying in two tests was 31 and 33%, respectively, while the use of roach tape alone gave 100% reduction with Dursban, 92 to 98% reduction with Diazinon, and 86% reduction with Baygon.[856]

Anobiid beetles attack wooden portions of houses, turning joists, subflooring, and wall interiors into sawdust.[857] Although the economic loss is not anywhere near that attributable to termites, it is serious enough to warrant the study of controlled release techniques as a means of alleviating this problem. The black carpet beetle is a seasonally serious pest in parts of the eastern United States.[858] Hercon dispensers have shown long-term effectiveness against this pest.[859]

Earwigs, *Forficula auricularia* L., and silverfish, *Ctenolepisma longicauda*, are unwanted house pests of little economic significance. Neither respond well to baits where a food stuff is used as the attractant,[851] although the use of a Baygon bait in conjunction with spraying may have some merit in earwig control.[860]

ANNOYANCE INSECTS

A number of insects do not attack man or his artifacts, nor are they significant in the spread of disease, but they are annoyances that one would rather avoid. The chironimid midge is an example. Larvae are members of the aquatic biota and thus susecptible to control with various larvicides, Baytex and Abate being useful for this purpose.[644] Various piscine species can be utilized as a biological control.[861] Adulticiding is possible and effective – if the community is willing to accept a daily spraying routine for one midge generation.[862] Mechanical control of larvae is feasible under very specific conditions.[863]

It is believed that midge control is a fertile area for the application of the controlled release technology. Organotin/elastomers evaluated by the author have been observed to destroy midge larvae at pellet concentrations of 10 ppm and less. Since such materials provide long-term release, use may be feasible in water of low organic content. More likely, a controlled release organophosphorus material would be apropos.

The common housefly, *Musca domestica*, is simply controlled with readily available aerosol preparations of synergized pyrethrins.[864] Both the No-Pest strip and Hercon dispenser provide housefly control.[857] Likewise, Altosid insect growth regulator, which has been incorporated in controlled release systems for mosquito control, is active against the housefly.[865]

Another pest spending its larval stage in water is the horsefly, *Tabanus* spp. Although tabanids may conceivably vector anthrax,[866] their pestiferousness arises from a propensity to annoy humans and domestic animals, often through biting. Although controlled release techniques have not, so far as is known, been applied to this problem, the author has observed tabanid larvae succumb to organotin/elastomers and Abate/EPDM biocides at low concentrations under laboratory conditions.

CONTROL OF INSECTS DESTRUCTIVE TO FORESTS AND NONAGRICULTURAL PLANTS

The gypsy moth was introduced into the United States during the last century (1869) under the assumption that it could crossbreed with the silkworm and create a hybrid of economic value. Once loose it became a major threat to trees of all descriptions. The adult lays millions of eggs in the summer, each hatching into a caterpillar the following spring that commences to devour the foliage of shade trees, fruit trees, and, if nothing else is available, evergreens. One caterpillar will consume up to a square foot of leaf surface in a day. It has been estimated that nearly 2 million acres of forests and wooded residential areas were defoliated in 1974.[867]

Conventional control measures have relied greatly upon DDT; however, unacceptable environmental contamination led to its being phased out in 1968. A biological weapon, *Bacillus thuringiensis*, has been used with some effectiveness as Thuricide® 16B (Sandoz, Inc., Homestead, Fla.), a low volume aerial spray.[867-869] An improved strain of *B. thuringiensis* Dipel® (Abbot Laboratories, North Chicago, Ill.), is also available and appliable with hydraulic and mist sprayers, backpack garden sprayers, and garden hose attachments.

Insecticides in use, essentially as DDT substitutes, include Imidan®, *O,O*-dimethyl *S*-phthalimidemethyl phosphorodithioate,[870] and Dylox®,[871] dimethyl (2,2,2-trichloro-1-hydroxyethyl) phosphonate. Gypsy moth attack abated somewhat in the 1974 and 1975 seasons, but the mentioned control methods obviously were not totally successful since 750,000 acres of northeastern woodlands were lost in 1974.[872]

Controlled release technology is under investigation as a method of controlling the gypsy moth. The main thrust has been to utilize disparlure, cis-7,8-epoxy-2-methyloctadecane, a synthetic sex attractant,[873] in a dispensing system (Figure 6–4).[873,874] Mechanical traps utilizing a sticky interior substance[875] or an area-wide release of the attractant which would utterly confuse or exhaust the male in his attempt to locate the female and thus depress the reproduction rate[876,877] are two other approaches. The simple use of filter paper treated with disparlure and distributed by aircraft prevented released males from locating virgin-female-baited traps or 10-μg-disparlure-baited traps, thus confirming the potential value of the confusant system.[878]

The observed problem with mechanical trapping or simple release as a confusant arises from the lack of adequate longevity. Disparlure incorporated in Hercon dispensers appears to be a method of achieving the necessary long-term release,[879] at least 120 days being realized.[874]

The tussock moth has been doing for the Northwest what the gypsy moth has done for the Northeast. The problem became so critical, with millions of dollars worth of timber at risk, that the ban on DDT was lifted temporarily and spraying of 400,000 acres took place in 1974.[872] Pheromone trapping of northwestern moths is demonstrably feasible,[880] and undoubtedly controlled release dispensers will be found suitable in this area.

In another development, Allan et al. prepared

FIGURE 6–4. Gypsy moth eradication/control trap baited with Hercon®/departure wafer (visible inside left) and coated with sticky "Tack Trap" (bottom of trap, holding moth). (Photograph courtesy of the Health-Chem Corp., New York, N.Y.)

several controlled release materials using carbofuran (2,3-dihydro-2,2-dimethyl benzofuran-7-yl *N*-methyl-carbamate), methomyl (*S*-methyl *N*-[(methyl-carbamoyl)oxy]-2-thioacetimide, monocrotophos (3-[dimethoxy-phosphinyloxy]-*N*-methyl-cis-crotonamide, phosphamidon (2-chloro-2-diethylcarbamoyl-1-methyl-vinyl dimethyl phosphate), and isolan (1-isopropyl-3-methyl-5-pyrazoyl dimethylcarbamate) in a polyamide base.[881] The polymer used was Versamid® 940 (General Mills Chemicals, Inc., Kankakee, Ill.), due to its reportedly excellent solvency for insecticides. All materials were evaluated against the mahogany shoot borer, *Hypsipyla grandella.* The duration of efficacy was increased from that normally observed, and the carbofuran-polyamide formulation suppressed insect attack for nearly a year.

An interesting pest of west Louisiana and east Texas is the town ant, *Atta texana,* whose colonies may be measured in acres. This ant creates huge, intricate, subterranean towns that may extend to depths of 30 or more feet. The ant defoliates field and garden crops, shrubs, and young trees, with pine forests being especially hard hit.[882] Leaves are used to feed fungi that the ants cultivate for food. In order to establish pathways from town exits to food suppliers, the ant exudes an odor-trail substance, a pheromone, that is detected and followed by other members of the community.[882] *A. texana* ants also secrete an alarm pheromone, identified as 4-methyl-3-heptanone, that radically alters behavior.[883] Conventional control is achieved by fumigation with methyl bromide. Residual treatment in the past has made use of 2% dieldrin or 10% chlordane dust.

Both the trail pheromone and the alarm pheromone could probably be incorporated in an elastomeric, and possibly a plastic, controlled release formulation. To date, as far as is known, only sex attractant pheromones have been evaluated as noncontaminating specific control agents. Intriguing possibilities are envisioned. Can the area-wide slow release of the trail pheromone act as a confusant so that the individual town ant cannot locate food, or cannot find its way home to the nest? Can a slow release alarm pheromone maintain a steady, continuous defense posture among members of a colony (at which time foraging is minimal), or will the individual chemoreceptors fatigue so that a true alarm cannot be given, thus enabling predators to decimate the population? Granted, the last point is rather unlikely, but it must be agreed that the reaction of a colony in a state of continual alarm should be most interesting.

INSECTS ATTACKING LIVESTOCK

As in most pest control areas, essentially no effort has been made to utilize controlled release technology in protecting fowl and cattle from various insect pests. Stable fly control is presently achieved using naled (1,2-dibromo-2,2-dichloroethyl dimethyl phosphate)[885] or other insecticides. Experience has shown that naled can be incorporated in various polyacrylate film-forming polymers or probably in polyisobutylene or chlorinated polyisobutylene with a dramatic extension in longevity.[886] Controlled release naled formulations have not been evaluated for stable fly control.

Controlled release insecticidal or repellent gels and foams could be developed in formulations suitable for spraying on cattle, horses, and sheep. Using Carbopol (B. F. Goodrich Co., Cleveland, Oh.) as a gellant, gelled water mixtures of malathion, crotoxyphos, and methoxychlor were prepared and evaluated against horn flies, *Haematobia irritans* (L.), on cows. Materials performed as well but no better than the toxicant alone.[887] In this instance, proper methods of application are lacking, and formulations were very preliminary. Carbopol/Carboset film-forming *gels,* similar to the cercaria penetration-inhibiting lotions in the binding portion of the formulation, would likely have provided greater efficacy and lifetime. Stabilized foam formulations similar to those discussed by Weidman and Yates but of Carbopol/Carboset/toxicant compounds might also be meritorious.[888]

INSECTS OF AGRICULTURAL IMPORTANCE

The future thrust for the development and application of controlled release methodology will almost certainly lie in the area of agricultural insect and weed control. The absolute necessity of expanding food production, both in terms of total acreage and yield per acre, will make it imperative to exploit this emerging technology if we are to preserve a quality environment. Without agricultural expansion an unprecedented world-wide food

TABLE 6–4

Pheromone Bait Development

Pheromone	Insect	Chemical identity	Reference
Gossyplure	Bollworm	(Z,Z)- and (Z,E)-7,11-hexadeca-dienyl acetates	879
Disparlure	Gypsy moth	cis-7,8-Epoxy-2-methyloctadecane	889
Looplure	Cabbage looper	(Z)-7-dodecenyl acetate	889
Grandlure	Boll weevil	(Mixture of four materials)	890,892–894
Codlelure	Codling moth	(E,E)-8,10-dodecadiene-ol	
Cuelure	Melon fly	4-(*p*-hydroxyphenyl)-2-butanone acetate	886
Trimedlure	Mediterranean fruit fly	tert-Butyl 4-chloro-2-methylcyclohexane carboxylate	886
	Oriental fruit fly	Methyl eugenol	886

crisis looms whose net result will be, as history so aptly teaches, continuous warfare and socio-economic unrest. Agricultural expansion with increased use of conventional pest control agents and methods will lead to a serious threat to human life and welfare from contamination of the ecosphere.

So far, relatively little has been done in the application of controlled release technology to agricultural chemicals. Most of the activity has involved the development of bait formulations based upon the use of sex attractants to lure the target into a mechanical or toxicant trap. Table 6–4 lists most of the pheromones under study.

Gossyplure has been used in the Hercon dispenser as a controlled release bait for trapping the pink bollworm, and it has shown a four-fold increase in efficacy over the use of a simple cotton wick system.[879] Likewise, the incorporation of disparlure, looplure, and grandlure in the same dispensing mechanism has provided long-term attractiveness.[889,890]

The boll weevil, *Anthonomus grandis* Boheman, causes cotton crop losses estimated at $200 to $300 million per year, with about 25% of the total amount of agricultural insecticides used applied towards its control.[890,891] Grandlure, the male boll weevil pheromone, was known to work under field conditions, but had a life span of only 1 to 2 days.[892] Use of various formulations in cotton wicks did not extend the release life much beyond 4 days. Various gel formulations provided 2 weeks' field life, as did polyethylene glycol/grandlure formulations impregnated in cigarette filters.[893] Grandlure release from the Hercon dispenser proved effective for 2 to 3 weeks.[890]

The pheromones cited in the above studies were used in connection with mechanical trapping devices.

In the ongoing fruit fly investigations pursued by the Hawaiian Fruit Fly laboratory (U.S. Department of Agriculture), sex attractants are used in connection with an insecticide.[886] Cuelure/toxicant mixtures were impregnated in solid carriers such as cotton wicks, cigarette filters, etc., and in some cases 10 months' release life was observed. Both naled and malathion have been used as the toxic agent. Recently, formulations composed of a pheromone (Cuelure, trimedlure, and methyl eugenal) and a toxic agent, usually naled, incorporated in Carboset 525, carboxylated polyacrylate, or liquid polyisobutylene (LM-430, Exxon Chemical Co., Akron, Oh.), have shown considerable promise in field evaluations (Figures 6–5 and 6–6).[886]

Another area of increasing interest lies in the use of juvenile hormone mimics, JHM, or insect growth regulators, IGR, chemicals that interfere with the growth and development of insects. Such morphogenetic materials are fairly selective, affecting only one or a few related species, so their relatively low environmental impact renders them attractive as substitutes for insecticides. Menn and Beroza have ably described the structure and activity of these agents.[895]

Morphogenic agents can be applied to the infested environment in a manner similar to techniques used with insecticides. For instance, baits have been successfully used against cockroaches with over a year's effective control;[896] emulsifiable concentrates have shown efficacy against mealworms, cockroaches, and

FIGURE 6–5. Application of liquid lure/naled/Carboset® to guava leaf. (Photograph courtesy of Hawaiian Fruit Fly Laboratory, U.S. Department of Agriculture.)

FIGURE 6–6. A 1-week catch of melon flies by a leaf with 0.010 g of Cuelure/naled/Carboset® formulation on it. The leaf had been previously weathered for 3 weeks before picking and exposure. (Photograph courtesy of Hawaiian Fruit Fly Laboratory, U.S. Department of Agriculture.)

houseflies,[897] as well as other formulation types.[898] However, morphogenic agents possess many of the disadvantages associated with first- and second-generation insecticides, although usually to a lesser degree. Most materials are essentially nontoxic to mammals, but some have a relatively high fish toxicity; Ent. No. 70221, for instance, has a piscine acute LC_{50} of 3 to 5 ppm.[899] Proper formulation is necessary for adequate control activity.[900] Potentiation is possible using phosphonate esters,[901] and insecticide/JHM combinations can be synergetic, the mix providing superior activity over either used individually at comparative dosage rates.[902] Induced resistance has been encountered with mosquitoes,[903,904] houseflies,[905,906] and others.[907,908] Also, there are indications that plant secretions may control some insect mating activity.[909]

While only three naturally occurring juvenile hormones have been identified, literally several hundred mimics demonstrating morphogenic activity have been discovered. Only a few have reached the field test stage, however, so judgment must be reserved as to their respective merits. In general, persistence is low (a day or two at most), so that, as in the previously described success of Altosid SR-10, a controlled release dispensing mechanism will be needed.[898,908]

Relatively little work has been performed to date with nonbait-type formulations. Allan et al. have suggested the absorption of insecticides on lignocellulosic or proteinaceous waste materials (bark, sawdust, oat hulls, corn cobs, etc.) and their use as controlled release pesticides.[910] The suggestion is not new in concept, as formulators have long been using silicas, clays, charcoal, and so on as absorbants to reduce loss rate in soil applications. The use of the millions of tons of waste products generated by this nation in this respect is intriguing, but, unfortunately, such materials in general provide neither a long-term release nor a controlled one. However, Campbell, for one, has carried the idea a step further: He has absorbed various organophosphorus and other insecticides on carriers such as wood chips, corn cob dust, etc., and then coated the system with a microcrystalline wax emulsion.[911] In tests against the potato flea beetle, the wax-coated material outperformed a nonencapsulated formulation over an 80-day evaluation period.

The theory governing the slow release of agricultural chemicals in the soil and movement from the site of application to the target insect, nematode, fungi, or weed has seen some preliminary development by Chandrasekaran et al.[912] Assuming that diffusion is the controlling transport mechanism, the kinetics of continuous release have been investigated. Soil concentration at any given time is probably a function of particle size, diffusion coefficient, and partition coefficient. In the subject analysis it is assumed that (1) the controlled release particle is non-water degradable and (2) the agent movement through the polymeric matrix is via internal diffusion. Equations for movement within the controlled release granule have been previously considered. Chandrasekaran and colleagues advance the following equation as governing transport through the soil to the target, and apply the results of mathematical analysis of the controlled release system towards predicting delivery system configurations that would maximize the effectiveness of the agent in soil.

$$\frac{\partial C}{\partial t} = D_2 \left(\frac{\partial^2 C}{\partial r^2} + \frac{2}{r} \quad \frac{\partial C^2}{\partial r} \right) \quad - KC$$

where C = concentration of agent in soil, D = diffusion coefficient in soil, r = granule radius, and K = degradation rate constant for the given agent in soil.

Diffusion coefficients of pesticides in soil are not well known. Scott and Phillips, among others, have determined values for a few herbicides.[913]

The analysis depicts the soil as an ideal isotropic medium and neglects the effects of rainfall, agent evaporation, and chemical interactions with soil constituents. However, it does present a point of departure for experimental and theoretical studies aimed at the development of a comprehensive theory for controlled release soil systems.

Stokes et al. have combined aldicarb [2-methyl-2-(methylthio) propionaldehyde *O*-(methycarbamoyl) oxime] with polymers such as cellulose acetate, polyamide, polyvinyl chloride, polyester, polyurethane, and copolymers of urea and formaldehyde, and applied the resulting granule to the soil.[914]

FERTILIZERS

Although plant nutrients do not properly fall

within the scope of this text, they are a fit subject for controlled release dispersal, and knowledge gained in their development, evaluation, and application will obviously play an important role in the general development of controlled release pesticides. Fertilizers are used in great quantity by all agriculturally advanced nations. In order to feed the anticipated global population, the need for fertilizers can only increase unless more effective use is made of present-level production. The equally anticipated world-wide shortage of energy and the use of increasingly scarce petroleum and natural gas in fertilizer production indicate a looming crisis. Coeval with the desire of the United States to maintain peace and prosperity within its borders will be the use of American agricultural surplus as a chief export and bartering tool.[915]

The modern techniques of using fertilizers, somewhat as with herbicides, depends on the use of a massive one-time dose. The quantity of fertilizer used must be great enough to overcome environmental features that reduce the amount available for plant consumption, rainfall and ground run water being basic to this loss. Possibly 80% of the fertilizer used never reaches the intended growing plant, but ends up in water bodies or in the deep soil unreached by roots. The lost portion can be considered an environmental contaminant, leading especially to the poisoning of water bodies and concommitant loss of fish and shellfish life.[916]

A controlled release fertilizing material would be relatively unaffected by rainfall or ground water wash out in that the size of the polymeric granule would not be conducive to movement laterally or downward in soil. A downpour would, of course, likely wash away some portion of the released material, thus rendering it unavailable to the crop, but the dispensing pellet would continue to release its contents after each rainfall so that nutrient would always be available. The uniformity of fertilizer availability would also be important in that the purpose of massive conventional application, if hopes are realized, is to provide a long-term residual; since a residual would always be available from controlled release granules, the necessity for overusage could be avoided.

The use of less fertilizer would obviously reduce environmental pollution. However, controlled release can provide several other important values as well. Uniform release means no injury to plants arising from an overabundance of fertilizer at application time, and avoids too-rapid growth of foliage and so on at the expense of the fruit.

A few controlled release materials are commercially available for rather specialized, essentially nonagricultural, uses. Medicap® capsules containing ferric ammonium citrate are implanted in trees suffering from chlorosis and provide an adequate alleviation of this iron deficiency disease,[917] with up to 3 years' effectiveness reported. Jobe's Tree Spikes® containing impacted plant nutrients, are driven into the ground around trees and slowly release their ingredients over a period of time. Strictly speaking, these are probably not controlled release devices, as the spike disintegration rate likely depends on the moisture content of the soil at a given time. Isobutylene diurea, Ibdu®, granules for turf fertilization work on the same principle.[918]

Controlled availability fertilizers, usually depending on a surface coating of a wax, sulfur, or other material to retard solubilization, have been investigated and in some instances marketed.[919, 920] Sommer et al. in their seminal invention have described the incorporation of plant nutrients in a hydrophobic binding matrix such as natural or synthetic rubber, waxes, asphalt, and so on.[921] In this work four critical aspects of the mechanism are described. The nutrient, which apparently can be any fertilizer, must be finely divided and uniformly dispersed in a hydrophobic binding element. The material must be shaped as a cylinder and partially coated with a water-insoluble, water-impermeable coating, which may be the same as or different from the binding polymer. Lastly, a portion of the cylinder must remain uncoated so that fertilizer can be lost to the external environment; up to 50 weeks' longevity is claimed.

Another valuable aspect in considering controlled release fertilizers is that several nutrients can be bound in one granule[922] or other agents can be added as retardants or synergists.[923] Hydrophillic polymeric release systems can also be utilized.[718]

Chapter 7

OTHER CONTROLLED RELEASE PESTICIDE AREAS AND NEW DISPERSING SYSTEMS

RODENTICIDES

Rats and other rodents represent structural, agricultural, and health pests. Commensal rodents destroy or contaminate food products in all phases of food manufacture and use. Minor economic loss is caused by the gnawing proclivities of domestic rodents. Field rodents are especially nuisances by their gnawing damage to underground cables. Both human life and national defense have been affected by rodent interference in communication systems and destruction of wiring in ground vehicles and aircraft.

In many developing nations rat predation on growing and stored crops leads to a significant reduction in foodstuffs. Grain losses prior to harvest have been estimated to be as high as 25% in India.[924]

Implication in the transmission of disease renders the rodent an extremely serious health problem, ranking only behind the mosquito and snail. House rats and mice transmit plague, typhus, leptospirosis, rat bite fever, *Salmonella*, etc., while field rodents carry and pass to man spotted fevers, plague, typhus, tularemia, and other dread diseases. Bats transmit rabies, histoplasmosis, chagas disease, and other ailments affecting man and his domestic cattle and horses. Vampire bats, directly through transmission of rabies and indirectly through secondary infection of the bite wound, destroy a million or more head of cattle in an average year in South and Central America. During local outbreaks bats may literally kill nearly all cattle and horses in a given district. In protein-deficient nations, bat depredations are important factors in malnutrition.

The general approach to rat and mouse control lies in environmental restructuring through sanitation, improved food storage, ratproofing of dwellings, and so forth. Rodenticides are usually considered supplemental to such measures.[829]

Various poisonous substances have been used in rat baits since early times. During this century red squill (scilliroside glycoside), strychnine, and sodium fluoroacetate ("1080") were in widespread use, although they have now been supplanted by various coumarin-type chemicals. Coumarins inhibit vitamin K, necessary for the synthesis of prothrombin, which in turn is the precursor of thrombin, an essential element for blood coagulation.

Since the late 1950s the anticoagulant poisons have dominated the rodenticide field. Major materials in use are warfarin, 3-(α-acetonylbenzyl)-4-hydroxycoumarin; chlorophacinone, 2-[(*p*-chlorophenyl) phenylacetyl]-1,3-indandione; diphacinone, 2-diphenyl-acetyl-1,3-indandione; Fumarin®, 3-(α-acetonylfurfuryl)-4-hydroxycoumarin; Pival®, 2-pivalyl-1,3-indandione; and PMP, 2-isovaleryl-1,3-indandione. Anticoagulants can be used as tracking powders, but the major use is in bait preparations. Conventional baits use attractants such as sugar, molasses, or table oil to enhance acceptance of the cereal grains normally used as the holding element for the toxicant. Dye additives such as methylene blue are utilized to provide distinctive coloration to reduce the possibility of mistaken identity as a human or animal foodstuff.

Rodents can and do develop resistance to anticoagulants.[925,926] Also, baiting as a protective device for growing crops has not been very successful.[927] Rats may become bait shy, and there is little doubt that many individuals of a given population can detect the presence of small amounts of a toxic material in baits.

Nontoxic control methods such as the use of pathogens[928] and chemosterilants[929,930] do not appear practical at this time. Integrated control involving cultural manipulation of vegetation in crop-growing areas can aid in reducing rodent populations.[931]

Probably the best current approach to chemical rat control is the use of improved baits. A basic improvement lies in increasing acceptability by masking the presence of the toxic agent. Greaves et al. recognized the value of toxicant encapsulation as a method of masking,[932] and Brooks also favored encapsulation as a means of delaying the onset of symptoms so that the poisoned rodent would continue consuming until a lethal amount had been ingested.[933]

Warfarin has been encapsulated using the Wurster air suspension process.[700] Tests showed

FIGURE 7–1. Rodent consuming encapsulated warfarin bait. (Photograph courtesy of Wisconsin Alumni Research Foundation.)

that the encapsulated product was three to ten times more acceptable to laboratory rats.[934] Norbormide, a complex dicarboximide very specific to the Norway rat (*Rattus norvegicus*),[935] has been encapsulated, and bait studies are underway.[936] Figure 7–1 shows a rodent consuming encapsulated warfarin bait.

Control of the vampire bat is a complex problem due to the often inaccessible habitats, widespread populations, and general movement and feeding habits. Poisoning or trapping campaigns are sufficiently difficult in execution to render them impractical. Repellents have been evaluated; one, a controlled release tributyltin sulfide in an acrylic base, prepared at the University of Akron, performed well under laboratory conditions.[937] In this test the target preferred a nonprotected "blood bath" over a TBTS-protected unit by a factor of three to one.

Repulsion of bats from a likely victim such as a horse or cow, if possible, would only encourage movement and attack of nonprotected victims, one of which is man. Destruction rather than repellency seems to be the more meritorious approach. Present effort centers upon the use of a bat-specific anticoagulant. Topically treated bats carry the agent, diphenadione, back to the nest where others are infected by contact.[938] Also, cattle can be injected with the agent and the bat will become exposed during the ingestion of blood.[939]

Protection of seed from field rodents can be at least partially achieved by repellent coatings. Mestranol coated on Douglas fir seed reduced deer mice feeding by up to 66% over that observed on nontreated seed.[940] Mestranol is both nontoxic and nonpersistent, and therefore more environmentally acceptable than endrin and other toxic coatings commonly used.

NEMATICIDES

Nematode attack on field, fruit, and nut crops can result in a 20% or greater reduction in yield. This translates into a revenue loss of 1.5 billion dollars per annum or greater to the U.S. grower.[941] At this time there are approximately 25 nematicidal materials registered for use in the United States. Nine are soil fumigants, and, except for DBCP, 1,2-dibromo-3 chloropropane, are phytotoxic and thus restricted to preplant usage. Others are organophosphate and carbamate insecticides showing nematicidal activity. Presently, about 2,000,000 acres, mostly in tobacco, cotton, fruit crops, and ornamentals, are treated with nematicides.

The effects of nematode infections are insidious and difficult to detect. Damage symptoms develop slowly and are nonspecific, so the grower may be generally unaware of the problem. Nematodes are difficult to destroy in that the soil mass they inhabit serves both as a barrier to permeation by control agents and a chemical degradant. Control often requires the use of inordinately large quantities of the selected agent in order to thoroughly saturate the soil. Application of controlled release technology may provide the advantages of improved control with the use of smaller quantities of agent.

In the first known attempt to utilize controlled release in this area, DBCP was encapsulated in a water-insoluble polymer and accordingly evaluated against DBCP used in the conventional sense.[942] It was observed that DBCP from the capsular formulation became available to the subject saprophytic nematode at a slower rate. A longer life was indicated, although not necessarily an improved degree of control.[942] This study indicated that an encapsulated nematicide may remain in the soil for some time until moisture conditions trigger release.

In a later program, various elastomeric matrices were examined as to their ability to retain and slowly release ethoprop, *O*-ethyl *S,S*-dipropyl phosphorodithioate. Ethylene-propylene diene terpolymers and natural rubber were found efficacious in this respect.[943] Ethoprop, unlike most in-use nematicides, has a relatively high boiling point (86 to 91°C at 0.2 torr) and thus is more likely to be able to be processed in elastomer-based formulations. It is readily degraded in vivo through P-S bond cleavage and deethylation to liberate the mercaptan, which is then rapidly methylated and oxidized to the sulfide, sulfoxide, and sulfone.[944] The acute oral LD_{50} (rat) is 61.5 mg/kg and the acute dermal LD_{50} (rabbit) is 25.9 mg/kg.

Ethoprop was found to be incompatible with both gum butyl and liquid butyl rubber, styrene-butadiene, cis-polybutadiene, and solution cast polyacrylates.[943] Fortunately, this agent was found to be highly soluble in precured natural rubber and EPT without appreciable swelling or loss of physical properties. The diffusion-volatilization loss rate in air is moderately slow, so half-lives in excess of 30 days are predictable. Bioassay studies are presently underway at the Plant Protection Institute of the U.S. Department of Agriculture, Beltsville, Md.

The following formulation appears superior:

Ingredient	Formulation (parts)
EPT (Epcar 5465)	100
HAF carbon black	10
Zinc oxide	3
Stearic acid	0.5
Sulfur	1
Mercaptobenzothiazole	0.2
Oil (Circosol 4240)	5
Tetramethyl thiuram disulfide	1
Ethoprop	20–30

The theory underlying the very high retention of ethoprop by EPT is unknown. It is suspected that true solubility exists in EPT, but certainly not to 30%. The carbon black content likely aids in ethoprop retention through adsorption. Release is in vapor form, not liquid, suggesting the development of bubble formation arising from supersaturation and the low activation energy involved.[945,946]

VERTEBRATE CONTROL

Other than rodents, thousands of species of vertebrates create control problems for man. Some problems such as those arising from the activities of domesticated dogs, cats, and birds are trivial, while others are far more serious, especially in regard to the destruction of foodstuffs.

Birds are the leading offenders. The quelea (*Quelea quelea*), or "milleteater," by destroying crops in the field, has been implicated in famine throughout central Africa.[924] Fifty percent or more of millet, wheat, sorghum, and other field crops may be lost in various areas visited by flocks composed of millions of individual birds. In addition to the quelea, species such as the golden sparrow (West Africa), sparrow (Malagasy), parrots, sparrows, and crows (India), starlings (Tunisia), goldfinches (Israel), cowbirds and grackles (Mexico), dickcissels (Columbia and Venezuela), weaver birds (Philippines), and others cause huge crop losses. In the United States up to 21 bushels of corn per acre may be consumed by birds in localized areas,[947] although the national loss probably does not exceed a fraction of a bushel per acre.

Bird control is difficult. Nontoxic methods such as fright devices usually only induce refractoriness in the exposed bird or encourage movement to unprotected fields. Avian predators are too few in number, due often to man's depletion of populations through hunting, to have any appreciable control effect. Mechanical protective devices such as screening of blackberry bushes are too awkward and uneconomical to be practical on a commercial scale.

In general, repellency through chemical or physical devices only translocates the problem to another field crop or feedlot. Chemical bird repellents such as carbamates are being used with mixed results.[948] Such materials can be used on sprouting crops[948] or on seeds.[949]

Toxic baits have been used in both urban and rural settings to reduce pest bird populations,[950] with 4-aminopyridine one commonly used material. Even the ingestion of sublethal dosages gives rise to distress calls by the poisoned bird, thus frightening away the rest of the flock. Unfortunately, poisoned birds are often eaten by nontarget animals which could in turn lead to poisoning of domestic pets, predator birds such as hawks, and so on, although this possibility has been discounted through experimentation.[951] Organophosphorus insecticides are, in some cases, quite toxic to birds and could be used in baits.[952]

Baiting campaigns suffer from the problems of short-term effectiveness, hazards to nontargets, and refractoriness of the bird population through either the sensing of the toxic element or observation of poisoned animals. Possibly the encapsulation of the toxicant will sufficiently mask odor and taste and thus prevent avoidance, while slowing physiological reaction so that the onset of symptoms occurs hours after ingestion and distress is therefore not associated with the bait.

In a few instances repellents may be a satisfactory method of preventing bird damage to structural objects. Woodpecker damage to utility poles, fence posts, and other objects could probably be prevented by the use of repellent paints or polymeric coverings. In this instance the offender has an alternate, albeit less desirable, choice – returning to the forest trees to construct nests and roosting holes.

The author has observed that paints comprising high rosin content with bis(tri-*n*-butyltin) sulfide and selenium diethyl dithiocarbamate as repellents will at least temporarily stop pileated woodpecker entrance into protected poles under experimental conditions.

Vertebrate predators play an important ecological role in keeping wild animal populaces at manageable levels. For instance, the coyote, eagle, and other western predators consume great quantities of jack rabbits, thus protecting grasslands from overforage and allowing a sizable sheep industry to subsist. However, the coyote especially may also destroy lambs and young sheep, so selective control measures are indicated. Sheepmen suffer perhaps a 2% yearly economic loss due to coyote predation.[953]

Coyote control must be selective in that (1) only a very small proportion of the total coyote population, perhaps 1%, ever take sheep, and (2) wholesale destruction of this pest leads to an increase in rabbits and other small vertebrates that in turn may compete with sheep for the available forage. Control through exclusion, fright devices, chemosterilants, and adverse agents does not appear to be either practical or economically attractive. Recently the use of poisons, which are not only nonselective in nature but tend to destroy wildlife in general and nonoffending coyotes in particular, have been banned on federal lands. Sheep ranchers in turn claimed a sizable increase in coyote predation losses and the future bankruptcy of the industry.[954]

One control method under investigation uses a toxicant-containing collar attached to the neck of a sacrificial lamb tethered away from the flock, providing an apt coyote target. As a rule, the coyote grabs at the neck and kills through crushing the trachea, resulting in suffocation. Packets of sodium or potassium cyanide solution or gel that are affixed to the collar and punctured by the coyote during the downing of the victim make it possible to induce a sufficient quantity of toxicant into the predator to afford destruction. Such collars have been evaluated in pen tests with lethal results,[955] but they are not always effective and their use will almost certainly entail loss of the lamb used as bait as well as some degree of contamination due to cyanide spillage from the punctured collar. An obvious hazard to handlers also exists. Technical problems associated with toxic collar development yet to be solved include means of coloring toxicant solutions for easy location and detoxification of spills and lowering or masking alkalinity of NaCN solutions to reduce the adverse effect so that the predator will ingest a lethal dose.[956]

Toxic collars constitute a rather special form of controlled release, in that the toxic element is only released at the time of target attack and tends to be a "one-shot" control method, unlike baits or the continuous release associated with insect and plant control.

PEST AREAS OF FUTURE POTENTIAL FOR THE CONTROLLED RELEASE APPROACH

Several areas exist where the need for adequate control is obvious, but, to date, no known attempt has been made to apply controlled release technology. These problems are mentioned in passing, for the future will likely see such applications.

Lamprey eel control is presently achieved in a somewhat nebulous fashion by the use of various toxicants applied to infested watercourses. TFM (3-trifluoromethyl-4-nitrophenol) is used as a lampricide, often jointly with niclosamide. TFM usage at 9 ppm is necessary for adequate control, and at this level various nontarget plants and animals show severe toxicological effects.[957] The use of baits may well overcome adverse environmental effects.

Another pest control area that may be conducive to the use of baits is in starfish control. Recent years have seen a considerable threat to various Pacific islands arising from *Acanthaster planci*, the coral-eating starfish.[958] The offenders have been picked by hand and destroyed, in some cases by treating each animal with an injection of formalin or other toxicant. The reliability of this method is open to doubt when infestation is serious. Neither toxicants nor suitable attractants useful in baits are known, although a copper sulfate gel has been pumped into very localized infestation with success.[959] Small members of the *Patiria* and *Pisaster* species were examined in Y-tube-type bait chambers in the author's laboratory. Attraction to baits composed of either powdered clam, oyster, fish, or shrimp in an acrylic binder were observed. Whether the coral-destroying *A. planci* can be controlled in this manner remains unknown.

Pest marine life might be controllable through the use of repellents, the depredations arising from shark attack, especially on downed pilots, have been well documented.[960] Other than providing psychological comfort for one stranded in shark-infested waters, dyes now employed are probably useless as repellents. What does repel sharks, barracudas, and other man-attacking marine denizens remains a mystery to the scientific community. However, the author does wish to point out an observation well known to fishermen through the Caribbean: a shark will not take fish bait composed of well-rotted shark meat. This is a rather important fact for one wishing to catch only edible fish, since sharks will aggressively take any other bait thrust in their way. Fresh shark flesh is taken, rotted flesh is not; ergo, determine what degradation product is repellent and proceed to isolate and/or synthesize for evaluation.

Probably the most intriguing marine pest problem area lies in the determination of how one prevents entrance of venemous animals such as the elasmobranchs (stingrays) and Cnidaria (jelly fish) into coastal areas used for recreation or industrial purposes. Some of the hydrozoa such as the Spanish-man-of-war and scyphozoa, e.g., the six deadly *Cubomedusae* species (sea wasps), will kill man within minutes of attack.[961] Jellyfish infestations have resulted in the closing of many miles of ocean beaches along the mid-Atlantic coast and the initiation of various projects keyed to studying the ecology and biology of these menaces with an eye towards future control.[962] So far, no control method has been developed, other than the use of beach nets; these will prevent the entrance of the larger animals but not the smaller, and are overridden during heavy weather. A dismembered tentacle alone is dangerous, and can cause an extremely serious wound if from *Cyanea* or one of the *Cubomedusae*. It cannot be isolated via mechanical means, and cannot be poisoned. The firing of the nematocysts occurs upon contact, even though the pest, in the biological sense, is dead.

NEW MATERIALS OF POTENTIAL VALUE FOR CONTROLLED RELEASE

In ending this work the author would be remiss in not including a brief section on several new materials and pesticide-dispensing systems that promise much to the future of controlled release technology. Controlled-pore ceramics presently used as carriers for enzymes are one possibility.[963] Of greater versatility are the nitrile-acrylic copolymers and polyurethanes used to hold and slowly release lubricating oils.[964,965] One such material, Micro-well®, produced by the

Mohawk Data Sciences Corp., Newport, N.Y., is an ultraporous elastomeric foam with up to 70% void space that can be filled with any number of liquids. Present use is in ball bearing retainer rings and ink rollers.[966] An oil loss of 3% per year has been measured under environmentally harsh conditions, so a "safe" prediction of 10 or more years' life can be made.[967] The Micro-well materials readily adsorb liquid insecticides and pheromones to near 70% capacity by volume, and their value as dispensing systems for pest control agents is presently under evaluation.

Another new material is Poroplastic®, an ultramicroporous plastic consisting of a cellulose triacetate network with molecular size pores varying from 14 to 60 Å.[968] One form, Sustrelle®, microbeads of 10 to 200 μm, can carry up to 96% of its total weight as a liquid payload. In both forms the material is supplied water filled. In order to load with the active agent the water is replaced by any of a number of techniques. Release is diffusion controlled and thus significantly different from encapsulation systems. Release rates depend upon four controllable factors: (1) active agent concentration, (2) permeability within the polymer network, (3) surface area of the loaded material, and (4) the rate of released agent removed from the interface into the external medium.[969]

Conrel® controlled release vapor dispenser systems, manufactured by FRL, an Albany International Company (Dedham, Mass.), consist of a parallel array of hollow polymeric fibers fixed to an adhesive tape backing.[970] After filling with the *liquid* agent, the fibers are heat sealed at regular intervals along the length of the tape. Release is activated by cutting the tape. The air-exposed liquid surface has the diameter of the fiber, which is variable, and a predetermined reservoir depth depending on the length of the fiber section. As volatilization occurs the fiber reservoir gradually depletes. Effective release life is a function of length. Various pheromones and insecticides have been incorporated in Conrel, and evaluations are underway.[970] The versatility of this system is aptly demonstrated by noting that each of the three components of the elm bark beetle pheromone are placed in a different hollow fiber. The desired dose rate is then achieved by adjusting the number of fibers loaded with each component.

REFERENCES

1. **Anon.**, *Marine Fouling and Its Prevention,* United States Naval Institute, Annapolis, Md., 1952.
2. **Doochin, H. and Smith, F. G. W.**, Marine boring and fouling in relation to velocity of water currents, *Bull. Mar. Sci. Gulf Caribb.*, 1, 196, 1951.
3. **Dunn, P. and Oldfield, D.**, A Novel Concept in Marine Antifouling Elastomers, Report of the Defence Standards Laboratory, Dept. of Defence, Maribyrnong, Victoria, Australia, 1974.
4. **Cardarelli, N. F., King, T., and Walker, K. E.**, Marine Biological Adhesives, Natural Products Symp., Chemical Society of Washington, D.C., December 12, 1968.
5. **Bernard, F. and Lane, C. E.**, Early settlement and metamorphosis of the barnacle, *Balanus amphitrite niveus, J. Morphol.*, 110, 19, 1962.
6. **Barnes, H., Barnes, M., and Finlayson, D. M.**, The metabolism during starvation of *Balanus balanoides, J. Mar. Biol. Assoc.* (U.K.), 43, 213, 1963.
7. **Southwood, A. J.**, On the behavior of barnacles: III. Further observations on the influence of temperature and age on cirral activity, *J. Mar. Biol. Assoc.* (U.K.), 36, 323, 1957.
8. **Ganapati, P. N. and Prasada, R. D. G. V.**, Studies on the respiration of barnacles: oxygen uptake and metabolic rate in relation to body size in *Balanus amphitrite communis, J. Anim. Morphol. Physiol.*, 1, 27, 1960.
9. **Crisp, D. J. and Bhupendra, P.**, The interaction between breeding and growth rate in the barnacle *Eliminius modestus Darwin, J. Limnol. Oceanog.*, 6, 105, 1961.
10. **Cardarelli, N. F.**, *Barnacle Cement as a Dental Restorative Adhesive,* Pub. No. 151, National Institutes of Health, Bethesda, Md., May 1968.
11. **Miller, S.**, Hidden enemy in our harbors, *Sea Frontiers,* 5, 23, 1959.
12. **Muraoka, J. S.**, Biodeterioration of Materials: Part III, U.S. Naval Civil Engineering Laboratory, Port Hueneme, Cal., AD631078, Feb. 1966.
13. **Snoke, L. R. and Richards, A. P.**, Marine borer attack on lead cable sheath, *Science,* 124, 443, 1956.
14. **Chellis, R. D.**, *Pile Foundations,* 2nd ed., McGraw-Hill, New York, 1961.
15. **Muraoka, J. S.**, Effects of deep sea microorganism on rubber and plastic insulation, *Mater. Prot.*, 5(4), 35, 1966.
16. **Katzev, S. W. and Katzev, M. L.**, Last harbor for the oldest ship, *Nat. Geograph.*, 146, 618, 1974.
17. **McIness, J.**, Compositions for Coating the Bottoms of Iron Ships, British Patent 1356, 1854.
18. **Tarr, J. G. and Wonson, A. H.**, Paint for Ships' Bottoms, U.S. Patent 40515, 1863.
19. **Saroyan, J. R.**, The Word on Antifouling Paints, Marine Coatings Conference, 80th annual meeting, National Paint, Varnish and Lacquer Association, San Francisco, Nov. 1, 1967.
20. **Anon.**, Boat paints, *Consumer Rep.*, p. 223, April 1969.
21. **Bufkin, B. G., Bounds, R. D., and Thames, S. F.**, Antifouling coatings, *Paint Varn. Prod.*, 64, 25, 1974.
22. **Anon.**, The Effect of Water Flow on Leaching Rates of Antifouling Paints: Consideration of Mechanism and Preliminary Results of Laboratory Tests, ACC/F 129/63, H. M. Dockyard, Portsmouth, Engl., 1963.
23. **Butler, G. and Ison, H. C. K.**, *Corrosion and Its Prevention in Waters,* Reinhold, New York, 1966.
24. **Bishop, J. H., Silva, S. R., and Silva, V. M.**, A study of microfouling on antifouling coatings using electron microscopy, *J. Oil Colour Chem. Assoc.*, 57, 30, 1974.
25. **Jennings, D. M., Littover, E. L., and Woodhouse, J. R.**, Marine Fouling and Antifouling Techniques, Trans. 3rd Annual Marine Technol. Conf., San Diego, Ca., 1967.
26. **Turner, H. J.**, Marine fouling, *Geo-marine Technol.*, 3, 112, 1967.
27. **White, W. R.**, The Effect of Low-Level Chlorination on Mussels at Poole Power Station, RD/LN17/66, Central Electrical Research Laboratory, U.K., March 7, 1966.
28. **Anon.**, 10-year antifouling coatings under development at Calspan, *Sea Technol.*, p. 23, October 1973.
29. **Tuthill, A. H. and Fielding, S. A.**, Shipboard Piping Systems – Costs and Reliability, *Mar. Mater.*, 27(3), 7, 1974.
30. **Maclean, J. D.**, *Preservative Treatment of Wood by Pressure Methods,* Agriculture Handbook No. 40, U.S. Dept. of Agriculture, December 1952.
31. **Vervall, A. F.**, Preserving Wood by Brush, Dip and Short Soak Methods, Tech. Bull. No. 1334, U.S. Dept. of Agriculture Forest Service, Washington, D. C., May 1965.
32. **Anon.**, Standards for Water-Borne Preservatives, *Manual of Recommended Practice,* P5-64, American Wood Preservers Association, 1964.
33. **Ray, D. L. and VanNiel, C. B.**, Method for Preventing the Destruction of Wood by Wood-Boring Animals, U.S. Patent 3,158,496, 1964.
34. **Smith, F. G. W. et al.**, A new method for long term preservation of wood by chemical modification, *For. Prod. J.*, p. 340, September 1956.
35. **Anon.**, Piling wrap saves docks, *Plast. World,* p. 300, January 1967.
36. **Morgans, W. M.**, *Outlines of Paint Technology,* CRC Press, Cleveland, 1969.
37. **Fisk, P. M.**, *The Physical Chemistry of Paints,* Chemical Publishing, New York, 1965.
38. **Crisp, D. J. and Austin, A. P.**, The action of copper in antifouling paints, *Ann. App. Biol.*, 48, 787, 1960.
39. **Herbert, P. A., et al.**, Chlorinated rubber marine coatings – performance tests including antifouling and blends of CR with tar, *A.C.S. Meeting,* Los Angeles, 34, 594, 1974.

40. **Varvill, M.**, Protection of Structures in Sea Water Against Fouling by Marine Growths, British Patent 851,902, 1960.
41. **Miller, S.**, Antifouling Potentials of Pesticidal Materials, final report, Contract NOas 59-6182-C (USN), Marine Laboratory, Institute of Marine Science, University of Miami, March 1961.
42. **Robbins, J. and Updegraf, D. M.**, bis (Tributyltin) Adipate and Antifouling Materials Comprising Same, U.S. Patent 3,236,793, 1966.
43. Dainippon Paint Manufacturing Co., Japanese Patent 6,903,103, 1969; *C.A.*, 71,714243, 1969.
44. **Hechenbleikner, I. and Thompson, P. F.**, Carboxymercaptal Hydrocarbon Tin Salts with Antifouling Marine Paint or Coating Biocidal Activity, U.S. Patent 3,463,644, 1969.
45. Mitsui Shipbuilding and Engineering Co., Antifouling Paints, British Patent 1,211,768, 1970.
46. Wirth, H. O., Friedrich, H. H., and Mras, V., German Patent 1,937,307, 1971; *C.A.*, 74,143396, 1971.
47. **Anon.**, bioMET bis(Tri-*n*-butyltin) Oxide Ingredient for Shipbottom Paints, Tech. Data Sheet No. 6223, M&T Chemicals, Inc., Rahway, N. J., December 1965.
48. **Anon.**, bioMET Tri-*n*-butyltin Fluoride Ingredient for Shipbottom Paints, Tech. Data Sheet No. 259, M&T Chemicals, Inc., Rahway, N.J., 1970.
49. **Beers, R. W.**, Antifouling Coating Composition, U.S. Patent 3,801,534, 1974.
50. **Vind, H. P. and Hochman, H.**, *An Evaluation of Organotin Compounds as Preservatives for Marine Timbers,* American Wood Preservers' Association, 1962.
51. **Anon.**, Harbor Screening Tests of Marine Borer Inhibitors-IV, Tech. Rep. 184, U.S. Naval Civil Engineering Laboratory, Port Hueneme, Cal., 1962.
52. **Oldfield, D. and Sansom, G. F.**, Potential Antifouling Coatings for Timber, Tech. Note 319, Defense Standards Laboratory, Maribyrnong, Australia, November 1973.
53. **Lomasney, H. L. and Lomasney, W. J.**, Solventless Coal Tar Extended Antifouling Coating, U.S. Patent 3,717,606, 1973.
54. Nowacki, L. J., Dick, R. J., and Johnston, H. N., Antifouling Paint Formulations, Fin. Rep., International Lead Zinc Research Organization Project LC-81, Battelle Memorial Institute, Columbus, Oh., March 1970.
55. Dick, R. J. and Nowacki, L. J., Organolead compounds in antifouling paints, *J. Paint Technol.*, 42, 535, 1970.
56. Carr, D. S., Triphenyllead acetate, *Paint Varn. Prod.*, 64, 19, Dec. 1974.
57. Kronstein, M., Organolead antifoulants: zeroing in on marine fouling, *Paint Varn. Prod.*, 63, 47, 1973.
58. **Phillip, A. T.**, Modern trends in marine antifouling paint research, *Prog. Org. Coat.*, 2, 159, 1973/74.
59. Anon., Industrial Biocide DS-2747 Tech. Bull., Diamond Shamrock Corp., Painesville, Oh., 1970.
60. Cardarelli, N. F., Development of An Antifouling Paint Utilizing TCPN as the Active Agent, Final Report, Creative Biology Laboratory, Barberton, Oh., to Diamond Shamrock Corp., Painesville, Oh., May 30, 1971.
61. Cardarelli, N. F., Recent Studies of Marine Fouling, Report to the American Institute of Aeronautics and Astronautics, Brecksville, Oh., Nov. 16, 1967.
62. **Cardarelli, N. F.**, Antifouling Elastomeric Coverings, Proc. Symp. on Deep Submergence Propulsion and Marine Systems, U.S. Naval Ordnance Plant, Forest Park, Ill., Feb. 28–March 1, 1966.
63. **Wuerzer, D. R., Senderling, R. L., and Cardarelli, N. F.**, Protecting buoys with antifouling rubber coverings, *Rubber World,* 157, 177, 1967.
64. **Natwick, J. O.**, Investigation of Nofoul Antifouling Rubber as a Protective Coating for Steel Sonar Domes, NUSC Tech. Rep. 4649, Naval Underwater Systems Center, New London, Conn., Nov. 26, 1973.
65. **Bollinger, E. H.**, Controlled Release Antifouling Rubber Coating, Rep. No. 19, Proc. Controlled Release Pesticide Symp., University of Akron, Akron, Oh., Sept. 16–18, 1974.
66. **Craig, A. S.**, *Rubber Technology: A Basic Course,* Oliver and Boyd, London, 1963.
67. **Morton, M.**, Ed., *Introduction to Rubber Technology,* Van Nostrand Reinhold, New York, 1969.
68. **Cardarelli, N. F. and Neff, H. F.**, Compositions de Préservation Contre les Incrustations Sous-Marines, French Patent 1,506,704, 1966.
69. **Cardarelli, N. F. and Caprette, S. J.**, Antifouling Covering, U.S. Patent 3,426,473, 1969.
70. **Cardarelli, N. F. and Neff, H. F.**, Biocidal Elastomeric Compositions, U.S. Patent 3,639,583, 1972.
71. **Anon.**, "Nofoul" Elastomeric Sheet, B. F. Goodrich Co. Rep., Akron, Oh., Oct. 5, 1967.
72. **Anon.**, Nofoul Technical Background, Sect. VI, B. F. Goodrich Co., Akron, Oh., 1973.
73. **Cardarelli, N. F.**, Biologically Active Rubber, Elastomer and Plastics Group, New England Section, American Chemical Society, Cambridge, Mass., Jan. 14, 1969.
74. **Hueck, H. J.**, The Biodeterioration of Materials as a Part of Hylobiology, TNO Comm. No. 218, Centraal laboratorium, Organische Chemie Institute, T.N.O., Delft, Netherlands, 1965.
75. **Cardarelli, N. F.**, Biocidal Elastomeric Composition and Method for Dispersing Biocides Therewith, U.S. Patent 3,851,053, 1974.
76. **Woodford, J. M. D.**, Underwater Marine Coatings, Part II, Marine Biocidal Rubbers Containing Organotin Toxics, Rep. 496, Australian Defence Science Services Defence Standards Laboratory, Maribyrnong, Victoria, Australia, March 1972.
77. **Anon.**, Antifouling Systems Based on Organotins: Australian Navy Trials, Tin and its Uses, No. 96, p. 7, 1973.
78. **deForest, A., Pettis, R. W., and Phillip, A. T.**, Underwater Marine Coatings: A Detailed Examination of Elastomeric Antifouling Materials After Marine Immersion, Rep. 589, Australian Defence Science Services Defence Standards Laboratory, Maribyrnong, Victoria, Australia, May 1974.

79. **Anon.,** Goodyear Rubber Anti-Foulant Systems for Ship Hulls, Rep. GTRAP-115, Goodyear Tire and Rubber Co., Goodyear, Ariz., Oct. 21, 1971.
80. **Dunn, P. and Oldfield, D.,** A Novel Concept in Marine Antifouling Elastomers, Australian Defence Science Services Defence Standards Laboratory Maribyrnong, Victoria, Australia, presented at Australian Rubber Technology Convention, Terrigal, New South Wales, Sept. 18–21, 1974.
81. **Dyckman, E. J., Montemarano, J. A., and Fischer, E. C.,** Environmentally compatible antifouling materials: organicmetallic polymers, *Nav. Eng. J.,* 85, 33, 1973.
82. **Anon.,** Organometallic polymers targeted at barnacles, *Chem. Eng. News,* 53, 18, 1975.
83. **Montemarano, J. A. and Dyckman, E. J.,** Performance of organometallic polymers as antifouling materials, in *ACS Meeting,* Los Angeles, 34, 607, 1974.
84. **Montemarano, J. A. and Dyckman, E. J.,** Biologically Active Polymeric Materials Exhibiting Controlled Release Mechanisms for Fouling Prevention, Rep. No. 21, Proc. Controlled Release Pesticide Symp., University of Akron, Akron, Oh., Sept. 16–18, 1974.
85. **Van Der Kerk, G. J. M.,** New developments in organolead chemistry, *Ind. Eng. Chem.,* 58, 29, 1966.
86. **Dub. M.,** *Organometallic Compounds,* Vol. 2, Springer-Verlag, Berlin, 1961.
87. **Walden, C. C. and Allen, I. V. F.,** Organolead Compounds as Wood Preservatives, First Int. Conf. Organolead Chemistry and its Applications, U. S. Army Natick Laboratory, Natick, Mass., May 2, 1967.
88. **Kronstein, M.,** Study of released matter from organometallic toxicant – antifouling paints, *Controlled Release Molluscicide Newsl.* (University of Akron), 1(1), 8, 1975.
89. **DeWolf, P. and Van Londen, A. M.,** Antifouling compositions, *Nature,* 209, 272, 1966.
90. **Harris, F. W.,** Controlled release herbicides – polymer formulations for aquatic weed control, *Hyacinth Control J.,* 11, 61, 1973.
91. **Higachi, T.,** Mechanisms of sustained action medication, *J. Pharm. Sci.,* 52, 1145, 1963.
92. **Kanakkanatt, S. V.,** Schistosomiasis Control By Slow Release Molluscicides, Product Development Laboratory, University of Akron, Report to World Health Organization, Sept. 15, 1972.
93. **Gardiner, J. B.,** Measurement of curative diffusion between rubbers by microinterferometry, *Rubber Chem. Technol.,* 42, 1058, 1969.
94. **Kraus, G. and Gruver, J.,** Thermal expansion, free volume and molecular mobility in a carbon black-filled elastomer, *J. Polym. Sci., Part A-2,* 8, 571, 1970.
95. **Aboytes, A. and Voet, A.,** Accessibility of the carbon black particle surface to elastomers, *Rubber Chem. Technol.,* 43, 464, 1970.
96. **Smith, W. R. and Kasten, G. A.,** Porosity studies of some oil furnace blacks, *Rubber Chem. Technol.,* 43, 960, 1970.
97. **Sircar, A. K. and Voet, A.,** Immobilization of elastomers at the carbon black particle surface, *Rubber Chem. Technol.,* 43, 973, 1970.
98. **Kraus, G. and Gruver, J. T.,** Molecular weight effects in adsorption of rubbers on carbon black, *Rubber Chem. Technol.,* 41, 1256,1968.
99. **Riven, D., Aron, J., and Medalia, A. I.,** Bonding of rubber to carbon black by sulfur vulcanization, *Rubber Chem. Technol.,* 41, 330, 1968.
100. **Horns, R. A.,** The Structure of Liquid Water, Arthur D. Little, Inc., report to the Office of Naval Research Nonr-4424(00), AD 666579, Cambridge, Mass., March 15, 1968.
101. **Riley, J. P. and Skirrow, G.,** *Chemical Oceanography,* Vol. I, Academic Press, New York, 1965.
102. **Senderling, R.,** Nofoul Rubber Application Procedure (Hydrolock-11, 21, 30 System), B.F. Goodrich Co., Aerospace and Defense Products, Akron, Oh., 1971.
103. **Van Der Schalie, H.,** People and their snail borne diseases, *Mich. Q. Rev.,* 2(2), 106, 1963.
104. **Muirhead-Thomson, R. C.,** *Pesticides and Freshwater Fauna,* Academic Press, New York, 1971.
105. **Jordan, P., Unrav, G. O., and Dalton, P.,** Preliminary Results of a Scheme to Control Transmission of *Schistosoma mansoni* through Provision of Water, WHO Rep. Ser. WHO/SCHISTO/73.28, 1973.
106. **Pellegrino, J. and Katz, N.,** Experimental chemotherapy of schistosomiasis mansoni, in *Advances in Parasitology,* Vol. 6, Academic Press, New York, 1968, 233.
107. **Bruce, J. I. et al.,** Schistosomiasis Drug Test Systems, WHO Inf. Rep., WHO/SCHISTO/73.30, 1973.
108. **Jarockij, L. S., Eyakuze, V. M., and McCullough, F. S.,** An Exploratory Trial of Schistosomiasis Mass Treatment with Niridazole in a Rural Population at Misungwi, Tanzania, WHO Inf. Rep., WHO/SCHISTO/71.16, 1971.
109. **Hillyer, G. V. et al.,** An attempt to induce resistance in mice to *Schistosoma mansoni* infection using millipore diffusion chambers, *Rev. Inst. Med. Trop. Sao Paulo,* 12(2), 149, 1970.
110. **Felsenfeld, O.,** *The Epidemiology of Tropical Diseases,* Charles C Thomas, Springfield, Ill., 1966.
111. **Warren, K. S. and Newill, V. A.,** *Schistosomiasis: A Bibliography of the World's Literature from 1852 to 1962, Western Reserve University Press, Cleveland, 1967.*
112. **Hoeffler, D. F.,** Cercarial dermatitis, *Arch. Environ. Health,* 29, 225, 1974.
113. **Chiriboga, J., de Leon, D., and Ritchie, L. S.,** Fasciola Hepatica Program in Puerto Rico, Report No. 156, Puerto Rico Nuclear Center, San Juan, December 1971.
114. **Liddel, J. S.,** Fascioliasis and cattle production, in *Proc. Control of Fascioliasis Symp.,* Shell International Chem. Co., Sittingbourne, Sussex, U.K., Feb. 23–25, 1970.
115. **Martin, H., Ed.,** *Insecticide and Fungicide Handbook for Crop Protection,* 4th ed., Blackwell, London, 1972.

116. **Van Der Schalie, H.**, Man meddles with nature – Hawaiian style, *Biologist*, 51(4), 136, 1969.
117. **Sollers, P.**, Toxic Bait for Oyster Beds, U.S. Patent 2, 613, 474, 1950.
118. **Davis, A. and Plestina, R.**, Preliminary Report on Phase I of Clinical Trials of Metrifonate Against Schistosomiasis, WHO Rep. Ser. WHO/SCHISTO/73.31, 1973.
119. **Abdallah, A. et al.**, Report of a WHO Consultant Group on the Comparative Evaluation of New Schistosomocidal Drugs for Use in Treatment Campaigns, WHO Rep. Ser., WHO/SCHISTO/72.19, 1972.
120. **Abdallah, A. et al.**, Report of a WHO Consultant Group on Hycanthone, WHO Rep. Ser. WHO/SCHISTO/72.20, 1972.
121. **Katz, N. et al.**, Preliminary clinical trials with oxamniquine, a new antischistosomal agent, *Rev. Inst. Med. Trop. Sao Paulo*, 15(1), 25, 1973.
122. **Arfaa, F., Farahmandian, I., and Soleimani, M.**, Evaluation of the effect of mass chemotherapy with niridazole as a method of bilharziasis control in Iran, *Trans. R. Soc. Trop. Med. Hyg.*, 64, 130, 1970.
123. **Goldsmith, E. I. and Kean, B. H.**, Extracorporeal hemofiltration in urinary schistosomiasis, *Am. J. Trop. Med. Hyg.*, 18, 382, 1969.
124. **Hunter, G. W. et al.**, Studies on schistosomiasis. V. Testing protective ointments against schistosome dermatitis producing cercaria, *Trans. R. Soc. Trop. Med. Hyg.*, 46, 201, 1952.
125. **Moon, A. P. and Hunter, G. W.**, Studies on schistosomiasis. VIII. Failure of copper oleate to afford complete protection to mice against *Schistosoma mansoni, J. Exp. Parasitol.*, 4, 165, 1955.
126. **Kemp, H. A. et al.**, Studies on schistosomiasis. XI. Some ointments examined for protection against *Schistosoma mansoni*, cercariae in preliminary tests, *Mil. Med.*, 119, 1, 1956.
127. **Radke, M. G.**, Studies on schistosomiasis. IX. Results of screening some ointments for protection against *Schistosoma mansoni, Mil. Surg.*, 115, 401, 1954.
128. **Hunter, G. W. et al.**, Studies on schistosomiasis. XII. Some ointments protecting mice against the cercariae of *Schistosoma mansoni, Am. J. Trop. Med. Hyg.*, 5, 713, 1956.
129. **Cardarelli, N. F. and Mansdorf, S. Z.**, Dermal Protective Film, U.S. Patent 3, 749, 772, 1973.
130. **Quinn, S. A., Walker, K. E., and Cardarelli, N. F.**, Development of a Long Lasting Topical Lotion Effective in the Prevention of Cercaria Penetration of Human Skin, Annual Summary Report of the Creative Biology Laboratory to U.S. Army Medical Research and Development Command, DADA 17-71-G-9334, August 1971.
131. **Walker, K. E. et al.**, Protective Lotions for Military Personnel, final comprehensive report, Creative Biology Laboratory to U. S. Army Medical Research and Development Command, DADA 17-73-C-3031, May 1, 1974.
132. **Jobin, W. R.**, A Rationale for Selection of Molluscicides for Schistosomiasis Control Programmes, WHO Rep. Ser., PD/MOL/69.2, 1969.
133. **Jobin, W. R.**, Cost/Benefit Analysis for Application of Molluscicides in the Prevention of Schistosomiasis, WHO Rep. Ser., WHO/SCHISTO/73.24, 1973.
134. **Amin, M. A.**, The control of schistosomiasis in the Gezira, Sudan, *Sud. Med. J.*, 10, 75, 1972.
135. **Reimer, D. N. and Toth, S. J.**, Absorption of copper by clay materials, humic acids and bottom muds, *J. Am. Water Works Assoc.*, 62, 195, 1970.
136. **Toth, S. J. and Reimer, D. N.**, Algae control in farm ponds, *Agric. Chem.*, 16(5), 44, 1968.
137. **Chu, K. Y., Massoud, J., and Arfaa, F.**, Comparative studies of the molluscicidal effect of cuprous chloride and copper sulfate in Iran, *Bull. WHO*, 39, 320, 1968.
138. **Ritchie, L. S. and Malek, E. A.**, Molluscicides: Status of Their Evaluation, Formulations and Methods of Applications, WHO Rep. Ser. PD/MOL/69.1, 1969.
139. **Harry, H. W. and Aldrich, D. V.**, The distress syndrome in *Taphius glabratus* (Say) as a reaction to toxic concentrations of inorganic ions, *Malacology*, 1, 283, 1963.
140. **Nastel, G. K. and Albright, F. R.**, Chemistry of Copper in Water, Annual Report, Lancaster Laboratory to International Copper Research Association, INCRA Proj. 177A, Aug. 9, 1973.
141. **Cheng, T. C. et al.**, Biological Studies on Copper Containing Molluscicides, Institute of Pathiobiology, Lehigh University, Annual Report to International Copper Research Association, INCRA Proj. 193, June 30, 1973.
142. **Kapkov, V. T.**, Toxicity of copper complexes to freshwater molluscs, *Vest. Mosk. Univ. Biol. Pochvoved*, 26(2), 34, 1971.
143. **Barbosa, F. S.**, Insoluble or slightly soluble chemicals as molluscicides, *Bull. WHO*, 25, 710, 1961.
144. **Frick, L. P. et al.**, Molluscicidal properties of copper protoxide (Cu_2O) as revealed by tests on stages of *Australorbis glabratus, Bull. WHO*, 30, 295, 1964.
145. **Puma, R. and Zagustin, T.**, Molusquicidas en Uso en la Compaña de Lucha Contra La Bilharziasis, II. Reunion del Caribe pro Investigacion de la Bilharziasis, Maracay, Venezuela, Feb. 13–14, 1969.
146. **Olivier, L. and Hoskins, W. T.**, The effects of low concentrations of sodium pentachlorophenate on the fecundity and egg viability of *Australorbis glabratus, Am. J. Trop. Med. Hyg.*, 9, 199, 1960.
147. **Wright, W. H., Dobrovolny, C. G., and Berry, E. G.**, Field trials of various molluscicides (chiefly sodium pentachlorophenate) for the control of aquatic intermediate hosts of human bilharziasis, *Bull. WHO*, 18, 963, 1958.
148. **Pesigan, T. P. et al.**, Screening of Chemical Compounds for Molluscicidal Activity Against *Oncomelania quadrasi*, WHO Rep. Ser. MOL/INF/67.23, 1967.
149. **Ayad, N. et al.**, Molluscicides: Second Report of the Expert Committee on Bilharziasis, WHO Tech. Rep. Ser. No. 214, 1961.

150. **Yokogawa, M.**, The Decline of *Schistosoma japonicum* Infections in Japan, WHO Rep. Ser., WHO/SCHISTO/74.34, 1974.
151. **Jobin, W. R. and Unrau, G. O.**, Chemical control of *Australorbis glabratus, Public Health Rep.*, 82, 63, 1967.
152. **Anon.**, Molluscicide screening and evaluation, *Bull. WHO*, 33, 567, 1965.
153. **Gonnert, R. and Schraufstatter, E.**, A new molluscicide, Bayer 73, in *Proc. 6th Int. Congr. Trop. Med. Malar.*, London, 2, 197, 1959.
154. **Duncan, J.**, A review of the development and application of molluscicides in schistosomiasis control, in Cheng, T. C., Ed., *Molluscicides in Schistosomiasis Control*, Academic Press, New York, 1974.
155. **Ritchie, L. S. et al.**, Molluscicidal qualities of Bayluscide revealed by 6-hour and 24-hour exposures against representative stages and sizes of *Australorbis glabratus, Bull. WHO*, 29, 81, 1963.
156. **Shiff, C. J. et al.**, Molluscicide for the Control of Schistosomiasis in Irrigation Schemes, *Bull. WHO*, 48, 299, 1973.
157. **Dawood, I. K. et al.**, Field trials in Egypt with IC 24223, Bayluscide and sodium pentachlorophenate molluscicides, *Bull. WHO*, 32, 261, 1965.
158. **Anon.**, Bayluscide, Chemagro Corp., Kansas City, Mo., Technical Specifications, January 1970.
159. **Sturrock, R. F., Barnish, G., and Upatham, E. S.**, Snail findings from an experimental mollusciciding program to control *Schistosoma mansoni* transmission on St. Lucia, *Int. J. Parasitol.*, 4, 231, 1974.
160. **Sturrock, R. F. and Barnish, G.**, The aerial application of molluscicides with special reference to schistosomiasis control, *Bull. WHO*, 49, 283, 1973.
161. **Newton, M. E. and Fetterolf, C. M.**, Experimental Use of Bayluscide for the Control of Swimmers Itch, Report of the State of Michigan Water Resources Commission, Dept. of Conservation, Lansing, June 1967.
162. **Etges, F. J. and Gilbertson, D. E.**, Repellent action of some chemical molluscicides on schistosome vector snails, *Am. J. Trop. Med. Hyg.*, 15, 618, 1966.
163. **Sturrock, R. F.**, Persistence of the molluscicide Bayluscide (clonitralide) emulsifiable concentrate on mud surfaces in the tropics, *Ann. Trop. Med. Parasitol.*, 68, 427, 1974.
164. **Ritchie, L. S. and Fox, I.**, Comprehensive laboratory evaluations of the molluscicide 2-5-dichloro-4-nitro-3-phenylsalicylanilide (PSA); time-concentration relationships and susceptibilities of stage-size arrays of snails, *Bull. WHO*, 39, 312, 1968.
165. **Hopf, H. S. et al.**, Laboratory Assessment of the Molluscicidal Activity of Two Halogenated Salicylanilides, WHO Rep. Ser. PD/MOL/69.3, 1969.
166. **Etges, F. J., Bell, E. J., and Ivins, B. E.**, A field survey of molluscicide-degrading organisms in the Caribbean area, *Am. J. Trop. Med. Hyg.*, 18, 472, 1969.
167. **Meredith, R. and Meridith, G. C.**, The Assay of 5,2-Dichloro-4-nitrosalicylicanalide-Ethanolamine Salt Using the Ciliate Protozoan *Spirostomum ambiguum* Ehrenberg var. *Minor*, WHO Rep. Ser. WHO/SCHISTO/71.11, 1971.
168. **Meredith, R.**, The Dependence on pH of the Molluscicidal Activity of the Ethanolamine Salt of 5,2-dichloro-4-nitro-salicylicanilide in Solution in 10° Artificial Hard Water, WHO Rep. Ser. WHO/SCHISTO/71.12, 1971.
169. **Meredith, R.**, The Adsorption of Ethanolamine-Niclosamide by River Muds – The Effect of the Organic, Clay and Mineral Content of the Mud, WHO Rep. Ser. WHO/SCHISTO/71.15, 1971.
170. **Marking, L. L. and Hogan, J. W.**, Toxicity of Bayer-73 to fish, *Invest. Fish Control*, 19, 1, 1967.
171. **Paulini, E., Chaia, G., and de Freitas, J. R.**, Trials with the molluscicides, Rhodiacide and Bayer-73, *Bull. WHO*, 25, 706, 1961.
172. **Harrison, A. D.**, The effects of Bayluscide on gastropod snails and other aquatic fauna, *Hydrobiologia*, 28, 371, 1966.
173. **Shiff, C. J. and Garnett, B.**, The short term effects of three molluscicides on the microflora and microfauna of small, biologically stable ponds in southern Rhodesia, *Bull. WHO*, 25, 543, 1961.
174. **Shiff, C. J.**, Trials with *N*-tritylmorpholine (Shell WL8008) as a molluscicide in southern Rhodesia, *Bull. WHO*, 35, 203, 1966.
175. **Shiff, C. J., Crossland, N. O., and Millar, D. R.**, The susceptibilities of various species of fish to the molluscicide, *N*-tritylmorpholine, *Bull. WHO*, 36, 500, 1967.
176. **Boyce, C. B. C., Tyssul-Jones, T. W., and Van Tongeren, W. A.**, The molluscicidal activity of *N*-tritylmorpholine, *Bull. WHO*, 37, 1, 1967.
177. **Yasuraoka, K., Hosaka, Y., and Komiya, Y.**, Laboratory Trials with the New Frescon (WL8008) Formulations Against *Oncomelania nosophora*, WHO Rep. Ser. PD/MOL/69.5, 1969.
178. **Fenwick, A. and Lidgate, H. J.**, Attempts to eradicate snails from impounded water by the use of *N*-tritylmorpholine, *Bull. WHO*, 42, 581, 1970.
179. **Gilbert, B., Castleton, C., and Bulhoes, M. S.**, Molluscicidal activity of trifenmorph in field trials, *Bull. WHO*, 49, 377, 1973.
180. **Degremont, A. A., Geigy, R., and Perret, P.**, Preliminary results of the project for controlling and preventing schistosomiasis in the lower Mongoky (Malagasy Republic), *Acta Trop. Suppl.*, 29(2), 138, 1972.
181. **Crossland, N. O., Bennett, M. S., and Cawdery, M. J. H.**, Preliminary observations on the control of *Fasciola hepatica* with the molluscicide *N*-tritylmorpholine, *Vet. Rec.*, 84, 182, 1969.
182. **Crossland, N. O.**, Some practical aspects concerning the use of the molluscicide *N*-tritylmorpholine (Frescon) for the control of fascioliasis, *Helminthologia*, 10, 83, 1969.

183. **Ogersby, J. M.**, The Distribution of a Chemical Following Its Application at a Point Source in an Irrigation System, WHO Rep. Ser. PD/MOL/70.11, 1971.
184. **Meyling, A. H., Meyling, J., and Pitchford, R. J.**, Physical-chemical properties of substances used as molluscicides, *Bull. WHO*, 34, 141, 1966.
185. Fenwick, A., Snail Eradication with Frescon (*N*-tritylmorpholine) Using the Split Dosage Technique of Application (Part II), Misc. Rep. No. 677, Tropical Pesticides Research Institute, Arusha, Tanzania, 1970.
186. **Berrios-Duran, L. A., Ritchie, L. S., and Wessel, H. B.**, Field screening tests on molluscicides against *Biomphalaria glabrata* in flowing water, *Bull. WHO*, 39, 316, 1968.
187. Fenwick, A., **Beasley, J. S. S., and Lidgate, H. J.**, Snail Eradication Experiment With Frescon (*N*-tritylmorpholine) Using the Split Dosage Technique of Application, Misc. Rep. No. 641, Tropical Pesticides Research Institute, Arusha, Tanzania, May 1968.
188. **Lemma, A.**, Laboratory and field evaluation of the molluscicidal properties of endod (*Phytolacca dodecandra*), *Bull. WHO*, 42, 597, 1970.
189. **Unrau, G. O. et al.**, Field trials in Egypt with acrolein herbicide-molluscicide, *Bull. WHO*, 32, 249, 1965.
190. **Anon.**, Snail control in the prevention of bilharziasis, *WHO Monogr. Ser.*, No. 5, 1965.
191. **Cheng, T-H**, Schistosomiasis in Mainland China: A Review of Research and Control Programs Since 1949, WHO Rep. Ser. WHO/SCHISTO/73.27, 1973.
192. **Gaaboub, A., Abales, I. M., and El-Khodary, A. S.**, The Effectiveness of Certain Mosquito Larvicides in Comparison with the Molluscicide, Niclosamide, Against the Schistosome-Bearing Snail, *Biomphalaria alexandrina*, in the Surroundings of Alexandria, Egypt, WHO Rep. Ser., WHO/SCHISTO/72.18, 1972.
193. **Yasuraoka, K.**, Laboratory Evaluation of the Molluscicidal Properties of Triterpenoid Saponin, A Molluscicidal Fraction from Endod (*Phytolacca dodecandra*), Against *Oncomelania nosophora*, WHO Rep. Ser. WHO/SCHISTO/71.8, 1971.
194. **Hopf, H. S. et al.**, Molluscicidal properties of organotin and organolead compounds, *Bull. WHO*, 36, 955, 1967.
195. **Hira, P. R. and Webbe, G.**, The effect of sublethal concentrations of the molluscicide, triphenyllead acetate, on *Biomphalaria glabrata* (Say) and on the development of *Schistosoma mansoni* in the snail, *J. Helminthol.*, 46, 11, 1972.
196. **Shiff, C. J., Mitchell, L., and Coutts, W.**, Trials with triphenyllead acetate for the destruction of aquatic snails, WHO Rep. Ser., WHO/SCHISTO/73.32, 1973.
197. **Anon.**, Industrial Biocide DS-2787, Tech. Bull., Diamond Shamrock Corp., Cleveland, 1970.
198. **Cardarelli, N. F.**, Method of Controlling the Gastropod Vectors of Parasitic Flukes, U.S. Patent 3,767,809, 1973.
199. **Hopf, H. S. and Muller, R. L.**, Laboratory breeding and testing of *Australorbis glabratus* for molluscicidal screening, *Bull. WHO*, 27, 783, 1962.
200. **de Villers, J. P. and Mackenzie, J. G.**, Organo-tin and Organo-lead Molluscicides, WHO Rep. Ser. MOL/INF 13, 1963, 63.
201. **Paulini, E.**, Laboratory Experiments with Some Organotin Compounds, WHO Rep. Ser. MOL/INF/16, III/1, 1963.
202. Anon., Trials Carried Out Against *Q. quadrasi*, WHO Rep. Ser., MOL/INF./16, III 4, 1963.
203. **Seiffer, E. A. and Schoof, H. F.**, Tests of 15 experimental molluscicides against *Australorbis glabratus*, *Public Health Rep.*, 82, 833, 1967.
204. **Hopf, H. S. and Duncan. J.** The Molluscicidal Properties of Organotin and Organolead Compounds, WHO Rep. Ser., MOL/INF./66.21, 1966, VII/1.
205. **Ritchie, L. S., Berrios-Duran, L. A., Frick, L. P., and Fox, I.**, Molluscicidal time-concentration relationships of organotin compounds, *Bull. WHO*, 31, 147, 1964.
206. **Floch, H., Deschiens, R., and Floch, T.**, Sur le Proprietes Molluscicides de L'oxide et de L'acetate de tributyl-etain, *Bull. Soc. Pathol. Exot.*, 57, 454, 1964.
207. **Frick, L. P. and de Jimenez, W.**, Molluscicidal qualities of three organotin compounds revealed by 6h and 24h exposure against representative stages and sizes of *Australorbis glabratus*, *Bull. WHO*, 31, 429, 1964.
208. **Webbe, G. and Sturruck, R. F.**, Laboratory tests of some new molluscicides in Tanzania, *Ann. Trop. Med. Parasitol.*, 58, 234, 1964.
209. **Deschiens, R., Brottes, H., and Mrogo, L.**, Application sur le terrain au Cameroun das la prophylaxie des bilharzioses de l'action molluscicide de l'oxide de tributyletain, *Bull. Soc. Pathol. Exot.*, 59, 968, 1966.
210. **Freiberg, A. H.**, Tributyltin Molluscicides, U.S. Patent 3,439,098, 1969.
211. Anon., Epidemiology and Control of Schistosomiasis, WHO Tech. Rep. Ser. No. 372, 1967.
212. **Deschiens, R., Floch, H., and Floch, T.**, The molluscicidal properties of tributyltin oxide and acetate for prophylaxis of bilharziasis, *Bull. Soc. Pathol. Exot.*, 57, 454, 1964.
213. **Chernin, E., Michelson, E. H., and Augustine, D. L.**, Studies on the biological control of schistosome-bearing snails, II. The control of *Australorbis glabratus* populations by the leech, *Helobdella fusca*, under laboratory conditions, *Am. J. Trop. Med. Hyg.*, 5, 308, 1956.
214. **Chernin, E., Michelson, E. H., and Augustine, D. L.**, *Daubaylia potomaca*, a nematode parasite of *Helisoma trivolvis*, transmissable to *Australorbis glabratus*, *J. Parasitol.*, 46, 599, 1960.
215. **Michelson, E. H.**, Studies on the biological control of schistosome-bearing snails. Predators and parasites of fresh-water mollusca: a review of the literature, *J. Parasitol.*, 47, 413, 1957.

216. **Michelson, E. H.**, *Plistophora husseyi* sp. n., a microsporidian parasite of aquatic pulmonate snails, *J. Insect Pathol.*, 5, 28, 1963.
217. **Pan, C-T**, Studies on the biological control of schistosome-bearing snails: a preliminary report on pathogenic microorganisms found in *Australorbis glabratus*, *J. Parasitol.*, 42 (Sect. 2), 33, 1956.
218. **Michelson, E. H.**, An acid-fast pathogen of fresh-water snails, *Am. J. Trop. Med. Hyg.*, 10, 423, 1961.
219. **Chernin, E., Michelson, E. H., and Augustine, D. L.**, Studies on the biological control of schistosome-bearing snails. I. The control of *Australorbis glabratus* by the snail *Marisa cornuarietis* under laboratory conditions, *Am. J. Trop. Med. Hyg.*, 5, 297, 1956.
220. **Ruiz-Tiben, E., Palmer, J. R., and Ferguson, F. F.**, Biological control of *Biomphalaria glabrata* by *Marisa cornuarietis* in irrigation ponds in Puerto Rico, *Bull. WHO*, 41, 329, 1969.
221. **Webbe, G.**, Molluscicides in the control of schistosomiasis, in *Molluscicides in Schistosomiasis Control*, Cheng, T. C., Ed., Academic Press, New York, 1974, 41.
222. **Lawson, G. W., Petr, T., Biswas, S., Biswas, E. R. I., and Reynolds, J. D.**, Hydrobiological work on the Volta basin research project 1963–1968, *Bull. Inst. Fondam. Afr. Noire, Ser. A*, 31, 965, 1969.
223. **Obeng, L. E. et al.**, Fourth Annual Report, 1969–1970, Institute of Aquatic Biology, Ghana, 1971.
224. **Anon.**, Research on the Epidemiology and Methodology of Control of Schistosomiasis in Man-Made Lakes (UNDP/WHO Project REG-217/IR-0658): Study Proposal, WHO Doc. PD/72.3, 1972.
225. **Dazo, B. C. and Biles, J. E.**, Schistosomiasis in the Kainji Lake Area, Nigeria, WHO Rep. Ser., WHO/SCHISTO/72.21, 1972.
226. **Dazo, B. C. and Biles, J. E.**, Follow-up Studies on the Epidemiology of Schistosomiasis in the Kainji Lake Area, Nigeria (November-December 1971), WHO Rep. Ser., WHO/SCHISTO/73.29, 1973.
227. **Dazo, B. C. and Biles, J. E.**, The Schistosomiasis Situation in the Lake Nasser Area, The Arab Republic of Egypt, WHO Rep. Ser., WHO/SCHISTO/72.23, 1972.
228. **Anon.**, Annual Report 1973–1974, National Council for Research, Khartoum, Sudan, July 1974.
229. **Amin, M. A.**, The control of schistosomiasis in the Gezira, Sudan. The use of copper sulfate and mechanical barriers, *Sudan Med. J.*, 10, 75, 1972.
230. **Arfaa, F., Bijan, H., and Farahmandian, I.**, Present status of urinary bilharziasis in Iran, *Trans. R. Soc. Trop. Med. Hyg.*, 61, 358, 1967.
231. **Fenwick, A.**, The development of snail control methods on an irrigated sugar estate in northern Tanzania, *Bull. WHO*, 42, 589, 1970.
232. **Jobin, W. R. and Michelson, E. H.**, Operation of irrigation reservoirs for the control of snails, *Am. J. Trop. Med. Hyg.*, 18, 297, 1969.
233. **McCullough, F. S. and Eyakuse, V. M.**, World Health Organization Tanzania schistosomiasis pilot control and training project at Mwanza, Tanzania, *East Afr. Inst. Med. Res. Ann. Rep.*, 1971, 52, 1971.
234. **Shiff, C. J., Clarke, V. de V.**, The effect of snail surveillance in natural waterways on the transmission of *Schistosoma haematobium* in Rhodesia, *Cent. Afr. J. Med.*, 13(6), 133, 1967.
235. **Shiff, C. J.**, The value of incidence for the assessment of schistosomiasis control, *Bull. WHO*, 48, 409, 1973.
236. **Sturrock, R. F.**, Control of *Schistosoma mansoni* transmission: strategy for using molluscicides on St. Lucia, *Int. J. Parasitol.*, 3, 795, 1973.
237. **Duke, B. O. L. and Moore, P. J.**, The control of *Schistosoma haematobium* in West Cameroon, *Trans. R. Soc. Trop. Med. Hyg.*, 65, 841, 1971.
238. **Berrios-Duran, L. A. and Ritchie, L. S.**, Molluscicidal activity of bis (tri-*n*-butyltin) oxide formulated in rubber, *Bull. WHO*, 39, 310, 1968.
239. **Hopf, H. S. and Goll, P. H.**, Biocidal Rubber: A Two Year Assessment of the Organotin Containing Formulations, WHO Rep. Ser. PD/MOL/70.14, 1970.
240. **daSouza, C. P. and Paulini, E.**, Laboratory and Field Evaluation of Some Biocidal Rubber Formulations, WHO Rep. Ser. PD/MOL/69.9, 1969.
241. **Fenwick, A.**, Some Observations on Molluscicides in Connexion with the Schistosomiasis Pilot Control and Training Project at Misungwi (Tanzania), Addendum, WHO Rep. Ser. AFR/BILHARZ/14 Add. 1, May 21, 1969.
242. **Fenwick, A.**, Laboratory Evaluation of Biocidal Pellets 633B for Molluscicide Use, Misc. Rep. 676, Tropical Pesticide Research Institute, Arusha, Tanzania, 1968.
243. **Cardarelli, N. F.**, Method for Dispersing Toxicants to Kill Disease Spreading Water Spawned Larva, Trematodes, Mollusks and Similar Organisms, and Products Used in Such Method, U.S. Patent 3,417,181, 1968.
244. **Cardarelli, N. F.**, Slow release molluscicides and related materials, in *Molluscicides in Schistosomiasis Control*, Cheng, T. C., Ed., Academic Press, New York, 1974, 177.
245. **Cardarelli, N. F.**, personal communication, 1968.
246. **Struffe, R.**, Method of Colorimetric Determination for Molluscicide Bayer 73 in Field Test, Pflanzenschutz-Nachrichten Bayer, 15/1962, 1, p. 42, 1962 (WHO Rep. WHO/BILHARZ/59, 1960).
247. **Cardarelli, N. F.**, Development and Testing of Molluscicidal and Cercariacidal Formulations, Annual report, 1973, WHO Ref. B2/181/62, Jan. 30, 1974.
248. **Walker, K. E. and Cardarelli, N. F.**, Development of Slow Release Copper Sulfate as a Molluscicide, Proj. 203, Final Report, International Copper Research Association, New York, July 1, 1974.

249. **Walker, K. E. and Cardarelli, N. F.**, Development of Slow Release Copper Sulfate as a Molluscicide, Proj. 203, Annual Report, International Copper Research Association, New York, July 1, 1973.
250. **Ansari, N. et al.**, Meeting of Directors of Collaborating Laboratories on Molluscicide Testing and Evaluation, 31 Aug.–4 Sept. 1970, WHO Inf. Rep. WHO/SCHISTO/ 71.6, 1971.
251. **Campion, D. G. and Outram, I.**, Insecticidal and possible chemosterilant effects of certain organo-metal compounds against red bollworm, *Int. Pest Control J.*, 10, 21, 1968.
252. **Anon.**, Tin Chemicals for Industry, Pub. No. 477, Tin Research Institute, Utrecht, Netherlands, 1972.
253. **Guess, W. L. and Stetson, J. B.**, Tissue reactions to organotin-stabilized polyvinyl chloride (PVC) catheters, *JAMA*, 204(7), 118, 1968.
254. **Frieden, E.**, The chemical elements of life, *Sci. Am.*, 227(1), 51, 1972.
255. **Schwarz, K., Milne, D. B., and Vinyard, E.**, Growth effects of tin compounds in rats maintained in a trace element-controlled environment, *Biochem. Biophys. Res. Commun.*, 40(1), 22, 1970.
256. **Schwarz, K.**, Tin as an essential growth factor for rats, in *Newer Trace Elements in Nutrition*, Mertz, W. and Cornatzer, W. E., Eds., Marcel Dekker, New York, 1970.
257. **Ingham, R. K. et al.**, Physiological properties of organotin compounds, *Chem. Rev.*, 60, 459, 1960.
258. **Cremer, J. E.**, Selective inhibition of glucose oxidation by triethyltin in rat brain in vivo, *Biochem. J.*, 119, 95, 1970.
259. **Barnes, J. M. and Stoner, H. B.**, The toxicology of tin compounds, *Pharmacol. Rev.*, 11, 211, 1959.
260. **Cerny, E. and Pelikan, Z.**, Toxic effects of bis(tri-*n*-butyltin) oxide (TBTO) on the skin of rats, *Berufs-Dermatosen*, 17, 305, 1969.
261. **Elsa, J. and Paynter, V.**, Toxicological studies on bis(tri-*n*-butyltin) oxide, *AMA Arch. Ind. Health*, 18, 214, 1958.
262. **Pelikan, Z.**, Effects of bis(tri-*n*-butyltin) oxide on the eyes of rabbits, *Br. J. Ind. Med.*, 26, 165, 1969.
263. **Luijten, J. G. A.**, A Bibliography of Organotin Toxicity, Tin Pub. No. 436, Tin Research Institute, Utrecht, Netherlands, August 1971.
264. **Bollen, W. B. and Tu, C. M.**, Effects of an organotin on microbial activities in soil, *Tin Its Uses*, 94, 13, 1972.
265. **Cenci, P. and Cremonini, B.**, Thin-layer chromatography of two organotin pesticides, and their stability in various soils, *Ind. Sacc. Ital.*, 62(11-12), 313, 1969.
266. **Hocking, D. and White, P. J.**, Phytotoxicity of triphenyltin acetate to rice seed, *East Afr. Agric. For. J.*, 32(4), 380, 1967; *C.A.*, 67:63244.
267. **Katsumura, T.**, Japanese Patent 21360/65, 1965.
268. **Bruggeman, J. et al.**, Nutritional physiology, analytical and toxicological studies with the fungicide triphenyltin acetate, *Zentralbl. Veterinaermed. Reihe A*, 11(1), 1, 1964.
269. **Herok, J. and Gotte, H.**, Radiometric analysis of the metabolism of triphenyltin acetate in plants and animals, *Int. J. Appl. Radiat. Isot.*, 14, 461, 1963.
270. **Kroller, E.**, Investigations on the use of IPC as an inhibitor of stored potatoes, *Dtsch. Lebensm. Rundsch.*, 56, 190, 1960.
271. **Ascher, K. R. S. and Nissim, S.**, Organotin Compounds and Their Potential Use in Insect Control, Rep. No. 775-E, National Institute of Agriculture, Rehovat, Israel, 1964.
272. **Upatham, E. S., Engelhart, J. E., and Seeyave, J.**, Effect of Organotin Compounds on St. Lucian *Biomphalaria glabrata* and Bananas, Rep. No. 24, Proc. Controlled Release Pesticide Symp., University of Akron, Akron, Oh., Sept. 16–18, 1974.
273. **Sheldon, A. W.**, The effect of organotin anti-fouling coatings on man and his environment, in *A.C.S. Meeting*, Los Angeles, 34(1), 600, 1974.
274. **Bridges, J. W. et al.**, The fate of ethyltin and diethyltin derivatives in the rat, *Biochem. J.*, 105, 1261, 1967.
275. **Deschiens, R., Brottes, H., and Mvogo, L.**, Controle sur le Terrain des Proprietes Molluscicides de L'oxyde de Tributyl-Etain (Prophylaxie des Bilharzioses), *Bull. Soc. Pathol. Exot.*, 59, 231, 1966.
276. **Deschiens, R. and Floch, H. A.**, Comparative biological effects of six chemical molluscicides used for bilharziasis prophylaxis, *Bull. Soc. Pathol., Exot.*, 61, 640, 1968.
277. **Alabaster, J. S.**, Survival of fish in 164 herbicides, insecticides, fungicides, wetting agents and miscellaneous substances, *Int. Pest Control*, 11(2), 29, 1969.
278. **Cardarelli, N. F.**, Chronic Intoxication Effects of Ultralow Molluscicide Concentrations on *Lebistes reticulatus* and *Biomphalaria glabrata*, Schistosomiasis Progress Rep. 73–1, Creative Biology Laboratory, Barberton, Oh., 1973.
279. **Cardarelli, N. F.**, Effects of Ultra-low Molluscicide Dosages on *Biomphalaria glabrata* and *Lebistes reticulatus* in Micro-Ecological Systems, I, Engineering and Science Division Rep., University of Akron, 1973.
280. **Matthiessen, P.**, Some Effects of Slow Release bis(tri-*n*-butyltin) Oxide on the Tropical Food Fish, *Tilapia mossambica* Peters, Rep. No. 25, Proc. Controlled Release Pesticide Symp., University of Akron, Akron, Oh., Sept. 16–18, 1974.
281. **Steinberg, M.**, Hazard Evaluation of B. F. Goodrich Biocidal Rubber No. 443A, Proj. 33-23-87/66, U.S. Army Environmental Health Agency, Nov. 28, 1967.
282. **Patnode, P. P., Kanakkanatt, S. V., and Cardarelli, N. F.**, Organotin Kill Mechanisms, unpublished report, 1970.
283. **Hecht, G. and Gloxhuber, C.**, Toleration of 5,2-dichloro-4-nitrosalicylamide etanolamine, *Z. Tropenmed. Parasitol.*, 13, 1, 1962.

284. **Duhm, B., Maul, W., Medenwald, H., Patzschke, K., and Wegner, R.,** Radioaktive untersuchungen mit einem neuen molluscicid, *Z. Naturforsch. Teil B,* 16, 509, 1961.
285. **Carr, D. S.,** Organolead Antifouling Paints, Rep. No. 20, Proc. Controlled Release Pesticide Symp., University of Akron, Akron, Oh., September 16–18, 1974.
286. **Anon.,** Forschungs Produkt, Schering AG, Bergkamen/Westf., January 1969.
287. **Willemsens, L. C.,** Investigations on the Field of Organolead Chemistry, Organische Chemie Institute, T.N.O., Utrecht, Netherlands, 1965.
288. **Fowler, D. G.,** Toxic Rating of Organolead Compounds, According to the U.S. Government, Rep. Second Int. Conf. Organolead and Organozinc Chem., Utrecht, Neth., May 8–9, 1969.
289. Donaldson, D. J., Organolead Compounds as Cotton Preservatives, Rep. First Int. Conf. Organolead Chem. and Its Applications, U.S. Army Natick Laboratory, Natick, Mass., May 2, 1967.
290. Skinner, C. E., Biocides for Emulsion Systems, *Paint Varn. Prod.,* 64(7), 43, July 1974.
291. **Carr, D. S.,** Some market potentials for organolead compounds, *Chem. Ind.* (London), Nov. 4, 1967, 1854.
292. Anon., Environmental Impact Analysis Report, International Copper Research Association, New York, June 18, 1974.
293. Gleason, M. N. et al., *Clinical Toxicology of Commercial Products,* 3rd ed., Williams & Wilkins, Baltimore, 1969.
294. **Cheng, T. C. and Sullivan, J. T.,** Possible Toxic Mechanism of Copper in *Biomphalaria glabrata* (Mollusca: Pulmonata), Rep. No. 26, Proc. Controlled Release Pesticide Symp., University of Akron, Akron, Oh., Sept. 16–18, 1974.
295. Dieterich, W. H., Acute Aqueous Exposure – Fish: DAC-2787, Final rep., Hazleton Labs, Inc., Falls Church, Va., Sept. 29, 1965.
296. **Stallard, D. E.,** Residue of Chlorothalonil on Bananas, personal communication.
297. Kanakkanatt, S. V., Theoretical Aspects of Diffusion of Molluscicides in Rubber, Rep. No. 10, Proc. Controlled Release Pesticides Symp., University of Akron, Akron, Oh., Sept. 16–18, 1974.
298. Kanakkanatt, S. V., Cumulative Effects of Matrix Properties on the Diffusion Coefficients of Liquids in Rubber, IUPAC Int. Symp. Macromolecules, Aberdeen, Scotland, Sept. 10–14, 1973.
299. Kanakkanatt, S. V., The effect of matrix properties on the diffusion coefficients of liquids through rubber, *Macromolecules,* Reprint I, 614, 1971.
300. Medalia, A. I., Effective degree of immobilization of rubber occluded with carbon black particles, *Rubber Chem. Technol.,* 45, 1171, 1972.
301. **Carr, W.,** Pigment stability in organic coatings, in *ACS Symp.,* Chem. Soc. Div. Org. Coat. Plast. Chem. Symp., Atlantic City, N.J., Sept. 9–13, 34(2), 543, 1974.
302. **Bishop, J. H. and Silva, S. R.,** Antifouling paint film structure, with particular reference to cross sections, in *ACS Meeting,* Div. Org. Coat. Plast. Chem., Toronto, May 1970, 30(1), 364, 1970.
303. **Driscoll, C.,** X-ray fluorescence spectrometry in antifouling coating research, in *ACS Meeting,* Div. Org. Coat. Plast. Chem., New York, September 1969, 29(2), 1969.
304. **Rowe, P. E.,** Glass microballon spheres as a filler for plastics, in *ACS Meeting,* Div. Org. Coat. Plast. Chem., Chicago, 1973, 33(2), 67, 1973.
305. **Engelhart, J.,** *Slow Release Organotin Molluscicides – A New Approach to Schistosomiasis Control,* report, M&T Chemicals, Inc., Rahway, N.J., 1974.
306. **Anon.,** *Calorimetric Determination of Tin with Dithiol: Preparation of Calibration Curve,* standard test method, V-OR-47, M&T Chemicals, Inc., Rahway, N.J., May 2, 1966.
307. **Anon.,** *Determination of Small Amounts of Tin: Rapid Distillation Separation and Dithiol Colorimetric Measurement,* standard test method, V-OR-48, M&T Chemicals, Inc., Rahway, N.J., May 2, 1966.
308. **Anon.,** *Dithiol Color Development for Traces of Tin in the Microgram Range,* standard test method, V-OR-51, M&T Chemicals, Inc., Rahway, N.J., Aug. 15, 1966.
309. **Anon.,** *Treatment of Water Solutions for the Determination of Small Amounts of Tin,* standard test method, V-OR-69, M&T Chemicals, Inc., Rahway, N.J., April 1, 1971.
310. **Meyling, A. H. and Meyling, J.,** A modified method for the estimation of the molluscicide N-tritylmorpholine (Frescon) in field waters, *S. Afr. Chem. Process.,* p. 63, June–July 1969.
311. **Otto, P. Ph, H. L., Creemers, H. M. J. C., and Luijten, J. G. A.,** Synthesis of tributyltin-113 benzoate, *J. Labelled Compd.,* 2, 339, 1966.
312. **Bruce, J. I. and Radke, M. G.,** Culturing *Biomphalaria* and *Oncomelania* (Gastropoda) for Large Scale Studies of Schistosomiasis, Biomedical Rep. No. 19, Part I, 406th Medical Laboratory, U.S. Army Medical Department Activity, Japan, July 1971.
313. **Sturrock, R. F. and Sturrock, B. M.,** The influence of temperature on the biology of *Biomphalaria glabrata* (Say), intermediate host of *Schistosoma mansoni* on St. Lucia, West Indies, *Ann. Trop. Med. Parasitol.,* 66, 385, 1972.
314. **Michelson, E. H.,** The effects of temperature on growth and reproduction of *Australorbis glabratus* in the laboratory, *Am. J. Hyg.,* 73, 66, 1961.
315. **Santos, A. T. et al.,** Summary of results of laboratory screening of compound 443A against *O. quadrasi., Controlled Release Molluscicide Newsl.* (University of Akron), 1, 4, 1975.

316. **Mansouri, A.**, Brief study of MT-1E slow release molluscicide, *Controlled Release Molluscicide Newsl.* (University of Akron), 1, 2, 1975.
317. **Cardarelli, N. F.**, Microenvironmental Evaluation of Slow Release Molluscicides, Rep. No. 29, Proc. Controlled Release Pesticide Symp., University of Akron, Akron, Oh., Sept. 16–18, 1974.
318. **Castleton, C. W.**, Brazilian Field Trials of MT-1E, Progress Dept. Series, Centro de Pesquisas de Produtos Naturais, Rio de Janeiro, Brazil, February 1974.
319. **Castleton, C. W.**, Brazilian Field Trials of MT-1E, An Organotin Slow Release Formulation, Rep. No. 22, Controlled Release Pesticide Symp., University of Akron, Akron, Oh., Sept. 16–18, 1974.
320. **Upatham, E. S. and Christie, J. D.**, Laboratory trials of copper and TBTO slow release compounds, *Controlled Release Molluscicide Newsl.* (University of Akron), 1, 2, 1975.
321. **Shiff, C. J.**, Perspective in the Use of Molluscicides, Rep. No. 23, Proc. Controlled Release Pesticide Symp., University of Akron, Akron, Oh., Sept. 16–18, 1974.
322. **Shiff, C. J. and Yiannakis, C.**, Slow release molluscicides, *Controlled Release Molluscicide Newsl.* (University of Akron), 1, 3, 1975.
323. **Gilbert, B., Leme, L. A. P., Ferreira, A. M., Bulhoes, M. S., and Castleton, C.**, Field tests of hexabutyldistannoxane (TBTO) in slow-release formulations against *Biomphalaria* spp., *Bull. WHO*, 49, 633, 1973.
324. **Taras, M. J., Greenberg, A. E., Hoah, R. D., and Rand, M. C., Eds.**, *Standard Methods for the Examination of Water and Waste Water*, American Public Health Association, Washington, D.C., 1971.
325. **Cardarelli, N. F.**, Effects of Ultralow Molluscicide Dosages on *Biomphalaria glabrata* and *Lebistes reticulatus* in Microecological Systems, Rep. II, University of Akron, Akron, Oh., Product Development Laboratory, Community and Technical College, October 1973.
326. **Cardarelli, N. F.**, Controlled Release Molluscicide Activity Through an Intervening Soil Barrier, Rep. RG401, University of Akron, Akron, Oh., Jan. 2, 1975.
327. **Sturrock, R. F.**, Persistence of the molluscicide Bayluscide (clonitralide) emulsifiable concentrate on mud surfaces in the tropics, *Ann. Trop. Med. Parasitol.*, 68, 427, 1974.
328. **Amin, M. A.**, Evaluation of Incracide E-51 against *Biompharlaria pfeifferi* and *Bulinus truncatus*, *Controlled Release Molluscicide Newsl.* (University of Akron), 1, 5, 1975.
329. **Pellegrino, J., DeMaria, M., and DeMoura, M. F.**, Observations on the predatory of activity of *Lebistes reticulatus* on cercariae of *Schistosoma mansoni*, *Am. J. Trop. Med. Hyg.*, 15, 337, 1966.
330. **Cardarelli, N. F.**, Biologically Active Rubber: An Adventure in Science, Report to the American Chemical Society, Elastomer and Plastic Group, New England Section, A.D. Little Research Center, Cambridge, Mass., Jan. 14, 1969.
331. **Cardarelli, N. F. and Walker, K. E.**, Slow Release Cercariacides as a Means of Interrupting the Schistosomiasis Transmission Cycle, Report to World Health Organization, Creative Biology Laboratory, Barberton, Oh., unpublished, 1971.
332. **Cardarelli, N. F.**, New Approaches to Interrupting the Schistosomiasis Transmission Cycle, report to Joint Conference on Filiariasis and Schistosomiasis, U.S.–Japan Cooperative Medical Sciences Program, National Institutes of Health, Bethesda, Md., July 26–28, 1971.
333. **Ritchie, L. S., Lopez, V. A., and Cora, J. M.**, Prolonged applications of an organotin against *Biomphalaria glabrata* and *Schistosoma mansoni*, in *Molluscicides in Schistosomiasis Control*, Cheng, T. C., Ed., Academic Press, New York, 1974, 77.
334. **Lopez, V. A. and Ritchie, L. S.**, The Release of TBTO and Other Toxicants from Elastomers as Measured by a Bioassay, Rep. No. 38, Proc. Controlled Release Pesticide Symp., University of Akron, Akron, Oh., Sept. 16–18, 1974.
335. **Cardarelli, N. F.**, Cercariacidal Potency of Slow Release Molluscicides, Rep. No. 39, Proc. Controlled Release Pesticide Symp., University of Akron, Akron, Oh., Sept. 16–18, 1974.
336. **Sandt, D. G.**, Laboratory comparison of four cercaria recovery techniques, *Bull. WHO*, 48, 35, 1973.
337. **Sandt, D. G.**, Direct filtration for recovery of *Schistosoma mansoni* cercariae in the field, *Bull. WHO*, 48, 27, 1973.
338. **Gilbert, B., deSouza, J. P., Fortes, C. C., Santos, D., do Prado Seabra, F. A., Kitagawa, M., and Pelligrino, J.**, Chemoprophylactic agents in schistosomiasis: active and inactive terpenes, *J. Parasitol.*, 56, 397, 1970.
339. **Smith, J. H., Reynolds, E. S., and von Lichtenberg, F.**, The integument of *Schistosoma mansoni*, *Am. J. Trop. Med. Hyg.*, 18, 28, 1969.
340. **Pan, C.**, Studies on the host-parasite relationship between *Schistosoma mansoni* and the snail *Australorbis glabratus*, *Am. J. Trop. Med. Hyg.*, 14, 931, 1965.
341. **Upatham, E. S.**, The effect of water temperature on the penetration and development of St. Lucian *Schistosoma mansoni* miracidia in local *Biomphalaria glabrata*, *Southeast Asian J. Trop. Med. Public Health*, 4, 367, 1973.
342. **Sturrock, B. M. and Sturrock, R. F.**, Laboratory studies of the host-parasite relationship of *Schistosoma mansoni* and *Biomphalaria glabrata* from St. Lucia, West Indies, *Ann. Trop. Med. Parasitol.*, 64, 357, 1970.
343. **Upatham, E. S.**, Effects of some physico-chemical factors on the infection of *Biomphalaria glabrata* (Say) by Miracidia of *Schistosoma mansoni* Sambon in St. Lucia, West Indies, *J. Helminthol.*, 46, 307, 1972.
344. **Upatham, E. S.**, Effect of water depth on the infection of *Biomphalaria glabrata* by miracidia of St. Lucian *Schistosoma mansoni* under laboratory and field conditions, *J. Helminthol.*, 46, 317, 1972.

345. **Upatham, E. S.,** Location of *Biomphalaria glabrata* (Say) by miracidia of *Schistosoma mansoni* Sambon in natural standing and running waters on the West Indian Island of St. Lucia, *Int. J. Parasitol.*, 3, 289, 1973.
346. **Stirewalt, M. A., Minnick, D. R., and Fregeau, W. A.,** Definition and collection in quantity of schistosomules of *Schistosoma mansoni, Trans. R. Soc. Trop. Med. Hyg.*, 60, 352, 1966.
347. **Upatham, E. S.,** Exposure of caged *Biomphalaria glabrata* (Say) to investigate dispersion of *Miracidia* of *Schistosoma mansoni* Sambon in outdoor habitats in St. Lucia, *J. Helminthol.*, 46, 297, 1972.
348. **Upatham, E. S.,** Rapidity and duration of hatching of St. Lucian *Schistosoma mansoni* egg in outdoor habitats, *J. Helminthol.*, 46, 271, 1972.
349. **Koltnow, M., Walker, K. E., and Cardarelli, N. F.,** Destruction of Aquatic Snails by Means of a Bait Concept, WHO Rep. Ser. PD/MOL/69.6, 1969.
350. **Michelson, E. H.,** Chemoreception in the snail *Australorbis glabratus, Am. J. Trop. Med. Hyg.*, 9, 480, 1960.
351. **Quinn, S. A., Romig, P., and Cardarelli, N. F.,** A Method of Destroying Snail Vectors of Schistosomiasis, final report, University of Akron to World Health Organization (B2/181/57) and the Rockfeller Foundation (GA MNS 6927), Nov. 10, 1969.
352. **Cardarelli, N. F. and Walker, K. E.,** Schistosomiasis: Its Potential Eradication with Biocidal Elastomers, American Institute of Chemistry Meeting, Kent, Oh., unpublished, May 20, 1969.
353. **Quinn, S. A., Walker, K. E., Mansdorf, S. Z., and Cardarelli, N. F.,** Pollution Control by Means of Target Biocides, *Proc. Inst. Environ. Sci.*, Mt. Prospect, Ill., p. 324, 1970.
354. **Butler, P. A.,** The southern oyster drill, *J. Nat. Shellfish Assoc.*, 43, 174, 1958.
355. **Anderson, W. F.,** personal communication, 1969.
356. **Cooley, N. R.,** Studies on *Parochis acanthus* (Trematoda: Diginea) as a biological control for the southern oyster drill, *Thais haemastoma, Fishery Bull.*, 62, 77, 1962.
357. **Mobius, K.,** An oyster bank is a bioconose or a social community, in *Readings in Ecology*, Kormondy, E. J., Ed., Prentice-Hall, Englewood Cliffs, N. J., 1965, 21.
358. **Walker, K. E., Quinn, S. A., Bille, S. M., and Cardarelli, N. F.,** Development of a Toxic Bait for Destruction of *Biomphalaria glabrata* and Other Snail Vectors of Schistosomiasis, final report to the National Institutes of Health, NIH-70-2273, University of Akron, July 1, 1971.
359. **Walker, K. E., Quinn, S. A., and Bille, S. M.,** Development of a Toxic Bait for Destruction of *Biomphalaria glabrata* and Other Snail Vectors of Schistosomiasis, National Institutes of Health, NIH-70-2273, University of Akron, Dec. 20, 1971.
360. **Bille, S. M., Walker, K. E., and Cardarelli, N. F.,** Field Tests of Molluscicidal Baits, final report, DADA 17-72-G-9351, U.S. Army Medical Research and Development Command, Washington, D.C., 1 Oct. 1971–19 Dec. 1971.
361. **Bille, S. M., Walker, K. E., and Cardarelli, N. F.,** A Bait Method of Destroying *Oncomelania* and *Biomphalaria glabrata*, Snail Hosts of Schistosomiasis and Related Studies, DADA 17-69-C-9166, Annual Report. II, U.S. Army Medical Research and Development Command, Washington, D.C., 1 July 1970–30 June 1971.
362. **Quinn, S. A.,** A Method of Destroying *Oncomelania*, the Amphibious Snail Vector of Schistosomiasis, unpublished report, University of Akron, June 6, 1970.
363. **Van der Schalie, H. and Davis, G. W.,** Culturing *Oncomelania* snails (Prosobranchia: Hydrobidae) for studies of oriental schistosomiasis, *Malacology*, 6, 321, 1968.
364. **Upatham, E. S.,** Interference by unsusceptible aquatic animals with the capacity of the miracidia of *Schistosoma mansoni* Sambon to infect *Biomphalaria glabrata* (Say) under field-simulated conditions in St. Lucia, West Indies, *J. Helminthol.*, 46, 277, 1972.
365. **Upatham, E. S. and Sturrock, R. F.,** Field investigations on the effect of other aquatic animals on the infection of *Biomphalaria glabrata* by *Schistosoma mansoni* miracidia, *J. Parasitol.*, 59, 448, 1973.
366. **Wright, C. A.,** Host location by trematode miracidia, *Ann. Trop. Med. Parasitol.*, 53, 288, 1959.
367. **MacInnes, A. J.,** Responses of *Schistosoma mansoni* miracidia to chemical attractants, *J. Parasitol.*, 51, 731, 1965.
368. **Etges, F. J. and Decker, C. L.,** Chemosensitivity of the miracidium of *Schistosoma mansoni* to *Australorbis glabratus* and other snails, *J. Parasitol.*, 49, 114, 1963.
369. **Abdel-Malek, E.,** Susceptibility of the snail *Biomphalaria boissyi* to infection with certain strains of *Schistosoma mansoni, Am. J. Trop. Med.*, 30, 887, 1950.
370. **Barbosa, F. A. S.,** Alguns aspectos das relacoes hospedeiro-parasito entre as fases larvarias do trematodeo *Schistosoma mansoni* ci o molusco *Australorbis glabratus, Bol. Fund. Goncalo Moniz*, 15, 1, 1960.
371. **Chernan, E. and Dunavin, C. A.,** The influence of host-parasite dispersion upon the capacity of *Schistosoma mansoni* miracidia to infect *Australorbis glabratus, Am. J. Trop. Med. Hyg.*, 11, 455, 1962.
372. **Chernan, E.,** Behavorial responses of miracidia of *Schistosoma mansoni* and other trematodes to substances emitted by snails, *J. Parasitol.*, 56, 287, 1970.
373. **Michelson, E. H.,** Miracidia-immobilizing substances in extracts prepared from snails infected with *Schistosoma mansoni, Am. J. Trop. Med. Hyg.*, 13, 36, 1964.
374. **Shiff, C. J.,** Location of *Bulinus* (physopsis) *globosus* by miracidia of *Schistosoma haematobium, J. Parasitol.*, 54, 1133, 1968.

375. **Chernin, E. and Perlstein, J. M.**, Further studies on interference with the host-finding capacity of *Schistosoma mansoni* miracidia, *J. Parasitol.*, 55, 500, 1969.
376. **Pappas, J. and Carman, G.**, Field screening tests with various materials against European brown snails on citrus in California, *J. Econ. Entomol.*, 48, 698, 1955.
377. **Getzin, J.**, Control of gray garden slug with bait formulation of a carbamate molluscicide, *J. Econ. Entomol.*, 58, 158, 1965.
378. **Pappas, J. and Carman, G.**, Control of European brown snails in citrus groves in southern California with guthion and metaldehyde sprays, *J. Econ. Entomol.*, 54, 152, 1961.
379. **Stephenson, J. W.**, A culture method for slugs, *Proc. Malacol. Soc. London*, 35, 43, 1961.
380. **Blackburn, R. D. and Willey, C. W.**, Integration of Biological and Chemical Control of Alligator Weed, Progress Report, Fiscal Year 1971, ARS Agricultural Research Center, U.S. Dept. of Agriculture, Ft. Lauderdale, Fla., 1971.
381. **Gangstad, E. O. et al.**, Integrated Control of Alligator Weed and Hyacinth in Texas, APCP Tech. Rep. 9, U.S. Army Engineers Waterways Experimental Station, Vicksburg, Miss., March 1975.
382. **Gangstad, E. O., Parkman, M. F., and Thompson, W. E.**, Control of Obnoxious Aquatic Plants of Louisiana and the Gulf Coast Area, Summary of Field Operations and Review of the Research Program, AD887531-L, Ser. No. 43, U.S. Army Corps of Engineers, Washington, D.C., July 1971.
383. **Barnes, W. W.**, Dissipation of residues of phenoxy herbicides applied for water milfoil control in large reservoirs, in Proc. Res. Planning Conf. Integrated Systems of Aquatic Plant Control, Oct. 29–30, 1973, U.S. Army, Office of the Chief Engineer, Vicksburg, Miss., August 1974, p. 37.
384. **Smith, G. E. and Ison, B. G.**, Investigations of effects of large-scale applications of 2,4-D on aquatic fauna and water quality, *Pestic. Monit. J.*, 1, 3, 1967.
385. **Anon.**, Hydrilla in Iowa, *Weeds Trees Turf*, 11 (10), 14, 1972.
386. **Rhett, J. T.**, Reconnaissance Report on Investigations of Aquatic Plants in Stoner Creek Vicinity of Paris, Kentucky, U.S. Army Engineering District Corps of Engineers, Louisville, Ky., Mar. 19, 1971.
387. **Gangstad, E. O., Seaman, D. E., and Nelson, M. L.**, Potential Growth of Aquatic Plants in the Republic of the Philippines and Projected Methods of Control, DAEN-CWP-V, U.S. Army, Office of the Chief Engineer, Washington, D.C., June 1972.
388. **Gangstad, E. O., Seaman, D. E., and Nelson, M. L.**, Potential growth of aquatic plants of the lower Mekong river basin, Laos-Thailand, *Hyacinth Control J.*, 10, 4, 1972.
389. **Terry, P. J. and Minto, S. D.**, Floating Islands in Lake Victoria, Report of the Tropical Pesticides Research Institute, Arusha, Tanzania, June 20, 1970.
390. **Palm, C. E. et al.**, *Principles of Plant and Animal Control*, Vol. 2, Weed Control, National Academy of Sciences, Washington, D.C., 1961.
391. **Shaw, W.**, Need for Controlled Release Technology in the Use of Agricultural Chemicals, keynote address, Proc. Controlled Release Pesticide Symp., University of Akron, Akron, Oh., Sept. 16–18, 1974.
392. **Stodola, J.**, *Encyclopedia of Water Plants*, T. F. H. Publications, Jersey City, N.J., 1967.
393. **Weldon, L. W., Blackburn, R. D., and Harrison, D. S.**, *Common Aquatic Weeds*, Agricultural Handbook No. 352, U.S. Dept. of Agriculture Research Service, Washington, D.C., January 1967.
394. **Lopinot, A. C.**, Aquatic Weeds, Their Identification and Methods of Control, Fishery Bull. No. 4, Department of Conservation, State of Illinois, Springfield, 1971.
395. **Weldon, L. W., DeRigo, H. T., and Blackburn, R. D.**, Control of alligator weed with dichlobenil, *Proc. South. Weed Conf.*, p. 287, 1968.
396. **Harrison, D. S., Blackburn, R. D., Weldon, L. W., Orsenigo, J. R., and Ryan, G. F.**, Aquatic Weed Control, Circ. 219B, Agriculture Extension Service, University of Florida, Gainesville, Fla., August 1966.
397. **Zur Burg, F. W., Foret, J. A., Solymosy, S. L., Averitt, W. K., and Barry, J. R.**, Chemical Control of Alligator Weed, annual report, University of Southwestern Louisiana, Lafayette, June 30, 1970.
398. **DeVaney, T. E.**, Chemical Vegetation Control Manual for Fish and Wildlife Management Programs, Bureau of Sport Fisheries and Wildlife, U.S. Dept. of the Interior, Washington, D.C., 1967.
399. **Anon.**, Final Environmental Statement – Aquatic Plant Control and Eradication Program – State of Texas, U.S. Army Engineering District, Galveston, Tex., May 30, 1972.
400. **Weldon, L. W. and Blackburn, R. D.**, *Herbicidal Treatment Effect on Carbohydrate Levels of Alligatorweed*, Jour. Ser. 2920, Florida Agricultural Experimental Station, Ft. Lauderdale, Fla., 1968.
401. **Zimmerman, P. W., Hitchcock, A. E., Kirkpatrick, H., and Earle, T. T.**, Control of water hyacinth, *Agric. Chem.*, 5(1), 45, 1950.
402. **Thompson, W. R., Parkman, F. W., and Gangstad, E. O.**, Operational Applications of Amine Aquatic Herbicide, Weed Science Society of America Meeting, St. Louis, Mo., Feb. 7–10, 1972.
403. **Brezny, O., Mehta, I., and Sharma, R. K.**, Studies on evapotranspiration of some aquatic weeds, *Weed Sci.*, 21, 197, 1973.
404. **Diem, J. R. and Davis, D. E.**, Effect of 2,4-D on ametryne toxicity, *Weed Sci.*, 22, 285, 1974.
405. **Dnyansagar, V. R. and Dharurkar, R. D.**, The effect of herbicides on the germination of water hyacinth seeds, *Hyacinth Control J.*, 10, 25, 1972.

406. **Elliston, R. E. and Steward, K. K.**, The response of Eurasian watermilfoil to various concentrations and exposure periods of 2,4-D, *Hyacinth Control J.*, 10, 38, 1972.
407. **Barnes, W. W.**, Dissipation of residues of phenoxy herbicides applied for watermilfoil control in large reservoirs, in Proc. Res. Plan. Conf. Integrated Systems of Aquatic Plant Control, U.S. Army Engineers Waterways Experimental Station, Vicksburg, Miss., 29–30 Oct. 1973, p. 37, Aug. 1974.
408. **Watson, J. D.**, Application for Residue Tolerance for Endothall in Water, Sect. C, Pennwalt Corp. report, Tacoma, Wash., Dec. 1970.
409. **Anon.**, Aquathol and Aquathol Plus for Aquatic Weed Control, Bull. No. A-35, Pennsalt Corp., Tacoma, Wash.
410. **Anon.**, Ortho Diquat, Experimental Data Sheet, Chevron Chem. Co., San Francisco, July 1969.
411. **Foret, J. A., Barry, J. R., Langlinais, S. J., Solymosy, S. L., and Viator, D. P.**, Testing of Herbicides in Pools and Field Tests for Control of *Elodea densa*, Rep. S. W., Louisiana University to Corps of Engineers, DACW 73-71-C-0029, 1973.
412. **Phillippy, C. L.**, Florida *Elodea* Control with Sulfuric Acid, U.S. Army Corps of Engineers Rep. AD881-260-L, Ser. 46, Append. J., February 1971.
413. **Mobley, G. S., Zeiger, C. F., Raynes, J. J., and Gangstad, E. O.**, Technical Report on Operational Large-Scale Field Test; Submersed Weed – Hydrilla, U.S. Army Corps of Engineers, May 1971.
414. **Jones, D.**, Dawn attack on hydrilla, *Weeds Trees Turf*, 13, 36, 1974.
415. **Anon.**, Casoron, Technical Information Bulletin, Vol. 4, Thompson, Hayward Chemical Co., Kansas City, Mo., December 1963.
416. **Anon.**, Aquathol, Tech. Info. Bull. A-39a, Pennsalt Chemical Corp.
417. **Day, B. E. et al.**, *Principles of Plant and Animal Pest Control*, Vol. 2, Weed Control, National Academy of Sciences, Washington, D. C., 1968.
418. **Anon.**, *Farm Chemicals Handbook*, Sect. D, Meister, Willoughby, Oh., 1974.
419. **Gangstad, E. O., Parkman, M. F., and Thompson, W. E.**, Control of Obnoxious Aquatic Plants of Louisiana and the Gulf Coast Area, Summary of Field Operations and Review of the Research Program, Interagency Research Advisory Committe, Aquatic Plant Control Program, U.S. Army Corps of Engineers, July 1971.
420. **Couch, R. W.**, Response of Aquatic Weeds to Laser Radiation, Tech. Rep. 2, Aquatic Plant Control Program, U.S. Army Engineers Waterways Experimental Station, Vicksburg, Miss., Dec. 1972.
421. **Gangstad, E. O.**, Aquatic Use Pattern for Silvex, Tech. Rep. 5, Aquatic Plant Control Program, U.S. Army Engineers Waterways Experimental Station, Vicksburg, Miss., October 1973.
422. **Ganstad, E. O., Raynes, J. J., Mobley, G. S., and Zeiger, C. F.**, Technical Report of the Potential Growth of Aquatic Plants of the Cross Florida Barge Canal, Review of the Aquatic Plant Control Research Program and Summary of the Research Area Development Operations in Florida, U.S. Army Office of the Chief Engineer, Washington, D.C., February 1971.
423. **Gangstad, E. O., Ed.**, Herbivorous Fish for Aquatic Plant Control, Tech. Rep. 4, Aquatic Plant Control Program, U.S. Army Waterways Experimental Station, Vicksburg, Miss., January 1973.
424. **Gangstad, E. O., Ed.**, Biological Control of Water Hyacinth with Insect Enemies, Tech. Rep. 6, Aquatic Plant Control Program, U.S. Army Engineers Waterways Experimental Station, Vicksburg, Miss., January 1974.
425. **Gordon, R. D. and Coulson, J. R.**, Report on Field Observations of Arthropods on Water Hyacinth in Florida, Louisiana and Texas, July 1969, AD881-260L, Append. F., U.S. Army Corps of Engineers Meeting, New Orleans, February 1971.
426. **Maddox, D. M. and Hambric, R. N.**, Use of alligatorweed flea beetle in Texas; an exercise in environmental biology, *Proc. Southern Weed Soc.*, 23, 283, 1970.
427. **Coulson, J. R.**, Biological Control of Alligator Weed, 1959–1972, report summary, Beltsville Agricultural Research Center, Agricultural Research Service, Beltsville, Md., August 1974.
428. **Spencer, N. R.**, Biological control of water milfoil, in Proc. Res. Plan. Conf. Integrated Systems of Aquatic Plant Control, Aquatic Plant Control Program, U.S. Army Engineers Waterways Experimental Station, Vicksburg, Miss., August 1974, p. 74.
429. **Goeden, R. D., Andres, L. A., Freeman, T. E., Harris, P., Pienkowski, R. L., and Walker, C. R.**, Present status of projects on the biological control of weeds with insects and plant pathogens in the United States and Canada, *Weed Sci.*, 22, 490, 1974.
430. **Freeman, T. E., Zettler, F. W., and Charudatten, R.**, Utilization of Phytopathogens as Biocontrols for Aquatic Weeds, in Proc. Res. Plan. Conf. on Integrated Systems of Aquatic Plant Control, U.S. Army Engineers Waterways Experimental Station, Vicksburg, Miss., August 1974, p. 97.
431. **Blackburn, R. D., Steward, K. K., and Sutton, D. L.**, Summary Semi-Annual Report on Chemical and Biological Control of Aquatic Vegetation, July 1970–Jan. 1971, U.S. Dept. of Agriculture, Agriculture Research Service, Ft. Lauderdale, Fla., January 1971.
432. **Ingersoll, J. M.**, *The Water Hyacinth*, OCD-OS-62-218, Dept. of the Army, Office of Civil Defense, Sept. 1, 1964.
433. **Gangstad, E. O.**, Aquatic Weed Control with Pathogens, Tech. Rep. 8, APCP, U.S. Army Eng. Waterways Exp. Sta., Vicksburg, Miss., Nov. 1974.
434. **Guerra, L. V.**, Integrated control of noxious aquatic plants in Proc. Res. Plan. Conf. on Integrated Systems of Aquatic Plant Control, U.S. Army Engineers Waterways Experimental Station, Vicksburg, Miss., August 1974, p. 85.

435. **Foret, J. A., Spencer, N. R., and Gangstad, E. O.**, Towards Integrated Control of Alligator Weed, Append. H in Aquatic Use Patterns for 2,4-D Dimethylamine and Integrated Control, Rep. 7, Aquatic Plant Control Program, U.S. Army Engineers Waterways Experimental Station, Vicksburg, Miss., November 1974.
436. **Hestand, R. S., May, B. E., Schultz, D. P., and Walker, C. R.**, Ecological implications of water levels on plant growth in a shallow water reservoir, *Hyacinth Control J.*, 11, 54, 1973.
437. **Meeks, R. L.**, The effect of drawdown rate on wetland plant succession, *J. Wildl. Manage.*, 33, 88, 1969.
438. **Timmons, F. L.**, Herbicides as Antipollutants, Append. F3, Integrated Control of Alligator Weed and Water Hyacinth in Texas, Rep. 9, Aquatic Plant Control Program, U.S. Army Engineers Waterways Experimental Station, Vicksburg, Miss., March 1975.
439. **Whitney, E. W., Esters, R. D., Smitherman, R. O., and Gangstad, E. O.**, Effects of Silvex on Aquatic Biota, Cooperative Study, Corps of Engineers, Division of Fishery Service, FWPCA, and Louisiana University Aquatic Plant Control Program, Office of the Chief Engineer, Washington, D.C., 1968.
440. **Anon.**, Expanded Project for Aquatic Plant Control: Field Test Areas Fishery Management Program, final report, U.S. Dept. of the Interior, Division of Fishery Services, Atlanta, Oct. 17, 1969.
441. **May, B. E., Hestand, R. S., and VanDyke, J. M.**, Comparative effects of diquat plus copper sulfate on aquatic organisms, *Weed Sci.*, 21, 249, 1973.
442. **Walker, C. R.**, Herbicide Chemicals and Their Effect on the Aquatic Environment, paper presented at the 24th Annual Meeting of the Southern Weed Science Society, Memphis, Tenn., Jan. 19–21, 1971.
443. **Norris, L. E.**, Dissipation of residues of phenoxy herbicides applied to the watershed, in Proc. Res. Plan. Conf. Integrated Systems of Aquatic Plant Control, Oct. 29–30, 1973, U.S. Army Engineers Waterways Experimental Station, Vicksburg, Miss., August 1974, p. E5.
444. **Knoll, J. E.**, Environmental Considerations in the Use of Herbicides on Reclamation Projects, paper presented at the Irrigation and Drainage Division Specialty Conference, American Society of Civil Engineers, Lincoln, Neb., Oct. 6–8, 1971.
445. **Hiltibran, R. C.**, Aquatic herbicides in irrigation waters, *Weeds Trees Turf*, 14(2), 48, 1975.
446. **Hummel, R. J. and Cummings, J. C.**, Registration of herbicides for aquatic use, in Proc. Res. Plan. Conf. Integrated Systems of Aquatic Plant Control, Oct. 29–30, 1973, U.S. Army Engineers Waterways Experimental Station, Vicksburg, Miss., August 1974, p. 33.
447. **House, W. B., Goodman, L. H., Gadberry, H. M., and Dockter, K. W.**, Phenoxy herbicide residues and their persistence, in Integrated Control of Alligator Weed and Water Hyacinth in Texas, Gangstad, E. O., Ed., Rep. 9, Aquatic Plant Control Program, U.S. Army Engineers Waterways Experimental Station, Vicksburg, Miss., March 1975.
448. **Hestand, R. S. and May, B. E.**, Effects of a diquat plus cutrine combination on invertebrates in Inglis reservoir, *Hyacinth Control J.*, 10, 40, 1972.
449. **Buchanan, G.**, Weed losses too high, *Weeds Today*, 6, 4, 1975.
450. **Bendixon, L. E.**, Weeds cost – everybody pays, *Weeds Today*, 3, 5, 1972.
451. **Behrens, R.**, Corn and weeds, *Weeds Today*, 6, 15, 1975.
452. **DeOng, E. R., Bishop, D., and Bishop, J. L.**, *Insect, Disease and Weed Control*, Chemical Publishing, New York, 1972.
453. **Anon.**, *Weed Control Manual and Herbicide Guide*, Meister, Willoughby, Oh., 1974.
454. **Anon.**, *Farm Chemicals Handbook*, Meister, Willoughby, Oh., 1974.
455. **Unger, P. W., Allen, R. E., and Wiese, A. F.**, Tillage and herbicides for surface residue maintenance, weed control and water conservation, *J. Soil Water Conserv.*, 26, 147, 1971.
456. **Penner, D.**, The effectiveness of pesticide interactions, *Weeds Today*, 5(3), 13, 1974.
457. **Schery, R. W.**, Cool weather weed control, *Weeds Trees Turf*, 13(9), 15, 1974.
458. **Husted, R. F. and Radke, R. O.**, Herbicidal Compositions, U.S. Patent 3,834,891, 1974.
459. **Turgeon, A. J., Beard, J. B., Martin, D. P., and Meggitt, W. F.**, Effects of successive applications of preemergent herbicides in turf, *Weed Sci.*, 22, 349, 1974.
460. **Harvey, W. A.**, Effects of weed control on the environment, *Weeds Trees Turf*, 11(1), 14, 1972.
461. **Davis, D. E.**, The 2,4,5-T story, *Weeds Today*, 5 (1), 12, 1974.
462. **Edwards, W. E. and Harrold, L. L.**, Agricultural pollution of waterbodies, *Ohio J. Sci.*, 70(1), 56, 1970.
463. **Anon.**, Biological weed control proposed by scientists, *Weeds Trees Turf*, 13 (9), 28, 1974.
464. **Bille, S., Mansdorf, S. Z., and Cardarelli, N. F.**, Development of Slow Release Herbicide Materials for Controlling Aquatic Plants, final report, November 1969-July 1971, DACW 73-70-C-0030, Dept. of the Army, Office of the Chief Engineer, Washington, D. C., July 1971.
465. **Steward, K. K.**, Weed Investigations – Aquatic and Non-Crop Areas, annual report fiscal 1970, U.S. Dept. of Agriculture, Agricultural Research Center, Plantation Field Laboratory, Ft. Lauderdale, Fla., June 30, 1970.
466. **Steward, K. K.**, Weed Investigations – Aquatic and Non-Crop Areas, annual report fiscal 1971, U.S. Dept. of Agriculture, ARS, Agriculture Research Center, Ft. Lauderdale, Fla., June 30, 1971.
467. **Janes, G. A. and Cardarelli, N. F.**, Aquatic Herbicides Chronicity Study, final report, Cont. No. DACW 73-73-C-0042, Dept. of the Army, Office of the Chief Engineer, Washington, D.C., Feb. 15, 1974.

468. **Bartley, T. R.**, *Control of Submersed Aquatic Weeds in Irrigation*, American Society of Civil Engineers Meeting on Water Resources Engineering, New Orleans, Preprint 794, Feb. 3–7, 1969.
469. **Bartley, T. R.**, Progress Report on Evaluation of Copper for Aquatic Weed Control and Herbicide Residues on Irrigation Systems, Rep. WC-32, U.S. Dept. of the Interior Bureau of Reclamation, October 1967.
470. **Cardarelli, N. F.**, Evaluation of Incracide E-51 Against Select Aquatic Weeds and Algae, final report, C.B.L. Proj. 104, International Copper Research Association, New York, April 5, 1975.
471. **Mansdorf, S. Z.**, Slow Release: Development of Aquatic Herbicide Carrier Systems, Rep. No. 12, Proc. Controlled Release Pesticides Symp., University of Akron, Akron, Oh., Sept. 16–18, 1974.
472. **Janes, G. A. and Cardarelli, N. F.**, Aquatic Herbicides Chronicity Study, final rep., Nov. 1, 1972–Oct. 31, 1973, Cont. No. DACW 73-73-C-0042, Dept. of the Army, Office of the Chief Engineer, Washington, D.C., Feb. 15, 1974.
473. **Janes, G. A.**, Chronicity Phenomenon, Proc. Controlled Release Pesticides Symp., Rep. No. 14, University of Akron, Akron, Ohio, Sept. 16–18, 1974.
474. **Ries, S. K.**, Herbicides as growth boosters, *Weeds Today*, 2(1), 4, 1971.
475. **Cardarelli, N. F. and Kanakkanatt, S. V.**, A Hypothesis on Chronic Intoxication by Herbicides in Ultralow Concentrations, American Chemical Society, 30th Northwest Regional Meeting, University of Hawaii, Honolulu, June 11–13, 1975.
476. **Anon.**,Plants have nerves says Cornell biologist, *Weeds Trees Turf*, 13, 83, 1974.
477. **Crafts, A. S.**, *The Chemistry and Mode of Action of Herbicides*, Interscience, New York, 1961.
478. **Levitt, J.**, *Responses of Plants to Environmental Stresses*, Academic Press, New York, 1972.
479. **Cardarelli, N. F.**, A Hypothesis Concerning the Mechanism of the Chronicity Phenomenon in Aquatic Plants, 2nd International Controlled Release Pesticide Symp., Wright State University, Dayton, Oh., Sept. 8–10, 1975.
480. **Dinman, B. D.**, "Non-concept" of "no-threshold": chemicals in the environment, *Science*, 175, 495, 1972.
481. **Thompson, W. E.**, Field Tests of Slow Release Herbicides, Rep. No. 15, Proc. Controlled Release Pesticide Symp., University of Akron, Akron, Oh., Sept. 16–18, 1974.
482. **Anon.**, Preliminary Evaluation of Controlled Release Herbicide Formulations for Aquatic Weeds, Oct. 1968-Aug. 1969, Special Study No. 31-004-69/70, Dept. of the Army, U.S. Army Environmental Hygiene Agency, Edgewood Arsenal, Md., Dec. 30, 1969.
483. **Young, W. W. and Nelson, L. L.**, Evaluation of Controlled Release Herbicide Formulations for Aquatic Weeds, Special Study No. 31-004-69/71, July 1969–July 1970, Dept. of the Army, U.S. Army Environmental Hygiene Agency, Edgewood Arsenal, Md., Oct. 23, 1970.
484. **Quacenbos, H. M.**, Plasticizers in vinyl chloride resins, *Ind. Eng. Chem.*, 46, 1335, 1954.
485. **Nelson, L. L.**, Evaluation of Slow Release Herbicide Formulations for Aquatic Weed Control, Special Study No. 31-004-69/72, Dept. of the Army, U.S. Army Environmental Hygiene Agency, Aberdeen Proving Ground, Md., July 1970–July 1971.
486. **Steward, K. K.**, Evaluation of Controlled Release Herbicidal Formulations for Aquatic Weed Control in Controlled Release Herbicides and Field Reports, technical report, Aquatic Plant Control Program, Dept. of the Army, Office of the Chief Engineer, Washington, D.C., December 1970.
487. **Harris, F. W., Norris, S. O., and Post, L. K.**, Investigation of Factors Influencing Release of Herbicides from Polymer Matrices: The Polyethylene Matrix, Quart. Rep. II, October 1971–December 1971, Cont. DACW 73-71-C-0053, Dept. of the Army, Office of the Chief Engineer, Washington, D.C., December 1971.
488. **Harris, F. W., Post, L. K., and Field, W. A.**, Investigation of Factors Influencing Release of Herbicides from Polymer Matrices, Quart. Rep. II, October 1972–December 1972, Cont. DACW 73-72-C-0064, Dept. of the Army, Off. Chf. Eng., Washington, D.C., December 1972.
489. **Beasley, M. L. and Rhode, N. G.**, Water-decomposable Urea Polymers, German Patent 2,238,912, 1973.
490. **Cardarelli, N. F., Mansdorf, S. Z., Bille, S. M., and Lancaster, L. L.**, Development of Slow Release Herbicide Material for Controlling Aquatic Plants, Quart. Rep. VI, Cont. DACW 73-70-C-0030, Dept. of the Army, Office of the Chief Engineer, Washington, D.C., January 1971–March 1971.
491. **Mehltretter, C. L., Roth, W. B., Weakley, F. B., McQuire, T. A., and Russell, C. R.**, Potential controlled-release herbicides from 2,4-D esters of starches, *Weed Sci.*, 22, 415, 1974.
492. **Harris, F. W., Feld, W. A., and Post, L. K.**, Investigation of Factors Influencing Release of Herbicides from Polymers Containing Herbicides as Pendent Side Chains, annual report, Cont. DACW 73-74-C-0001, Dept. of the Army, Office of the Chief Engineer, Washington, D.C., July 1973–June 1974.
493. **Harris, F. W. and Post, L. K.**, Synthesis of Polymers Containing Herbicides as Pendent Substituents, Rep. No. 17, Proc. Controlled Release Pesticide Symp., University of Akron, Akron, Oh., Sept. 16–18, 1974.
494. **Allan, G. G., Chopra, C. S., Neogi, A. N., and Wilkins, R. M.**, Design and synthesis of controlled release pesticide-polymer combinations, *Nature*, 234, 349, 1971.
495. **Beasley, M. L. and Collins, R. L.**, Water degradable polymers: controlled release for herbicides and other agents, *Science*, 169, 769, 1970.
496. **Collins, R. L.**, Optimum efficiency in application of biologically active substances, in Gangstad, E. O., Nailon, W. T., and Novosad, C., Environmental Impacts and Research Review of the Aquatic Plant Control Program, Texas, Louisiana, Florida and Related Areas, Technical Advisory Committee Meeting, Aquatic Plant Control Program, U.S. Army Engineers District, Galveston, Tex., January 1973.

497. **Collins, R. L.**, A Theoretical Foundation for Controlled Release, Rep. No. 9, Proc. Controlled Release Pesticide Symp., University of Akron, Akron, Oh., Sept. 16–18, 1974.
498. **Harris, F. W., Norris, S. V., and Post, L. K.**, Factors influencing release of Fenac from polyethylene matrices, *Weed Sci.*, 21, 318, 1973.
499. **Desai, S. J., Simonelli, A. P., and Higuchi, W. I.**, Influencing release of solid drug dispersed in inert matrices, *J. Pharm. Sci.*, 54, 1459, 1965.
500. **Liebhafsky, H. A., Marshal, A. L., and Verhoek, F. W.**, Loss of plasticizers from polyvinyl chloride plastics in vacuum, *Ind. Eng. Chem.*, 34, 704, 1942.
501. **Harris, F. W.**, Theoretical Aspects of Controlled Release, Rep. No. 8, Proc. Controlled Release Pesticide Symp., University of Akron, Akron, Oh., Sept. 16–18, 1974.
502. **Daynes, H. A.**, Gas analysis by measurement of thermal conductivity, *Proc. R. Soc. London, Ser. A*, 97, 286, 1920.
503. **Barrer, R. M.**, Diffusion in and through solids, *Trans. Faraday Soc.*, 35, 628, 1939.
504. **Baker, R. W. and Lonsdale, H. K.**, Controlled release of biologically active agents, in *Advances in Experimental Medicine*, Tanquary, A. C. and Lacey, R. E., Eds., Plenum, New York, 1974.
505. **Crank, J.**, *Mathematics of Diffusion*, Oxford University Press, London, 1970.
506. **Roseman, T. J. and Higuchi, T.**, Release of methoxyprogesterone acetate from a silicone polymer, *J. Pharm. Sci.*, 59, 353, 1970.
507. **Collins, R. L., Doglia, S., Mazak, R. A., and Samulski, E. T.**, Controlled release of herbicides – theory, *Weed Sci.*, 21, 1, 1973.
508. **Audus, L. J.**, Herbicide behavior in the soil, in *The Physiology and Biochemistry of Herbicides*, Audus, L. J., Ed., Academic Press, New York, 1964.
509. **Valentine, J. P. and Bingham, S. W.**, Influence of several algae on 2,4-D residues in water, *Weed Sci.*, 22, 358, 1974.
510. **Crafts, A. S.**, Herbicide behavior in the plant, in *The Physiology and Biochemistry of Herbicides*, Audus, L. J., Ed., Academic Press, New York, 1964.
511. **Gangstad, E. O. and Averitt, W. K.**, Dissipation of 2,4-D Residues in Ponds, Lakes, Bayous, and Other Quiescent or Slowly Moving Bodies of Water, Appendix G in Control of Obnoxious Aquatic Plants of Louisiana and the Gulf Coast Area, AD887531-L, Interagency Research and Advisory Committee, Aquatic Plant Control Program, U.S. Army Corps of Engineers, New Orleans, July 1971.
512. **Novosad, C. J. and Nailon, W. T.**, Aquatic Plant Control Program in Texas, in Proc. Res. Planning Conf. on Aquatic Plant Control, U.S. Army Engineers Waterways Experimental Station, Vicksburg, Miss., Jan. 12, 1972, 19.
513. **Roosenburg, W. H.**, Greening and copper accumulation in the American oyster, *Crassostrea virginica*, in the vicinity of a stream electric generating station, *Chesapeake Sci.*, 10, 241, 1969.
514. **Sprague, J. B.**, Lethal concentrations of copper and zinc for young Atlantic salmon, *J. Fish Res. Board Can.*, 21, 17, 1964.
515. **Mount, D. I.**, Chronic toxicity of copper to fathead minnows (*Pimephales promelas* Rafinesque), *Water Res.*, 2, 215, 1968.
516. **Lin, C. K., Nemo, A. J., and Vedder, D. L.**, Copper residue in rainbow trout, *Hyacinth Control J.*, 10, 43, 1972.
517. **Burkhalter, A.**, *Guidelines for Aquatic Weed Control*, Florida Dept. of Natural Resources, Tallahassee, 1972.
518. **Sutton, D. L., Weldon, L. W., and Blackburn, R. D.**, Effect of diquat on uptake of copper in aquatic plants, *Weed Sci.*, 18, 703, 1970.
519. **Sutton, D. L., Haller, W. T., Steward, K. K., and Blackburn, R. D.**, Effect of copper on uptake of diquat-14_C by *Hydrilla*, *Weed Sci.*, 20, 581, 1972.
520. **Sutton, D. L., Blackburn, R. D., and Barlowe, W.**, Response of aquatic plants to combinations of endothall and copper, *Weed Sci.*, 19, 643, 1971.
521. **Stanley, R. A.**, Effect of 2,4-D and various salts on Eurasian watermilfoil, *Weed Sci.*, 22, 591, 1974.
522. **Cardarelli, N. F.**, Slow release herbicides, *Weeds Trees Turf* 11(5), 16, 1972.
523. **Hill, W. G. and Donley, R. G.**, Alligatorweed eradication, *Weeds Trees Turf*, 12(5), 14, 1973.
524. **Metcalf, R. L.**, Chemistry and biology of pesticides, in *Pesticides in the Environment*, White-Stevens, R., Ed., Marcel Dekker, New York, 1971.
525. **Loeffler, J. E. and Van Overbeek, J.**, Metabolism of herbicides, in *Pesticides in the Environment*, White-Stevens, R., Ed., Marcel Dekker, New York, 1971.
526. **Crafts, A. S. and Robbins, W. W.**, *Weed Control*, 3rd ed., McGraw-Hill, New York, 1962.
527. **Anon.**, *Herbicide Handbook*, 2nd ed., W. F. Humphrey Press, Geneva, New York, 1970.
528. **Harris, K. F. and Bath, J. E.**, Regurgitation by *Myzus persicae* during membrane feeding: its likely function in the transmission of non-persistent plant viruses, *Ann. Entomol. Soc. Am.*, 66, 793, 1973.
529. **Kovacs, M. F. and Parochetti, J. V.**, Dhurrin: an allelopath identified in Johnsongrass rhizome exudate in *Proc. Weed Sci. Soc. Meeting*, Washington, D.C., Feb. 4–7, 1975, Weed Science Society of America, Champaign, Ill., 77.
530. **Shaw, W. C. and Swanson, C. R.**, Techniques and equipment used in evaluating chemicals for their herbicidal properties, *Weeds*, 2, 352, 1950.
531. **Que Hee, S. S. and Sutherland, R. G.**, Volatilization of formulated butyl esters of 2,4-D from Pyrex and leaves, *Weed Sci.*, 23, 119, 1975.

532. **MaKinlay, K. S., Brandt, S, A., Morse, P., and Ashford, R.**, Droplet size and phytotoxicity of herbicides, *Weed Sci.*, 20, 450, 1972.
533. **Que Hee, S. S. and Sutherland, R. G.**, Penetration of amine salt formulations of 2,4-D into sunflower, *Weed Sci.*, 21, 115, 1973.
534. **Bauer, J. R., Bovey, R. W., and Riley, I.**, Effect of pH on foliar uptake of 2,4,5-T-^{14}C^{1}, *Weed Sci.*, 22, 481, 1974.
535. **Hess, F. D., Bayer, F. E., and Falk, R. H.**, Herbicide dispersal patterns: I, as a function of leaf surface, *Weed Sci.*, 22, 394, 1974.
536. **Templeton, A. R.**, The Effect of Artificial Rainfall on the Herbicidal Activity of 2,4-D in Combination with Various Spray Additives, Weed Sci. Soc. Am. Meeting, Washington, D.C., Abstracts, February 1975, 51.
537. **Bukovac, M. J.**, Foliar Penetration of Pesticide Chemicals, Report, Pesticide Research Center, Michigan State University, Lansing, 1974.
538. **Chandrasekaran, S. K., Baker, R. W., Buckles, R. G., and Michaels, A. S.**, A Mathematical Analysis for Controlled Delivery of Agrichemicals, Rep. No. 11, Proc. Controlled Release Pesticide Symp., University of Akron, Akron, Ohio., Sept. 16–18, 1974.
539. **Collins, R. L. and Doglia, S.**, Concentration of pesticides slowly released by diffusion, *Weed Sci.*, 21, 343, 1973.
540. **Savage, K.** Adsorption and degradation of chlorbromuron in soil, *Weed, Sci.*, 21, 416, 1973.
541. **Ritter, W. F., Johnson, H. P., and Lovely, W. G.**, Diffusion of atrazine, propachlor and Diazinon in silt loam soil, *Weed Sci.*, 21, 381, 1973.
542. **Bode, L. E., Day, C. L., Gebhardt, M. R., and Goering, C. E.**, Prediction of trifluralin diffusion coefficients, *Weed Sci.*, 21, 485, 1973.
543. **Wilson, R. G. and Cheng, H. H.**, Breakdown and Movement of 2,4-D Compounds in Soil-Water Systems, Weed Sci. Soc. Am. Meeting, Washington, D.C., Abstracts, February 1975, 77.
544. **Ehlers, W., Farmer, W. J., Spencer, W. F., and Letey, J.**, Lindane diffusion: II. Water content, bulk density and temperature effects, *Soil Sci. Soc. Am. Proc.*, 33, 505, 1969.
545. **Lavy, T. L.**, Diffusion of three chloro-s-triazines in soil, *Weed Sci.*, 18, 53, 1970.
546. **Gray, R. A.**, Thiocarbamate Herbicides – Their History, Properties and Behavior in Plants, Animals, Soils and Microorganisms, Weed Sci. Soc. Am. Meeting, Washington, D.C., Abstracts, February 1975, 53.
547. **Danielson, L. L. and Campbell, T. A.**, Evaluation of a Latex Based Herbicide Formulation, Res. Note 41-A.1, Proc. Controlled Release Pesticide Symp., University of Akron, Akron, Oh., Sept. 16–18, 1975.
548. **Akobundo, I. O., Sweet, R. D., and Duke, W. B.**, A method of evaluating herbicide combinations and determining herbicide synergism, *Weed Sci.*, 23, 20, 1975.
549. **Rummens, F. H. A.**, An improved definition of synergistic and antagonistic effects, *Weed Sci.*, 23, 4, 1975.
550. **Colby, S. R.**, Calculating synergistic and antagonistic responses of herbicide combinations, *Weeds*, 15, 20, 1967.
551. **Baur, J. R.** and **Bovey, R. W.**, Herbicidal Effects of Tebuthiuron and Glyphosate, Weed Sci. Soc. Am. Meeting, Washington, D.C., Abstracts, February 1975, 26.
552. **Klepper, L. A.**, Nitrite Accumulation: A Basis for Explaining Synergistic Combinations of Herbicides, Weed Sci. Soc. Am. Meeting, Abstracts, p. 57, Washington, D.C., Feb. 4–7, 1975.
553. **Jordan, T. N. and McWhorter, C. G.**, Environmental Effects on the Toxicity of Dalapon to Johnsongrass, Weed Sci. Soc. Am. Meeting, Abstracts, Washington, D. C., Feb. 4–7, 1975.
554. **Treshow, M.**, *Environment and Plant Response*, McGraw-Hill, New York, 1970.
555. **Allan, G. G., Chopra, C. S., Friedhoff, J. F., Gara, R. J., Maggi, M. W., Neogi, A. N., Roberts, S. C., and Wilkins, R. M.**, Pesticides, pollution and polymers, *Chem. Technol.*, 4, 171, 1973.
556. **Palmer, C. M.**, *Algae in Water Supplies*, PHS Pub. No. 657, Public Health Service, Washington, D.C., 1962.
557. **Davis, R.**, *The Marine and Fresh Water Plankton*, Michigan State University Press, East Lansing, 1955.
558. **Chapman, V. J. and Chapman, D. J.**, *The Algae*, Macmillan, London, 1973.
559. **Hanson, L. P.**, *Plant Growth Regulators*, Noyes Data Corp., Park Ridge, N.J., 1973.
560. **Anon.**, 3M Brand Aquatic Herbicide System M., Tech. Data Sheet, 3M Co., St. Paul, Minn., March 1970.
561. **Walsh, G. E.**, Effects of herbicides on photosynthesis and growth of marine unicellular algae, *Hyacinth Control J.*, 10, 45, 1972.
562. **Bartley, T. R. and Otto, N. E.**, Progress Report on Antifouling Materials for Algae Prevention, Rep. WC-30, U.S. Dept. of the Interior, Bureau of Reclamation, Denver, March 3, 1967.
563. **Taft, C. E.**, *Water and Algae – World Problems*, Educational Publishing, Chicago, 1965.
564. **Anon.**, Cutrine Fact Sheet, Tech. Svc. Rep. 020-271, Applied Biochemist Inc., Mequon, Wisc., July 1971.
565. **Herble, R. A.**, Acrolein: An Effective Cooling Water Slimicide, Tech. Paper 209, Betz Laboratories, Inc., Trevose, Pa., 1968.
566. **Anon.**, A New Slime Control Agent of Proven Effectiveness, report, M & T Chemicals Corp., Rahway, N.J.
567. **Sanborn, J. R.**, *Slime Control in the Pulp and Paper Industry*, Lockwood Trade Journal Co., New York, 1965.
568. **Fitzgerald, G. P. and Faust, F. L.**, Factors affecting the algicidal and algistatic properties of copper, *Appl. Microbiol.*, 11, 345, 1963.
569. **Lewin, R. A., Ed.**, *Physiology and Biochemistry of Algae*, Academic Press, New York, 1962.
570. **James, G. V.**, *Water Treatment*, 3rd ed., Technical Press, London, 1965.

571. **Cedeno-Maldonado, A. and Swader, J. A.,** Studies on the mechanism of copper toxicity in *Chlorella, Weed Sci.,* 22, 443, 1974.
572. **Gross, R. E., Pugno, P., and Dugger, W. M.,** Observations on the mechanism of copper damage in *Chlorella, Plant Physiol.,* 46, 183, 1972.
573. **Sargent, D. F. and Taylor, C. P. S.,** The effect of cupric and fluoride ion on the respiration of *Chlorella, Can. J. Bot.,* 50, 905, 1972.
574. **Anon.,** Envirocap Controlled Release Algicide, technical bulletin, 3M Co., St. Paul, Minn., 1973.
575. **Volf, L. A., Kotetskic, V. V., Khokhlova, V. A., Konev, Yu. E., and Meos, A. J.,** Preparation of germicidal polymer materials by ion exchange reactions with antiseptics, *Zh. Prikl. Khim. (Moscow),* 39, 2608, 1966.
576. **Edwards, L. D.,** Infections and use of microbials in an 800 bed hospital, *Public Health Rep.,* 84, 453, 1969.
577. **Rankin, B. F.,** Bactericidal Contact Lens Solution, U.S. Patent 3,755,561, 1973.
578. **Michaels, A. S.,** Drug Therapy and Novel Drug Delivery Systems, Symp. Battelle Seattle Res. Center, unpublished report, April 1974.
579. **Hueck, H. J., Adema, D. M. M., and Wiegmann, J. R.,** Bacteriostatic, fungistatic and algistatic activity of fatty nitrogen compounds, *Appl. Microbiol.,* 14, 308, 1966.
580. **Major, C. J. and Cardarelli, N. F.,** Biocidal Rubber for Water Reclamation Systems, Rep. AMRL-TR-69-17, Aerospace Medical Research Laboratory U.S. Air Force Systems Command, Wright-Patterson AFB, Dayton, Oh., 1969.
581. **Anon.,** Tree injection systems, *Weeds Trees Turf,* 12(4), 14, 1973.
582. **Anon.,** Dutch elm disease, *Weeds Trees Turf,* 11(4), 13, 1972.
583. **Smiley, R. W.,** Systemic fungicides – their role in turf disease control, *Weeds Trees Turf,* 14(1), 30, 1975.
584. **Kemp, H. T., Cooper, C. W., and Kell, R. M.,** Determining effectiveness of biocidal additives in coatings, *J. Paint Technol.,* 38, 363, 1966.
585. **Ugi, I. and Fetzer, U.,** Methods for Controlling Pests and Fungi with Polyisonitriles, U.S. Patent 3,278,371, 1966.
586. **Carpenter, S. J. and LaCasse, W. J.,** *Mosquitoes of North America,* University of California Press, Los Angeles, 1955.
587. **Gerberg, E. J.,** *Manual for Mosquito Rearing and Experimental Techniques,* AMCA Bull. No. 5, American Mosquito Control Association, Albany, N.Y., January 1970.
588. **James, M. T. and Harwood, R. F.,** *Medical Entomology,* 6th ed., Macmillan, New York, 1969.
589. **Benenson, A. S.,** *Control of Communicable Diseases in Man,* American Public Health Association, Washington, D.C., 1970.
590. **Pratt, H. D., Barnes, R. C., and Litlig, K. S.,** *Mosquitoes of Public Health Importance and Their Control,* Public Health Service Communicable Disease Center, Atlanta, 1963.
591. **Cohn, E. J.,** Assessing the costs and benefits of anti-malaria programs: the Indian experience, *Am. J. Public Health,* 63, 1086, 1973.
592. **Gonzalez-Valdivieso, F. E.,** Aedes Aegyptic Eradication in Venezuela, report, Division of Rural Endemic Diseases, Ministry of Health and Public Welfare, Venezuela, December 1969.
593. **Sollers-Riedel, H.,** The 1963 world-wide mosquito situation, in *Proc. New Jersey Mosq. Extermin. Assoc.,* 51st Annual Meeting, New Brunswick, N. J., 1964, 18.
594. **Johnson, H. E. and Pecor, C.,** Coho salmon mortality and DDT in Lake Michigan, *Trans. North Am. Wildl. Nat. Resour. Conf.,* p. 159, 1969.
595. **Malone, C. R. and Blaylock, B. G.,** Toxicity of insecticide formulations to carp embryos reared in vitro, *J. Wildl. Manage.,* 34, 460, 1970.
596. **Schouboe, P. J.,** Pesticides in the aquatic environment, *Hyacinth Control J.,* 11, 7, 1973.
597. **Dindal, D. L.,** Accumulation and excretion of Cl^{36} DDT in mallard and lesser scaup ducks, *J. Wildl. Manage.,* 34, 74, 1970.
598. **Hill, E. F.,** Toxicity of selected mosquito larvicides to some common avian species, *J. Wildl. Manage.,* 35, 757, 1971.
599. **Gill, J. A. and Verts, B. J.,** Tolerances of two populations of ring-necked pheasants to DDT, *J. Wildl. Manage.,* 34, 630, 1970.
600. **Neave, D. J. and Wright, B. S.,** The effects of weather and DDT spraying on a ruffed grouse population, *J. Wildl. Manage.,* 33, 1015, 1969.
601. **Kury, C. R.,** Pesticide residues in a marine population of double crested cormorants, *J. Wildl. Manage.,* 33, 91, 1969.
602. **Hunt, L. B. and Sacho, R. J.,** Response of robins to DDT and methoxychlor, *J. Wildl. Manage.,* 33, 336, 1969.
603. **Dimond, J. B. and Sherburne, J. A.,** Persistence of DDT in wild populations of small animals, *Nature,* 221, 486, 1969.
604. **Sherburne, J. A. and Dimond, J. B.,** DDT persistence in wild hares and mink, *J. Wildl. Manage.,* 33, 944, 1969.
605. **Brown, N. J. and Brown, A. W. A.,** Biological fate of DDT in a sub-arctic environment, *J. Wildl. Manage.,* 34, 929, 1970.
606. **Woodwell, G. M., Wurster, C. F., and Isaacson, P. A.,** DDT residues in an east coast estuary: a case of biological concentration of a persistent insecticide, *Science,* 156, 821, 1967.
607. **Derbeneva-Uhova, V. P. and Drobozin, V. P.,** The Development of DDT Resistance in *Photophormia terrae-novae,* WHO Rep. WHO/Vector Control 164.65, 1965.

608. **Abedi, Z. H., Duffy, J. R., and Brown, A. W. A.,** Dehydrochlorination and DDT resistance in *Aedes aegypti, J. Econ. Entomol.,* 56, 511, 1963.
609. **Sutherland, D. J., Siewierski, M., Marei, A. H., and Helrich, K.,** The effect on mosquitoes of sublethal exposure to insecticides. II. DDT metabolism, *Mosq. News,* 30, 8, 1970.
610. **Sutherland, D. J., Beam, F. D., and Gupta, A. P.,** The effects on mosquitoes of sublethal exposure to insecticides. I. DDT, dieldrin, Malathion and the basal follicles of *Aedes aegypti* (L.), *Mosq. News,* 27, 316, 1967.
611. **Davidson, G. and Zahar, A. R.,** The practical implications of resistance of malaria vectors to insecticides, *Bull. WHO,* 49, 475, 1973.
612. **Hadjinicolaou, J. and Betzios, B.,** Resurgence of *Anopheles sacharovi* following malaria eradication, *Bull. WHO,* 48, 699, 1973.
613. **Breslow, L.,** Current status of malaria in California, *Calif. Vector Views,* 14, 46, 1967.
614. **Monath, T. P.,** The 1970 yellow fever epidemic in Okwago district, Benue Plateau State, Nigeria, *Bull. WHO,* 49, 113, 1973.
615. **Carver, H. D.,** St. Louis Encephalitis Outbreak – Memphis, Tennessee, in Proc. New Jersey Mosq. Control Assoc. Meeting, Atlantic City, N. J., 1975, to be published.
616. **Sudia, W. D. and Newhouse, V. F.,** Venezuelan equine encephalitis outbreak in Texas 1971, informational report, *Mosq. News,* 31, 350, 1971.
617. **Kessel, J. F.,** Filariasis as a world problem, *Mosq. News,* 26, 490, 1966.
618. **Sasa, M.,** Epidemiology of Filariasis and Schistosomiasis in Asia and the Pacific: A Review, abstract, Joint Conference on Filiariasis Filariasis and Schistosomiasis, U.S.-Japan and Schistosomiasis, United States-Japan Cooperative Medical Science Program Meeting, National Institutes of Health, Bethesda, Md., July 26–28, 1971, 70.
619. **Sutherland, D. J., Jobbins, D. M., and Hagman, L. E.,** Source reduction through water management, in *Proc. 35th Annual Conf. Calif. Mosq. Control Assoc.,* San Fransciso, Feb. 5–8, 1967, 31.
620. **Carreker, J. R.,** Mosquito potential in flood detention reservoirs, *Trans Am. Soc. Agric. Eng.,* 8(2), 161, 1965.
621. **Furlow, B. M. and Hays, K. L.,** Some influences of aquatic vegetation on the species and number of culicidae (diptera) in small pools of water, *Mosq. News,* 32, 595, 1972.
622. **McDonald, J. L.,** Mosquitoes and agriculture on Okinawa, *Mosq. News,* 32, 466, 1972.
623. **Owens, J. C., Ward, C. R., Huddleston, E. W., and Ashdown, D.,** Non-chemical methods of mosquito control for playa lakes in west Texas, *Mosq. News,* 30, 571, 1970.
624. **Hayes, G. R. and Ritter, A. B.,** The diminution of *Aedes aegypti* infestations in Louisiana, *Mosq. News,* 26, 381, 1966.
625. **Knapp, F. W.,** Aerial Application of Baygon for Adult Mosquito Control, progress report to Chemagro Co., Kansas City; University of Kentucky, Lexington, Oct. 3, 1974.
626. **Murray, W. D.,** Oil larvicides applied by aircraft in an area characterized by resistance to organophosphorous larvicides, *Mosq. News,* 28, 233, 1968.
627. **Norman, W. F., Fentem, B., and Mulhern, T. D.,** A safe and efficient mixing and loading plant for insecticides used in aerial larviciding, in *Proc. 35th Annual Conf. Calif. Mosq. Control Assoc.,* San Francisco, Feb. 5–8, 1967, 129.
628. **Bay, E. C.,** Potential for naturalistic control of mosquitoes, in *Proc. 35th Annual Conf. Calif. Mosq. Control Assoc.,* San Francisco, Feb. 5–8, 1967, 35.
629. **Sasa, M., Wada, Y., Sato, H., and Kurihara, T.,** The Use of American Poeciliid Fishes in the Control of Filariasis Vectors and for Improvements of Water Pollution, report, Dept. of Parasitology, Institute of Medical Science, University of Tokyo, 1971.
630. **Hoy, J. B. and Reed, D. E.,** The efficacy of mosquito fish for control of *Culex tarsalis* in California rice fields, *Mosq. News,* 31, 567, 1971.
631. **Bay, E. C.,** Mosquito fish, *Pest Control,* 40 (12), 32, 1972.
632. **Fisher, H. J., Smith, L. W., and Enns, W. R.,** The fate of fish introduced as predators on insects in Missouri oxidation lagoons, *Mosq. News,* 30, 523, 1970.
633. **Darwazeh, H. A. and Mulla, M. S.,** Toxicity of herbicides and mosquito larvicides to the mosquito fish *Gambusia affinis, Mosq. News,* 34, 214, 1974.
634. **Burke, W. D. and Ferguson, D. E.,** Toxicities of four insecticides to resistant and susceptible mosquito fish in static and flowing solutions, *Mosq. News,* 29, 96, 1969.
635. **Ferguson, D. E., Gardner, D. T., and Lindley, A. L.,** Toxicity of Dursban to three species of fish, *Mosq. News,* 26, 80, 1966.
636. **Petersen, J. J., Chapman, H. C., and Woodard, D. B.,** Preliminary observations on the incidence and biology of a mermithid nematode of *Aedes sollicitans* in Louisiana, in *Proc. 35th Annual Calif. Mosq. Control Assoc.,* San Francisco, Feb. 5–8, 1967, 95.
637. **Kalucy, E. C.,** Parasitism of *Anopheles annulipes* Walker by a mermithid nematode, *Mosq. News,* 32, 582, 1972.
638. **Savage, K. E. and Petersen, J. J.,** Observation of mermithid nematodes in Florida mosquitoes, *Mosq. News,* 31, 218, 1971.
639. **Petersen, J. J. and Willis, O. R.,** Experimental release of a mermithid nematode to control *Anopheles* mosquitoes in Louisiana, *Mosq. News,* 34, 316, 1974.

640. **Smith, R. F.**, Considerations on the safety of certain biological agents for arthropod control, *Bull. WHO*, 48, 685, 1973.
641. **Lowe, R. E. and Kennel, E. W.**, Pathogenicity of the fungus *Entomorphthora coronata* in *Culex pipiens quinquefasciatus* and *Aedes taeniorhynchus*, *Mosq. News*, 32, 614, 1972.
642. **Clark, T. B., Kellen, W. R., Fukuda, T., and Lindegren, J. E.**, Experiments on the biological control of mosquitoes with the fungus *Beauveria bassiana* (Bals.) Vuill., in *Proc. 35th Annual Conf. Calif. Mosq. Control Assoc.*, San Francisco, Feb. 5–8, 1967, 99.
643. **Dame, D. A.**, The Sterile Insect Approach to Mosquito Control, paper presented to the American Mosquito Control Association Meeting, Atlantic City, N.J., Mar. 12, 1975.
644. **Patterson, R. S., Sharma, V. P., Singh, K. R. P., La Brecque, G. C., Seetheram, P. L., and Grover, K. K.**, Use of radiosterilized males to control indigenous populations of *Culex pipiens quinquefasciatus* Say: laboratory and field studies, *Mosq. News*, 35, 1, 1975.
645. **Spielman, A. and Williams, C. M.**, Lethal effects of synthetic hormone or larvae of the yellow fever mosquito, *Aedes aegypti*, *Science*, 154, 1043, 1966.
646. **Wheeler, C. M. and Thebault, M.**, Efficacy of synthetic substances with hormonal activity against IV instar larvae of *Culex pipiens quinquefasciatus* Say in an outdoor insectary, *Mosq. News*, 31, 170, 1971.
647. **Schaefer, C. M. and Wilder, W. H.**, Insect development inhibitors: a practical evaluation as mosquito control agents, *J. Econ. Entomol.*, 65, 1066, 1972.
648. **Hoppe, T., Isler, H., and Vogel, W.**, Biological activity of juvenile hormone analogues against larvae of *Culex pipiens pipiens* tested in small scale field trials, *Mosq. News*, 34, 293, 1974.
649. **Anon.**, Altosid SR-10, Mosquito Growth Regulator, Z8-06 Zoecon Corp., Palo Alto, Cal., March 1975.
650. **Jakob, W. L.**, Additional studies with juvenile hormone-type compounds against mosquito larvae, *Mosq. News*, 32, 592, 1972.
651. **Rathburn, C. B. and Boike, J. R.**, Laboratory and Small Plot Field Tests of Altosid and Dimilin for the Control of Mosquito Larvae, report of the West Florida Arthropod Laboratory, Panama City, Fla., 1975.
652. **Smith, C. N. and Roos, J.**, Results of Tests with Two Formulations from Aerospace and Defense Products Division, B. F. Goodrich Co., Rep. USDA, Insects Affecting Man and Animal, Research Branch, Florida Station, Gainesville, Fla., Dec. 7, 1966.
653. **Simmons, I. H.**, personal communication.
654. **Schultz, H. A. and Webb, A. B.**, Laboratory bioassay of pesticide-impregnated rubber as a mosquito larvicide, *Mosq. News*, 29, 38, 1969.
655. **Boike, A. H. and Rathburn, C. B.**, An evaluation of several toxic rubber compounds as mosquito larvicides, *Mosq. News*, 33, 501, 1973.
656. **Stockman, C. H., Bollinger, E. H., and Naranch, R. N.**, Slow-release polymeric compositions for mosquito larviciding, *Mosq. News*, 30, 584, 1970.
657. **Cardarelli, N. F. and Jackson, D. L.**, Slow Release Organotin Larvicides, University of Akron report to the Entomological Society of America Meeting, Dallas, Tex., Dec. 2–5, 1968, unpublished.
658. **Whitlaw, J. T. and Evans, E. S.**, Selected plastic formulations for use as mosquito larvicides, *J. Econ. Entomol.*, 61, 889, 1968.
659. **Cardarelli, N. F.**, Slow Release of Pesticides from Elastomeric Matrices, American Chemical Society Symp., Fertilizer/Pesticide Combinations, Washington, D.C., September 1971.
660. **Cardarelli, N. F.**, Larvicide, Rhodesian Patent 009452, 1967.
661. **Cardarelli, N. F.**, Floating Larvicide, U.S. Patent 3,590,119, 1971.
662. **Eto, M.**, *Organophosphorus Insecticides: Organic and Biological Chemistry*, CRC Press, Cleveland, 1974.
663. **Bowman, J. S. and Orloski, E. J.**, Abate insecticide residues in streams and ponds treated for control of mosquito larvae, *Mosq. News*, 26, 557, 1966.
664. **Fitzpatrick, G. and Sutherland, D. J.**, Impact of Abate and Dursban Larvicidal Formulations on the Salt Marsh Pulmonate Snail *Melampus bidentatus* Say, report to the annual meeting, American Mosquito Control Association, Atlantic City, N.J., Mar. 13, 1975.
665. **Bond, H. A. and Fay, R. W.**, Factors influencing *Aedes aegypti* occurence in containers, *Mosq. News*, 29, 113, 1969.
666. **Brooks, G. D., Schoof, H. F., and Smith, E. A.**, Evaluations of five formulations of Abate against *Aedes aegypti*, Savannah, Georgia, 1965, *Mosq. News*, 26, 580, 1966.
667. **Glancey, B. M., Moussa, M. A., Scanlon, J. E., and Lofgren, C. S.**, Abate and Dursban against *Aedes aegypti* (L.) breeding in concrete water jars in Bangkok, Thailand, *Mosq. News*, 28, 205, 1968.
668. **McDonald, J. L.**, Evaluations of Abate for mosquito control in polluted water, *Mosq. News*, 32, 627, 1972.
669. **Sacher, R. M.**, A mosquito larvicide with favorable environmental properties, *Mosq. News*, 31, 513, 1971.
670. **Dunn, R. L. and Strong, F. E.**, Control of catch-basin mosquitoes using Zoecon ZR_{515} formulated in a slow release polymer – a preliminary report, *Mosq. News*, 33, 110, 1973.
671. **Anon.**, EPA grants limited use registration to new mosquito control pesticide, *Weeds Trees Turf*, 14(5), 30, 1975.
672. **Brown, A. W. A., Brown, T. M., and Hooper, G. H. S.**, Resistance Hazard of Juvenile Hormone Mimics, annual report, Pesticide Research Center, Michigan State University, East Lansing, 1964, 29.
673. **Barnes, W. W., Webb, A. B., and Savage, L. B.**, Laboratory evaluation of Abate and Malathion insecticidal briquettes, *Mosq. News*, 27, 488, 1967.

674. **Barnes, W. W. and Webb, A. B.**, A field evaluation of Abate briquettes in woodland pools, *Mosq. News*, 28, 458, 1968.
675. **Whitlaw, J. T. and Evans, E. S.**, Selected plastic formulations for use as mosquito larvicides, *J. Econ. Entomol.*, 61, 889, 1968.
676. **Nelson, J. H., Evans, E. S., Pennington, N. E., and Young, W. W.**, The U.S. Army Environmental Hygiene Agency Activities in the Area of Slow Release Insecticides, Rep. No. 30, Proc. Controlled Release Pesticides Symp., University of Akron, Akron, Oh., Sept. 16–18, 1974.
677. **McDonald, J. L. and Dickens, T. H.**, Field evaluation of Dursban insecticide briquettes when used as mosquito larvicide materials, *Mosq. News*, 30, 563, 1970.
678. **Wilkinson, R. N., Barnes, W. W., Gillogly, A. R., and Minnemeyer, C. D.**, Field evaluation of slow release mosquito larvicides, *J. Econ. Entomol.*, 64, 1, 1971.
679. **Parker, T. A.**, Effectiveness of 9.9 Percent Dursban in Polyethylene Applied as a Pre-Season Larvicide, U.S. Army Environmental Hygiene Agency Special Study No. 31-005-71, AD 725573, 1970.
680. **Lewis, L. F., Christenson, D. M., and Eddy, G. W.**, Results of tests with Dursban and fenthion for the control of mosquito larvae in log ponds of western Oregon, *Mosq. News*, 26, 579, 1966.
681. **Georghiou, G. P., Metcalf, R. L., and Milhollin, R.**, Comparative resistance of Dursban and its dimethyl analog (OMS 1155) to insecticide-susceptible and resistant *Culex* and *Anopheles* mosquitoes, *Mosq. News*, 29, 92, 1969.
682. **McNeill, J. C., Miller, W. O., and Wleczyk, C. M.**, Evaluation of Dursban as a larvicide in septic ditches, *Mosq. News*, 28, 160, 1968.
683. **Mathis, W. and Schoof, H. F.**, Residual activity of several new insecticides to adult *Anopheles quadrimaculatus* in adobe-lined experimental huts, *Mosq. News*, 28, 412, 1968.
684. **Ferguson, D. E., Gardner, D. T., and Lindley, A. L.**, Toxicity of Dursban to three species of fish, *Mosq. News*, 26, 80, 1966.
685. **Miller, T., Nelson, L. L., Young, W. W., Roberts, L. W., Roberts, D. R., and Wilkinson, R. N.**, Polymer formulations of mosquito larvicides: I. Effectiveness of polyethylene and polyvinyl chloride formulations of chlorpyrifos applied to artificial field pools, *Mosq. News*, 33, 148, 1973.
686. **Roberts, L. W., Roberts, D. R., Miller, T. A., Nelson, L. L., and Young, W. W.**, Polymer formulations of mosquito larvicides: II. Effects of a polyethylene formulation of chlorpyrifos on *Culex* populations naturally infesting artificial field pools, *Mosq. News*, 33, 155, 1973.
687. **Roberts, D. R., Roberts, L. W., Miller, T. A., Nelson, L. L., and Young, W. W.**, Polymer formulations of mosquito larvicides: III. Effects of a polyethylene formulation of chlorpyrifos on non-target populations naturally infesting artificial field pools, *Mosq. News*, 33, 165, 1973.
688. **Miller, T. A., Nelson, L. L., and Young, W. W.**, Polymer formulations of mosquito larvicides: IV. Larvicidal effectiveness of polyethylene and polyvinyl-chloride formulations of chlorpyrifos during an 18-month field test, *Mosq. News*, 33, 172, 1973.
689. **Nelson, L. L., Miller, T. A., and Young, W. W.**, Polymer formulations of mosquito larvicides: V. Effects of continuous low-level chlorpyrifos residues on the development of *Culex pipiens quinquefasciatus* Say populations in the laboratory, *Mosq. News*, 33, 396, 1973.
690. **Nelson, L. L., Miller, T. A., Whitlaw, J. T., Young, W. W., and Lawson, M. A.**, Polymer formulations of mosquito larvicides: VI. Laboratory and field evaluation of selected polyethylene formulations of chlorpyrifos, *Mosq. News*, 33, 403, 1973.
691. **Lawson, M. A., Miller, T. A., Oakleaf, R. J., and Young, W. W.**, Polymer formulations of mosquito larvicides: VIII. Laboratory evaluations of selected polyethylene formulations of chlorpyrifos, *Mosq. News*, 33, 561, 1973.
692. **Fanger, G.**, Microencapsulation, a brief history and introduction, in *A. C. S. Div. Org. Coatings Plast. Chem. Meeting*, Chicago, 33(2), 533, 1973.
693. **Macaulay, N.**, Recording Paper Coated With Microscopic Capsules of Coloring Material, Capsules and Methods of Making Same, U.S. Patent 3,016,308, 1962.
694. **Grevenstuk, A. B. and Hoogesteger, F. J. J.**, Method of Forming Free Flowing Particles Containing a Biologically Active Substance, U.S. Patent 3,202,731, 1965.
695. **Evans, R. B. and Herbst, W.**, Method of Encapsulating Water Insoluble Oils and Products Thereof, U.S. Patent 3,159,585, 1964.
696. **Sachsel, G. F. and Nack, H.**, Method of Encapsulating Liquids, U.S. Patent 3,202,533, 1965.
697. **Nack, H.**, Fluidized Bed Coating of Fragile Bodies, U.S. Patent 3,382,093, 1968.
698. **Grass, G. M. and Robinson, M. J.**, Apparatus for Coating Discrete Solids, U.S. Patent 3,237,596, 1966.
699. **Wurster, D. E.**, Method of Applying Coatings to Edible Tablets or the Like, U.S. Patent 2,648,609, 1953.
700. **Anon.**, The Wurster Air Suspension Process for Encapsulation, Wisconsin Alumni Research Foundation. Coating Laboratory Report, Middleton, Wisc., 1974.
701. **Somerville, G. R.**, Method for Mass Producing Small Spherical Particles, U.S. Patent 3,389,194, 1968.
702. **Langer, G. and Yamate, G.**, Apparatus for Electrostatic Encapsulation, U.S. Patent 3,159,874, 1964.
703. **Langer, G. and Yamate, G.**, Encapsulation of Liquid and Solid Particles to Form Dry Powders, *J. Colloid Interface Sci.*, 29, 3, 1969.
704. **Berger, B. B., Miller, C. D., and Langer, G.**, Electrostatic Encapsulation, U.S. Patent 3,208,951, 1965.

705. **Herman, D. F., Kruse, U., and Brancato, J. J.**, Polyethylene encapsulated cellulose, *J. Polym. Sci. Part C*, 3(11), 75, 1965.
706. **Gorham, F.**, Films of polymeric p-xylene on substrate particles, *J. Polym. Sci.*, 4, 3027, 1966.
707. **Vandegaer, J. E., Ed.**, *Microencapsulation: Processes and Applications*, Plenum, New York, 1974.
708. **Morgan, P. W. and Kwolek, S. L.**, Interfacial polycondensation II. Fundamentals of polymer formation at liquid interfaces, *J. Polym. Sci.*, 40, 299, 1959.
709. **Wurzburg, O. B. and Herbst, W.**, Encapsulating Agent with Controlled Water Repellency, U.S. Patent 3,091, 567, 1963.
710. **Koff, A. and Widmer, P. F.**, Vitamin-containing Gelatin Beadlets and the Process of Preparing Them, U.S. Patent 3,143,475, 1964.
711. **Brynko, C.**, Oil Containing Capsules and Methods of Making Them, U.S. Patent 2,969,330, 1961.
712. **Sternberg, S., Bixler, H. J., and Michaels, A. S.**, Encapsulating Particles and Processes for Making Same, U.S. Patent 3,639,306, 1972.
713. **Brynko, C. and Olderman, G. M.**, Replacement of Capsules Contents by Diffusion, U.S. Patent 3,516,943, 1970.
714. **Hall, H. and Hinkes, T. M.**, Air suspension encapsulation of moisture-sensitive particles using aqueous system, in *ACS Div. Org. Coatings Plast. Chem. Meeting*, Chicago, 33 (2), 545, 1973.
715. **Anon.**, How drug delivery systems may recost drug therapy, *Chem. Eng. News*, 49, 20, 1971.
716. **Goldschmiedt, H.**, Microencapsulation, *Soap Cosmet. Chem. Spec.*, p. 46, September 1973.
717. **Anon.**, Microencapsulation: The Process and Its Capabilities, brochure, National Cash Register Co., Dayton, Oh., 1971.
718. **Sinclair, R. G.**, Slow release pesticide system, *Environ. Sci. Technol.*, 7, 955, 1973.
719. **Koestler, R. C. and Ivy, E. E.**, Controlled Release of Biologically Active Agents from Nylon-Type Microcapsules, Rep. No. 33, Proc. Controlled Release Pesticide Symp., University of Akron, Akron, Oh., Sept. 16–18, 1974.
720. **Miller, T. A., Lawson, M. A., Nelson, L. L., and Young, W. W.**, Polymer formulations of mosquito larvicides: VII. Laboratory and field evaluation of encapsulated formulations of Malathion and Abate®, *Mosq. News*, 33, 413, 1973.
721. **Anon.**, Penncap-M Technical Bulletin, Pennwalt Co., Monrovia, Calif., 1972.
722. **DeSavigny, C. B. and Ivy, E. E.**, Microencapsulated pesticides, in *ACS Div. Org. Coatings Plast. Chem. Meeting*, Chicago, 33 (2), 554, 1973.
723. **Rachesky, S.**, Mosquitoes and PCO's, *Pest Control*, 41(6), 26, 1973.
724. **Rogers, A. J. et al.**, Ground equipment and insecticides for mosquito control, *Am. Mosq. Control Assoc. Bull.*, No. 2, August 1968.
725. **Husted, S. R., Mulrennan, J. A., Frederick, R. M., Blair, D. A., and U'ren, H. J.**, Volume as a Factor in ULV Dispersal of Equal Dosages of Chlorpyrifos for Adult Mosquito Control, report, Disease Vector Ecology and Control Center, Naval Air Station, Alameda, Calif., 1975, to be published.
726. **Mount, G. A., Meisch, M. V., Lee, J. T., Pierce, N. W., and Baldwin, K. F.**, Ultralow volume ground aerosols of insecticides for control of rice field mosquitoes in Arkansas, *Mosq. News*, 32, 444, 1974.
727. **Preiss, F. J.**, Mosquito adulticiding with synergized pyrethrins, *Soap Cosmet. Chem. Spec.*, p. 72, May 1973.
728. **Fluno, J. A.**, Controlling Mosquitoes, House and Garden Bull. No. 84, U.S. Dept. of Agriculture, Washington, D.C., June 1965.
729. **Lofgren, C. S.**, Recent developments in methods of mosquito control, *Bull. WHO*, 50, 323, 1974.
730. **Wilson, H. G., Labrecque, G. C., and Thomas, J. A.**, Vapor toxicity of residual insecticides against mosquitoes, *Mosq. News*, 35, 15, 1975.
731. **Folckemer, F. B., Hanson, R. E., and Miller, A.**, Resin Compositions Comprising Organophosphorus Pesticides, U.S. Patent 3,318,769, 1967.
732. **Miles, J. W., Pearce, G. W., and Woehst, J. E.**, Stable formulations for the sustained release of DDVP, *J. Agric. Food Chem.*, 10, 244, 1962.
733. **Brooks, G. D., Elmore, C. M., Schoof, H. F., and Carmichael, G. T.**, Field evaluations of three types of DDVP dispensers for the control of *Culex pipiens quinquefasciatus* Say in catch basins, in *Proc. 50th Annual Meeting, New Jersey Mosquito Extermination Assoc. and 19th Annual Meeting, Am. Mosq. Control Assoc.*, American Mosquito Control Association, Albany, N.Y., 1963, 355.
734. **Soloway, S. B. and Morales, J. G.**, Volatile Insecticide, U.S. Patent 3,740,428, 1973.
735. **Preiss, F. J., Barefoot, H. L., Stryker, R. G., and Young, W. W.**, Effectiveness of DDVP as a killing agent in mosquito killing jars, *Mosq. News*, 30, 417, 1970.
736. **Witters, R. F., Gaines, T. B., Short, J. G., Sedlak, V. A., and Maddock, D. R.**, Studies on the safety of DDVP for disinfection of commercial aircraft, *Bull. WHO*, 24, 1000, 1961.
737. **Funckes, A. V., Miller, S., and Hayes, W. V.**, Initial field studies in Upper Volta with dichlorvos residual fumigant as a malaria eradication technique, *Bull. WHO*, 29, 243, 1963.
738. **Fukuto, T. R. and Sims, J. J.**, Metabolism of insecticides and fungicides, in *Pesticides in the Environment*, Vol. 1, Part 1, White-Stevens, R., Ed., Marcel Dekker, New York, 1971, 145.
739. **Hodgeson, E. and Casida, J. E.**, Mammalian enzymes involved in the degradation of 2,2-dichlorovinyl dimethyl phosphate, *J. Agric. Food Chem.*, 10, 208, 1962.

740. Gaines, T. B., Hayes, W. J., and Linder, R. E., Liver metabolism of anticholinesterase compounds in live rats: relation to toxicity, *Nature*, 209, 88, 1966.
741. Casida, J. E., MacBride, L., and Niedermeier, R. P., Metabolism of 2,2-dichlorovinyl dimethyl phosphate in relation to residues in milk and mammalian tissues, *J. Agric. Food Chem.*, 10, 370, 1962.
742. Gillett, J. W., Harr, J. R., Lindstrom, F. T., Mount, D. A., St. Clair, A. D., and Weber, L. J., Evaluation of human health hazards on use of dichlorvos (DDVP), especially in resin strips, *Residue Rev.*, 44, 115, 1972.
743. Cervoni, W. A., Oliver-Gonzales, J., Kaye, S., and Slomka, M. B., Dichlorvos as a single dose intestinal anthelmintic therapy for man, *Am. J. Trop. Med. Hyg.*, 18, 912, 1969.
744. Jacobs, D. E., Experiences with a broad spectrum anthelmintic, dichlorvos, in the adult pig, *Vet. Rec.*, 83, 160, 1968.
745. Brooks, G. D., Carmichael, G. T., and Schoof, H. F., Evaluation of large-scale treatment with dichlorvos for the control of *Culex pipiens quinquefasciatus* Say, *Mosq. News*, 25, 427, 1965.
746. Schoof, H. F., Pearce, G. W., and Willis, M., Dichlorvos as a residual fumigant in mud, plywood, and bamboo huts, *Bull. WHO*, 29, 227, 1963.
747. Kashin, K., Development of an Orally Effective Insect Repellent, Annual Progress Report, AD663818, ITT Research Institute, Chicago, November 1967.
748. Lopp, O. V. and Buchanan, W. J., How granular mosquito repellents performed in the field, *Pest Control*, 27(4), 25, 1959.
749. Horsfall, W. R., New use for an old formulation: spot suppression of mosquitoes, *Pest Control*, 27(4), 24, 1959.
750. Berry, R. A., Joseph, S. R., and Langford, G. S., The question of area mosquito repellency, in *Proc. New Jersey Mosq. Extermin. Assoc.*, 52nd Annual Meeting, New Brunswick, N.J., 1965.
751. Langford, G. S., Joseph, S. R., and Berry, R., Some observations on mosquito repellents, *Mosq. News*, 26, 399, 1966.
752. Ginsburg, J. M., Protection from Mosquito Bites in Outdoor Gatherings, *Science*, 82, 2134, 1935.
753. King, W. V., Chemicals Evaluated as Insecticides and Repellents at Orlando, Florida, Handbook No. 69, U.S. Dept. of Agriculture, Washington, D.C., 1954.
754. Gouck, H., McGovern, T. P., and Beroza, M., Chemicals tested as space repellents against yellow fever mosquitoes, I. Esters, *J. Econ. Entomol.*, 60, 1587, 1967.
755. Bar-Zeev, M. and Sternberg, S., Testing space repellents against mosquitoes, *Mosq. News*, 30, 27, 1970.
756. Gouck, H. K., Goodwin, D. R., Schreck, C. E., and Smith, N., Field tests with repellent-treated netting against black salt marsh mosquitoes, *J. Econ. Entomol.*, 60, 1451, 1967.
757. Bar-Zeev, M. and Ben-Tamar, D., Effectiveness of repellents on cloth as determined by oviposition of *Aedes aegypti* L., *Mosq. News*, 28, 396, 1968.
758. Gilbert, I. H., Gouck, H. K., and Smith, C. N., New insect repellent, *Soap Chem. Spec.*, 33(5), 129, 1957.
759. Altman, R. M., Repellent tests against *Anopheles albimanus* Wiedemann in the Panama Canal Zone, *Mosq. News*, 29, 110, 1969.
760. Bar-Zeev, M. and Ben-Tamar, D., Evaluation of mosquito repellents, *Mosq. News*, 31, 56, 1971.
761. Gorham, J. R., Tests of mosquito repellents in Alaska, *Mosq. News*, 34, 409, 1974.
762. Gilbert, I. H., Evaluation of repellents against mosquitoes and deer flies in Oregon, *J. Econ. Entomol*, 50, 46, 1957.
763. Anon., Metadelphine®: The All Purpose Insect Repellent, Bull. AP-105 B, Hercules Co., Agricultural Chemicals Div., Wilmington, Del., 1969.
764. Travis, B. V., Smith, A. L., and Madden, A. H., Effectiveness of insect repellents against black flies, *J. Econ. Entomol.*, 44, 813, 1951.
765. Catts, E. P., DEET-impregnated net shirt repels biting flies, *J. Econ. Entomol.*, 61, 1765, 1968.
766. Travis, B. V. and Smith, A. L., Tests with repellents against the stable fly, *J. Am. Vet. Med. Assoc.*, 894, 214, 1951.
767. Gilbert, I. H. and Gouck, H. K., All-purpose repellent mixtures as clothing treatments against chiggers, *Fl. Entomol.*, 36, 45, 1953.
768. Gouck, H. K. and Gilbert, I. H., Field tests with new tick repellents in 1954, *J. Econ. Entomol.*, 48, 499, 1955.
769. Smith, C. N. and King, W. V., Field studies of tick repellents, *Am. J. Trop. Med.*, 30, 97, 1950.
770. Sjogren, R. D., An effective repellent for *Leptoconops kerteszi* Kieffer (Diptera: Ceratopogonidae) *Mosq. News*, 31, 115, 1971.
771. Anon., Insect Repellent, Bull. F-24, FMC Corp., Middleport, N. Y., 1967.
772. Tong, H. C., Skidmore, D., Maibach, H. I., and Skinner, W., A., Topical mosquito repellents, VIII: Substituted 2-thio-4-thiazolidineones and 2,4-thiazolidinediones, *Mosq. News*, 35, 76, 1975.
773. Kurtz, A. P., Logan, J. A., and Akers, W. A., Rep. No. 13, Letterman Army Institute of Research, San Francisco, May 1973.
774. Barton, L. J., Ointment, U. S. Patent 2,095,092, 1937.
775. Kemppe, A. J., Composition of Matter for Treating Skin Diseases, U.S. Patent 2,267,739, 1941.
776. Travis, B. V. and Jones, H. A., Insect Repellent Composition, U.S. Patent 2,420,389, 1947.
777. Rosenthall, M. W., Antiseptic Composition Containing Peroxide, Glycerol and Carboxypolymethylene Polymer, U.S. Patent 3,657,413, 1972.
778. Conrady, J. A. and Stockman, C. H., Long Lasting Insect Repellent Films for Skin and Other Substrates, U.S. Patent 3,590,118, 1971.

779. **Lange, P. M.,** Wash-resistant Skin Preparation, U.S. Patent 3,836,647, 1974.
780. **Iino, H.,** Cleaning Composition, U.S. Patent 3,790,488, 1974.
781. **Dickert, Y. J., Thill, B. P., and Steinhauer, A. F.,** Nongreasy Cosmetic Lotions, U.S. Patent 3,755,560, 1973.
782. **Blackford, E. A.,** Latex for High Gloss Floor Polish Formulations, U.S. Patent 3,766,112, 1973.
783. **Simmons, S. W.,** Some Current Developments in the Use of Economic Poisons, Proc. Annual Meeting, Southern Branch of the American Public Health Association, Birmingham, Ala., 1950.
784. **Gillies, P. A., Chambers, G. V., and Womeldorf, D. J.,** Evaluation of aerial applications of Flit® MLO for the control of organophosphorus-resistant *Aedes nigromaculis* and *Culex tarsalis, Mosq. News,* 31, 528, 1971.
785. **Glancey, B. M., Savage, K. E., and Lofgren, C. S.,** Laboratory evaluation of promising insecticides against adult black salt-marsh mosquitoes, *Aedes taeniorhynchus* (Wiedemann), *Mosq. News,* 26, 397, 1966.
786. **Gahan, J. B., Bertholf, J. H., Davis, A. N., and Smith, C. N.,** Field tests with two phosphorothioates against resistant salt-marsh mosquitoes, *Mosq. News,* 16, 91, 1956.
787. **Brown, A. W. A. and Pal, R.,** Insecticide resistance in arthropods, *WHO Monogr. Ser.,* No. 38, 1971.
788. **Dorval, C. and Brown, A. W. A.,** Inheritance of resistance to fenthion in *Culex pipiens fatigans* Wied., *Bull. WHO,* 43, 727, 1970.
789. **Womeldorf, D. J., Gillies, P. A., and White, K. E.,** Insecticide tolerance in *Culex tarsalis* larvae in California, in *Proc. 35th Annual Conf. Calif. Mosq. Control Assoc.,* San Francisco, Feb. 5–8, 1967, 110.
790. **Rathburn, C. B. and Boike, A. H.,** Studies of insecticide resistance in Florida mosquitoes, *Mosq. News,* 27, 377, 1967.
791. **Busvine, J. R.,** Insecticide resistance in mosquitos, *Pestic. Sci.,* 3, 483, 1972.
792. **Haise, H., McDuffie, W. C., Simms, D. H., Beattie, B., Gibbs, J. C., Ware, S. J., Yeager, L., Springer, P., Keesee, G., Hess, A. D., and Rainey, M. B.,** *Mosquito Prevention on Irrigated Farms,* Handbook No. 319, U.S. Dept. of Agriculture, Washington, D.C., 1967.
793. **Micks, D. W., Chambers, G., Jennings, J., and Rehmet, A.,** Mosquito control agents derived from petroleum hydrocarbons, I. Laboratory effectiveness, *J. Econ. Entomol.,* 60, 426, 1967.
794. **Dorer, R. E.,** Features and benefits of bulk Flit® MLO handling, *Mosq. News,* 32, 462, 1972.
795. **Abramitis, W. W.,** Mosquito Larvicide Compositions, U.S. Patent 3,755,597, 1973.
796. **Mulla, M. S., Darwazeh, H. A., and Majori, G.,** Field efficacy of some promising mosquito larvicides and their effects on nontarget organisms, *Mosq. News,* 35, 179, 1975.
797. **Pierce, N. W., Baldwin, K. F., Meisch, M. V., and Mount, G. A.,** Effectiveness of ten adulticides against *Psorophora confinnis* (Lynch Arribalzaga) in Arkansas, *Mosq. News,* 33, 418, 1973.
798. **Mulla, M. S., Darwazeh, H. A., and Peters, D. R.,** Mosquito control in sewage oxidation ponds with drip and pour-in larvicides, *Mosq. News,* 30, 456, 1970.
799. **Duncan, J.,** Post treatment effects of sublethal doses of dieldrin on the mosquito *Aedes aegypti* L., *Ann. Appl. Biol.,* 52, 1, 1963.
800. **Intermill, R. W.,** Development of *Dirofillaria immitis* in *Aedes triseriatus* Say, *Mosq. News,* 33, 176, 1973.
801. **Sanders, D. P., Riewe, M. E., and McNeill, J. C.,** Salt marsh mosquito control in relation to beef cattle production: a preliminary report, *Mosq. News,* 28, 311, 1968.
802. **Sanmartin, G.,** Annual Report, Cali Virus Laboratory, Cali., Colombia, 1967.
803. **Anderson, J. et al.,** Isolation of eastern encephalitis virus from diptera in Wisconsin, *Mosq. News,* 21, 244, 1961.
804. **Hamon, M. J. and Kartman, L.,** Onchocerchiasis, *Pest Control,* 40(6), 48, 1972.
805. **Anon.,** Black Flies in the Americas, WHO Doc. WHO/VBC 71.283, 1971.
806. **Waddy, B.,** Report on a Tour of Onchocerciasis Foci in Mexico, Guatemala, Venezuela and Colombia, WHO Doc. WHO/Oncho/67.60, 1967.
807. **Buck, A. A., Anderson, R. I., Kawata, K., and Hitchcock, J. C.,** Onchocerciasis: some new epidemiologic and clinical findings, *Am. J. Trop. Med. Hyg.,* 18, 217, 1969.
808. **Burton, G. J., Perkins, I. V., and Sodhi, H. S.,** Aerobic bacteria in the midgut of *Simulium damnosum* larvae, *Mosq. News,* 33, 115, 1973.
809. **Burton, G. J. and McRae, T. M.,** Phoretic attachment of *Simulium* larvae and pupae to mayfly and dragonfly nymphs, *Mosq. News,* 32, 436, 1972.
810. **McMahon, J. P.,** A review of the control of *Simulium* vectors of onchocerciasis, *Bull. WHO,* 37, 415, 1967.
811. **Carestia, R. R., Fromer, R. L., Vavra, R. W., and Loy, V. A.,** Field evaluation of black fly control – aerial applications, *Mosq. News,* 34, 330, 1974.
812. **Winston, D. C.,** Massive river spraying project planned, *Aviat. Week Space Technol.,* p. 88, Apr. 29, 1974.
813. **Swabey, Y. H., Schenk, C. F., and Parker, G. L.,** Evaluation of two organophosphorus compounds as black fly larvicides, *Mosq. News,* 27, 149, 1967.
814. **Travis, B. V. and Wilton, D. P.,** A progress report on simulated stream tests of black fly larvicides, *Mosq. News,* 25, 112, 1965.
815. **Jamnback, H. A., Duflo, T., and Marr, D.,** Aerial application of larvicides for control of *Simulium damnosum* in Ghana: a preliminary trial, *Bull. WHO,* 42, 826, 1970.
816. **Quelennec, G.,** Essais sur le terrain de nouvelles formulations d'insecticides, OMS-187, OMS-786 et OMS-971, contre les larves de simulies, *Bull. WHO,* 43, 313, 1970.

817. **Quelennec, G.**, Essais de portée de deux insecticides utilises contre les larves de *Simulium damnosum*, dans le Nord Dahomey, *Bull. WHO*, 27, 615, 1962.
818. **Metcalf, R. L., Kapoor, I., and Hirwe, A.**, Insecticidal Biodegradable Methylthio Analogues of DDT, U.S. Patent 3,855,313, 1974.
819. **Garris, G. I. and Adkins, T. R.**, The effects of Altosid®, an insect developmental inhibitor, on the last instar larva of *Simulium pictipes, Mosq. News*, 34, 335, 1974.
820. **Fairchild, G. B. and Barreda, E. A.**, DDT as a larvicide against *Simulium, J. Econ. Entomol.*, 38, 694, 1945.
821. **Waddy, B. B.**, Prospects for the control of onchocerciasis in Africa, *Bull. WHO*, 40, 843, 1969.
822. **Brown, A. W. A.**, A survey of *Simulium* control in Africa, *Bull. WHO*, 27, 511, 1962.
823. **Kanakkanatt, S. V. and Cardarelli, N. F.**, A New Concept in Black Fly Control, Proc. Int. Controlled Release Pesticide Symp., Wright State University, Dayton, Oh., Sept. 8–10, 1975.
824. **Arthur, D. R.**, *Ticks and Disease*, Row, Peterson & Co., Evanston, Ill., 1961.
825. **Anon.**, ZR-856 Miticide, technical bulletin, Zoecon Corp., Palo Alto, Calif., 1975.
826. **Gilbert, B.**, Slow Release Products in Tropical Disease Control, Proc. Controlled Release Pesticide Symp., University of Akron, Akron, Oh., Sept. 16–18, 1974.
827. **Turner, E. R., LeBreque, G. C., and Hoogstraal, H.**, Effectiveness of insecticides as residues and fogs for control of *Phlebotomus langeroni orientalis* parrot (Diptera: Psychodidae), *J. Egypt. Public Health Assoc.*, 40, 59, 1965.
828. **Hoogstraal, H. and Heyneman, D.**, Leishmaniasis in the Sudan Republic, *Am. J. Trop. Med. Hyg.*, 18(2), 1091, 1969.
829. **Anon.**, Public Health Pesticides, National Communicable Disease Center report, Savannah, Ga., 1967.
830. **Anon.**, Superbee, *Pest Control*, 4(10), 56, 1972.
831. **Alexander, S. D. and Norment, B. R.**, Effects of Mirex on populations of *Aedes triseriatus* Say, *Mosq. News*, 34, 112, 1974.
832. **Ruddock, J. C. and Rohe, D. L.**, A bait station system for controlling the ground nesting yellowjacket, *Vespula pensylvanica* (Saussure), *Calif. Vector Views*, 15(1), 3, 1968.
833. **Lyons, G. R. L.**, The control of quineaworm with Abate: a trial in a village of north-west Ghana, *Bull. WHO*, 49, 215, 1973.
834. **Bultman, J. D., Leonard, J. M., and Southwell, C. R.**, Termite Resistance of Polyvinyl Chloride Plastics, NRL Memo, Rep. 1704, Naval Research Laboratory, Washington, D.C., June 20, 1966.
835. **Bultman, J. D., Leonard, J. M., and Southwell, C. R.**, Termite Resistance of Polyvinyl Chloride Plastic – Two Years Exposure in the Tropics, NRL Rep. 6601, Naval Research Laboratory, Washington, D.C., Oct. 20, 1967.
836. **Beal, R. H. and Tynes, K. E.**, Testing of Wire Insulating Materials for Resistance to Termites, First Annual Field Progress Report, U.S. Depart. of Agriculture, Wood Products Insects Laboratory, Gulfport, Miss., Jan., 1967.
837. **Smythe, R. V. and Carter, F. L.**, Feeding responses to sound wood by the eastern subterranean termite, *Reticulitermes flavipes, Ann. Entomol. Soc. Am.*, 62, 335, 1969.
838. **Beal, R. H.**, Formosan invader, *Pest Control.*, 35(2), 13, 1967.
839. **Beal, R. H. and Smith, V. K.**, Carbamate or phosphate insecticides for subterranean termite control, *Pest Control*, 40(7), 20, 1972.
840. **Leger, R. L.**, Formosan termites, especially ferocious, posing peril to U.S., *Wall St. J.*, Oct. 18, 1968.
841. **Esenther, G. R. and Beal, R. H.**, Attractant-Mirex bait suppresses activity of *Reticulitermes* spp., *J. Econ. Entomol.*, 67, 85, 1974.
842. **Beard, R. L.**, Termite Biology and Bait-Block Method of Control, Bull. 748, Connecticut Agricultural Experimentation Station, New Haven, November 1974.
843. **Anon.**, Organotin-dieldrin combination protects wood against termites, *Tin Its Uses*, No. 97, p. 14, 1973.
844. **Anon.**, Cockroaches are linked to the disease parasitic toxoplasmosis, *Pest Control*, 45, 22, 1975.
845. **Mulrennan, J. A. and Burden, G. S.**, German cockroaches: evaluations for insecticide resistance in field strains, *Pest Control*, 42 (3), 21, 1974.
846. **Keller, J. C., Clark, P. H., and Lofgren, C. S.**, Susceptibility of insecticide-resistant cockroaches to pyrethrins, *Pest Control*, 24, 14, 1956.
847. **Collins, W. J.**, German cockroach resistance. 1. Resistance to Diazinon includes cross resistance to DDT, pyrethrins and propoxur in a laboratory colony, *J. Econ. Entomol.*, 66, 44, 1973.
848. **Gupta, A. P., Das, Y. T., Gusciora, W. R., Adam, D. S., and Jargowsky, L.**, Effectiveness of 3 spray-dust combinations, *Pest Control*, 43 (7), 28, 1975.
849. **Reierson, D. A.**, Field tests to control German cockroaches with ULV aerosol generators, *Pest Control*, 41 (1), 26, 1973.
850. **Grayson, J. M.**, Cockroach control research in 1974, *Pest Control*, 43 (4), 17, 1975.
851. **Ebeling, W. and Reierson, D. A.**, Bait trapping silverfish, cockroaches and earwigs, *Pest Control*, 42 (7), 24, 1974.
852. **Burden, G. S. and Madden, E. E.**, *Panchlora nivea*: comparative susceptibility to residuals, *Pest Control*, 43, 26, 1975.
853. **Shore, J.**, Accurate residue predictions help determine optimum dosages, *Pest Control*, 42 (12), 18, 1974.
854. **Kydonieus, A. F. and Smith, I. K.**, Efficacy of Hercon® Polymeric Controlled Release Dispensers (PCRD) in Cockroach Control, report, Herculite Protective Fabric Corp., New York, 1974.

855. **Kydonieus, A. F., Smith, I. K., and Hyman, S.,** A Polymeric Delivery System for the Controlled Release of Pesticides, Proc. Int. Controlled Release Pesticides Symp., Wright State University, Dayton, Oh., Sept. 8–10, 1975.
856. **Quisumbing, A. R., Lawatsch, D. J., and Kydonieus, A. F.,** Field Studies on the Control of the German Cockroach, *Blattella germanica* (L.) Using Hercon® "Roach Tape" and Standard Sprays, Proc. Int. Controlled Release Pesticides Symp., Wright State University, Dayton, Oh., Sept. 8–10, 1975.
857. **Williams, L. H.,** Anobiid beetles should be controlled, *Pest Control,* 41, (6), 18, 1973.
858. **Mampe, C. D.,** The relative importance of household insects in the continental United States, *Pest Control,* 40 (12), 24, 1972.
859. **Anon.,** Controlled release insecticides, *Soap Cosmet. Chem. Spec.,* 51, 44, 1975.
860. **Gupta, A. P. and Vasvary, L. M.,** European earwig and Baygon bait, *Pest Control,* 40 (7), 27, 1972.
861. **Kimball, J. H.,** Carp for chironomid midge control, *Mosq. News,* 28, 147, 1968.
862. **Nielson, E. J.,** A note on the control of the chironomid *Glyptotendipes paripes* Edwards, *Mosq. News,* 22, 1145, 1962.
863. **Spiller, D.,** Methods used for chironomid larva surveys of sewage oxidation ponds and natural waters at Auckland, New Zealand, *Calif. Vector Views,* 12 (3), 9, 1965.
864. **Winney, R.,** Pyrethrins, pyrethroids and mixtures, *Soap Cosmet. Chem. Spec.,* 50, 56, 1974.
865. **Anon.,** Altosid, technical bulletin, Zoecon Corp., Palo Alto, Calif., Jan. 1, 1974.
866. **Sirol, J., Gendron, Y., and Condat, M.,** La charbon humain en Afrique, *Bull. WHO,* 49, 143, 1973.
867. **Anon.,** One mistaken notion, *Weeds Trees Turf,* 14, 10, 1975.
868. **Anon.,** The Gulph Mills Story, Brochure, Sandoz, Inc., Homestead, Fla., 1974.
869. **Anon.,** Thuricide-16B, technical specifications, Sandoz, Inc., Homestead, Fla., 1974.
870. **Anon.,** Update: gypsy moth 1973, *Weeds Trees Turf,* 12, 24, 1973.
871. **Anon.,** Attack on two fronts: tussock moth and gypsy moth, *Weeds Trees Turf,* 12 (3), 17, 1973.
872. **Anon.,** Those nomad gluttons, *Weeds Trees Turf,* 13 (12), 50, 1974.
873. **Beroza, M., Bierl, B. A., Tardif, J. G. R., Cook, D. A., and Paszek, E. C.,** Activity and persistence of synthetic and natural sex attractants of the gypsy moth in laboratory and field trials, *J. Econ. Entomol.,* 64, 1499, 1971.
874. **Anon.,** Controlled-release pesticides attract interest, *Chem. Eng. News,* 52, (9), 20, 1974.
875. **Beroza, M.,** Insect sex attractant pheromones, a tool for reducing insecticide contamination in the environment, *Toxicol. Environ. Chem. Rev.,* 1, 109, 1972.
876. **Beroza, M.,** Insect sex attractants, *Am. Sci.,* 59, 320, 1971.
877. **Beroza, M. and Knipling, E. F.,** Gypsy moth control with the sex attractant pheromone, *Science,* 177, 19, 1972.
878. **Stevens, L. J. and Beroza, M.,** Mating-inhibition field tests using disparlure, the synthetic gypsy moth sex pheromone, *J. Econ. Entomol.,* 65, 1090, 1972.
879. **Brown, G. P.,** The Impact of Controlled Release on Future Product Development, Rep. No. 5, Proc. Controlled Release Pesticide Symp., University of Akron, Akron, Oh., Sept. 16–18, 1974.
880. **Riedl, H. and Croft, B. A.,** A study of pheromone trap catches in relation to coddling moth damage, *Can. J. Entomol.,* 106, 525, 1971.
881. **Allan, G. G., Gara, R. I., and Wilkins, R. M.,** Field testing of insecticide-polymer combinations for the long-term control of the mahogany shootborer, *Int. Pest Control,* reprint, Vol. 16, July–August 1974.
882. **Bennett, W. H.,** The Texas Leaf-Cutting Ant, Forest Pest Leaflet No. 23, U.S. Dept. of Agriculture, Washington, D.C., June 1958.
883. **Moser, J. C.,** Trials of the leaf cutters, *Nat. Hist.,* 76 (1), 32, 1967.
884. **Moser, J. C., Brownlee, R. C., and Silverstein, R.,** Alarm pheromones of the ant *Atta texana, J. Insect. Physiol.,* 14, 529, 1968.
885. **Clements, B. W. and Rogers, A. J.,** Some factors affecting kill of the stable fly, *Stomoxys calcitrans* (L.) with insecticidal thermal aerosols, in *Proc. 35th Annual Conf. Calif. Mosq. Control Assoc.,* San Francisco, Feb. 5–8, 1967, 109.
886. **Harris, E. J.,** Controlled Release Cue-Lure Formulations for the Detection and Control of the Melon Fly, Rep. No. 35, Proc. Controlled Release Pesticide Symp., University of Akron, Akron, Oh., Sept. 16–18, 1974.
887. **Blume, R. R.,** Results of Tests with Carbopol from the University of Akron, report, U.S. Dept. of Agriculture Livestock Insects Investigation, Kerrville, Tex., 1971.
888. **Weidman, V. W. and Yates, P. C.,** Alkali Metal Silicate Stabilized Pesticide Foams, U.S. Patent 3,689,245, 1972.
889. **Beroza, M., Paszek, E. C., Mitchell, E. R., Bierl, B. A., McLaughlin, J. R., Chambers, D. L., and Brown, G. P.,** A Hercon Dispenser for Controlled Emission of Attractants from Insect Traps, Rep. No. 37, Controlled Release Pesticide Symp., University of Akron, Akron, Oh., Sept. 16–18, 1974.
890. **Hardee, D. D.,** Slow Release Formulations for Grandlure, The Pheromone of the Boll Weevil, Rep. No. 36, Controlled Release Pesticide Symp., University of Akron, Akron, Oh., Sept. 16–18, 1974.
891. **Coker, R. R.,** Testimony at Hearings Before the Subcommittee on Cotton of the Committee on Agriculture, House of Representatives, 93rd Congress, Serial 93-HH, Nov. 6–7, 1973.
892. **Hardee, D. D., McKibben, G. H., Gueldner, R. C., Mitchell, E. B., Tumlinson, J. H., and Cross, W. W.,** Boll weevils in nature respond to grandlure, a synthetic pheromone, *J. Econ. Entomol.,* 65, 97, 1972.

893. **Bull, D. L., Coppedge, J. R., Ridgeway, R. L., Hardee, D. D., and Graves, T. M.,** Formulations for controlling the release of the synthetic pheromone (grandlure) of the boll weevil. 1. Analytical studies, *J. Environ. Entomol.,* 2, 829, 1973.
894. **Anon.,** Chemical Messengers, catalog, Story Chemical Corp., Willoughby, Oh., October 1973.
895. **Menn, J. J. and Beroza, M., Eds.,** *Insect Juvenile Hormones, Chemistry and Action,* Academic Press, New York, 1972.
896. **Cruickshank, P. A.,** Insect juvenile hormone analogues: effects of some terpenoid amide derivatives, *Bull. WHO,* 44, 395, 1971.
897. **Anon.,** Insect Growth Regulant CRD 9499, data sheet, FMC Corp., Princeton, N.J., 1972.
898. **Meltzer, Y. L.,** *Hormonal and Attractant Pesticide Technology,* Noyes Data Corp., Park Ridge, N.J., 1971.
899. **Anon.,** R-20458, technical information bulletin, Stauffer Chemicals, Inc., Mountain View, Calif., May 1972.
900. **Cruickshank, P. A. and Palmere, R. M.,** Terpenoid amides as insect juvenile hormones, *Nature,* 233, 488, 1971.
901. **Cruickshank, P. A.,** Some juvenile hormone analogs: a critical appraisal, *Mitt. Schweiz. Entomol. Ges.,* 44 (1, 2), 97, 1971.
902. **Chodnekar, M. S., Pfiffner, A., Rigassi, N., Schweiter, U., and Suchy, M.,** Insecticidal Pyrethrins in Combination with Juvenile Hormones, U.S. Patent 3,839,562, 1974.
903. **Schaefer, C. H. and Wilder, W. H.,** Insect development inhibitors. 2. Effects on target mosquito species, *J. Econ. Entomol.,* 66, 913, 1973.
904. **Brown, T. W. and Brown, A. W. A.,** Experimental induction of resistance to a juvenile hormone mimic, *J. Econ. Entomol.,* 67, 799, 1974.
905. **Plapp, F. W. and Vinson, S. B.,** Juvenile hormone analogues: toxicity and cross-resistance in the housefly, *Pestic. Biochem. Physiol.,* 3, 131, 1973.
906. **Cerf, D. C. and Georghiou, G. P.,** Evidence of cross-resistance to juvenile hormone analogue in some insecticide-resistant houseflies, *Nature,* 239, 401, 1972.
907. **Dyte, C. E.,** Resistance to a synthetic juvenile hormone in a strain of the flour beetle *Tribolium castaneum, Nature,* 238, 48, 1972.
908. **Menn, J. J. and Pallos, F. M.,** Development of Morphogenetic Agents in Insect Control, report, Stauffer Chem. Co., Mountain View, Calif., Aug. 13, 1974.
909. **Anon.,** Penn. State chemists' findings contradict current theories, *Weeds Trees Turf,* 14 (1), 64, 1975.
910. **Allan, G. G., Chopra, C. S., Neogi, A., and Wilkins, R. M.,** Concept of chemical combination of pesticides with solid waste macromolecules, *Int. Pest Control,* 13, 10, 1971.
911. **Campbell, C. C.,** Pesticidal Compositions, U.S. Patent 3, 740, 419, 1973.
912. **Chandrasekaran, S. K., Baker, R. W., Buckles, R. G., and Michaels, A. S.,** A Mathematical Analysis for Controlled Delivery of Agrichemicals, Rep. No. 11, Proc. Controlled Release Pesticide Symp., University of Akron, Akron, Oh., Sept. 16-18, 1974.
913. **Scott, H. D. and Phillips, R. E.,** Diffusion of selected herbicides in soil, *Soil Sci. Soc. Am. Proc.,* 36, 714, 1972.
914. **Stokes, R. A., Coppedge, J. R., Bull, D. L., and Ridgeway, R. L.,** Use of selected plastics in controlled release granular formulations of aldicarb and dimethoate, *J. Agric. Food Chem.,* 21, 103, 1973.
915. **Sutherland, G. L.,** Agriculture is our best bartering tool, *Farm Chem.,* 135, 43, 1972.
916. **Edwards, W. M. and Harold, L. L.,** Agricultural pollution of water bodies, *Ohio J. Sci.,* 70, 50, 1970.
917. **Wolfe, W.,** Creative cure for chlorosis, *Weeds Trees Turf,* 13, 44, 1974.
918. **Anon.,** Ibdu for Turf, Tech. Bull. No. 1, Swift Chemical Co., Chicago, 1975.
919. **Lunt, O. R.,** Controlled release fertilizers. Achievements and potential, *J. Agric. Food Chem.,* 19, 797, 1971.
920. **Hays, J. T. and Hewson, W. B.,** Controlled release fertilizers by chemical modification of urea, *J. Agric. Food Chem.,* 21, 498, 1973.
921. **Sommer, H. J., McLauren, G. A., Feichtmeir, E. F., Larsen, R. G., and Wittsell, L. E.,** Plant Nutrient Products, U.S. Patent 3, 748, 115, 1973.
922. **Philen, O. C., Silverberg, J., and Norton, M. M.,** Fertilizers Containing Micro and Macronutrients, U.S. Patent 3,520,651, 1970.
923. **Thorneburg, R. P. and Tweedy, J. A.,** A rapid procedure to evaluate the effect of pesticides on nitrification, *Weed Sci.,* 21, 397, 1973.
924. **Anon.,** Rodents, Bats and Birds, Damage to Food Supplies in the Diet-Deficient Regions of the World, report, Wildlife Research Laboratory, U.S. Dept. of the Interior, Denver, 1969.
925. **Drummond, D. C. and Rennison, B. D.,** The detection of rodent resistance to anticoagulants, *Bull. WHO,* 48, 239, 1973.
926. **Jackson, W. B., Brooks, J. E., Bowerman, A. M., and Kaukeinen, D. E.,** Anticoagulant resistance in Norway rats, *Pest Control,* 43 (4), 12, 1975.
927. **West, R. P., Fall, M. W., and Libay, J. L.,** Field Trial of Multiple Baiting with Zinc Sulfide to Protect Growing Rice from Damage by Rats, Proc. 3rd Annual Scientific Meeting, Crop Science Society, Philippines, May 15–17, 1972.
928. **Howard, W. E. and Marsh, R. E.,** Rodent control manual, *Pest Control,* 42 (8), Section D, 1974.
929. **Knipling, E. F. and McGuire, J. U.,** Potential Sale of Sterilization for Suppressing Rat Populations, Tech. Bull. No. 1455, Agriculture Research Service, U.S. Dept. of Agriculture, Washington, D.C., June 1972.
930. **Marsh, P. E., and Howard, W. E.,** Prospects of chemosterilant and genetic control of rodents, *Bull. WHO,* 48, 309, 1973.

931. **Tigner, J. R.**, Seasonal Food Habits of *Rattus rattus Mindanensis* (the Philippine Rice Field Rat) in Central Luzon, Ph.D. thesis, University of Colorado, Boulder, 1972.
932. **Greaves, J. H., Rowe, H. P., Redfern, R., and Ayres, P.**, *Nature*, 219, 402, 1968.
933. **Brooks, J. E.**, A review of comensal rodents and their control, *CRC Crit. Rev. Environ. Control*, 3 (4), 405, 1973.
934. **Abrams, J. and Hinkes, T. M.**, Acceptability and performance of encapsulated warfarin, *Pest Control*, 42 (5), 14, 1974.
935. **Roszkowski, A. P.**, The pharmacological properties of norbormide, a selective rat toxicant, *J. Pharmacol. Exp. Ther.*, 149, 288, 1965.
936. **Jackson, W. B.**, Use of Microencapsulation to Enhance Rodenticide Acceptance, Rep. 6, Proc. Controlled Release Pesticide Symp., University of Akron, Akron, Oh., Sept. 16–18, 1974.
937. **Linhart, S. B., Crespo, R. F., Burns, R. J., Mitchell, G. C., and Tigner, J. R.**, Vampire Bats: Rabies Transmission and Livestock Production in Latin America, 1969 annual report, Palo Alto, Mex., Field Station, Denver Wildlife Pest Center, 1969.
938. **Mitchell, G. C.**, Vampire Bat Control Programs in Latin America, Quarterly News Report, U.S. Agency for International Development, July 1974.
939. **Thompson, R. D., Mitchell, G. C., and Burns, R. J.**, Vampire bat control by systemic treatment of livestock with an anticoagulant, *Science*, 177, 806, 1972.
940. **Lindsey, G. D., Anthony, R. M., and Evans, J.**, Mestranol as a repellent to protect douglas-fir seed from deer mice, in *Proc. Sixth Vertebrate Pest Conf.*, University of California Press, Davis, 1974, 272.
941. **Feldmesser, J. et. al.**, Estimated crop losses due to plant-parasite nematodes in the United States, Soc. Nematologists, *J. Nematol.*, Spec. Publ. No. 1, Suppl., 1971.
942. **Feldmesser, J.**, Progress Report on Controlled Release of a Nematocide, Rep. 32, Proc. Controlled Release Pesticide Symp., University of Akron, Akron, Oh., Sept. 16–18, 1974.
943. **Cardarelli, N. F. and Feldmesser, J.**, Controlled Release Nematicide Development: Preliminary Investigations, Proc. Int. Controlled Release Pesticide Symp., Wright State University, Dayton, Oh., Sept. 8–10, 1975.
944. **Igbal, Z. M. and Menzer, R. E.**, Metabolism of O-ethyl S,S-dipropyl phosphorodithioate in rats and liver microsomal systems, *Biochem. Pharmacol.*, 21, 1569, 1972.
945. **Stewart, C. W.**, Nucleation and growth of bubbles in elastomers, *Rubber Chem. Technol.*, 44, 1363, 1971.
946. **Paul, D. R. and Dibenedetto, A. T.**, Diffusion in amorphous polymers, *Rubber Chem. Technol.*, 39, 1496, 1966.
947. **Stone, C. P. and Mott, D. F.**, Bird Damage to Ripening Field Corn in the United States, 1971, Wildlife Leaflet 505, U.S. Dept. of the Interior, Washington, D.C., October 1973.
948. **West, R. R., Brunton, R. B., and Cunningham, D. J.**, Repelling pheasants from sprouting corn with a carbamate insecticide, *J. Wildl. Manage.*, 33, 216, 1969.
949. **Besser, J. F.**, Protecting seeded rice from blackbirds with methiocarb, *Int. Rice Commission Newsl.*, 22(3), 9, 1973.
950. **Guarino, J. L. and Schafer, E. W.**, Magpie Reduction in an Urban Roost, Spec. Sci. Rep. Wildl. No. 104, U.S. Dept. of the Interior, Washington, D.C., June 1967.
951. **Schafer, E. W., Brunton, R. B., and Lockyer, N. F.**, Hazards to animals feeding on blackbirds killed with 4-aminopyridine baits, *J. Wildl. Manage.*, 38, 424, 1974.
952. **Schafer, E. W., Bunton, R. B., Lockyer, N. F., and DeGrazio, J. W.**, Comparative toxicity of seventeen pesticides to the *Quelea*, house sparrow and red-winged blackbird, *Toxicol. Appl. Pharmacol.*, 26, 154, 1973.
953. **Balser, D. S.**, A review of coyote research, in *Proc. Sixth Vertebrate Pest Conf.*, University of California Press, Davis, 1974, 171.
954. **Pringle, L.**, Each antagonist in coyote debate is partly correct, *Smithsonian Mag.*, 6, 74, 1974.
955. **Anon.**, Toxic Collar for Predator Damage Control, status report, Bureau of Sport Fisheries and Wildlife, U.S. Dept. of the Interior, Denver, May 1975.
956. **Cardarelli, N. F.**, Protective Collars for Sheep Wear, Quarterly Progress Rep. 14-16-0008-895, May–July 1975, Bureau of Sport Fisheries and Wildlife, U.S. Dept. of the Interior, Denver, Aug. 4, 1975.
957. **Maki, A. W., Geissel, L., and Johnson, H. E.**, The Influence of Larval Lampricide (TFM: 3-Trifluoromethyl-4-nitrophenol) on Growth and Reproduction of Two Species of Aquatic Macrophytes, *Elodea canadensis* and *Myriophyllum spicatum*, annual report, Michigan State University, East Lansing, 1974.
958. **Anon.**, Research shows starfish threat is extremely serious, *Ocean Ind.*, p. 52, February 1970.
959. **Suehiro, Y.**, Tokyo Aburatsubo Marine Park, personal communication.
960. **Ray, C. and Ciampi, E.**, *The Underwater Guide to Marine Life*, Barnes, New York, 1956.
961. **Russell, F. E.**, Marine toxins and venomous and poisonous marine animals, in *Advances in Marine Biology*, Vol. 3, Russell, F. S., Ed., Academic Press, New York, 1965, 256.
962. **Anon.**, Bureau of Commercial Fisheries Federal Aid Program, Circ. 293, U.S. Dept. of the Interior, Washington, D.C., June 1968.
963. **Messing, R. A.**, Controlled-pore ceramics, *Res. Dev.*, 25, 32, 1974.
964. **Christy, R. I.**, Evaluation of a Nitrile-Acrylic Copolymer as a Ball Bearing Retainer Material, 29th Annual Meeting, American Society of Lubrication Engineering, Preprint No. 74AM-5A-1, Cleveland, Apr. 28–May 2, 1974.
965. **Christy, R. I. and Meeks, C. R.**, Polyurethane Retainers for Ball Bearings, Int. Ball Bearing Symp., Charles Stark Draper Laboratory, Cambridge, Mass., June 5–7, 1973.

966. **Anon.,** It oils with super efficiency, *Bus. Week,* p. 70g, March 16, 1974.
967. **Anon.,** Advanced lubrication material may permit long-life parts, *Prod. Eng. (N.Y.),* 45, 27, 1974.
968. **Anon.,** *Sustrelle Laboratory Manual,* Moleculon Research Corp., Cambridge, Mass., June 2, 1975.
969. **Nichols, L. D.,** Poroplastic and Sustrelle: Controlled Release Vehicles Having Broad Compatibility with Dissolved and Precipitated Pesticides, Proc. Int. Controlled Release Pesticides Symp., Wright State University, Dayton, Oh., Sept. 8–10, 1975.
970. **Ashare, E., Brooks, T. W., and Swenson, D. W.,** Controlled Release from Hollow Fibers, Proc. Int. Controlled Release Pesticide Symp., Wright State University, Dayton, Oh., Sept. 8–10, 1975.

A

C

D

I

J

K

L

M

P

Q

R

S

T

U

V

W

Y

Z